Change of Time and Change of Measure

ADVANCED SERIES ON STATISTICAL SCIENCE & APPLIED PROBABILITY

Editor: Ole E. Barndorff-Nielsen

Published

Vol. 1 Random Walks of Infinitely Many Particles
by P. Révész

Vol. 2 Ruin Probabilities
by S. Asmussen

Vol. 3 Essentials of Stochastic Finance: Facts, Models, Theory
by Albert N. Shiryaev

Vol. 4 Principles of Statistical Inference from a Neo-Fisherian Perspective
by L. Pace and A. Salvan

Vol. 5 Local Stereology
by Eva B. Vedel Jensen

Vol. 6 Elementary Stochastic Calculus — With Finance in View
by T. Mikosch

Vol. 7 Stochastic Methods in Hydrology: Rain, Landforms and Floods
eds. O. E. Barndorff-Nielsen et al.

Vol. 8 Statistical Experiments and Decisions: Asymptotic Theory
by A. N. Shiryaev and V. G. Spokoiny

Vol. 9 Non-Gaussian Merton–Black–Scholes Theory
by S. I. Boyarchenko and S. Z. Levendorskiĭ

Vol. 10 Limit Theorems for Associated Random Fields and Related Systems
by A. Bulinski and A. Shashkin

Vol. 11 Stochastic Modeling of Electricity and Related Markets
by F. E. Benth, J. Šaltytė Benth and S. Koekebakker

Vol. 12 An Elementary Introduction to Stochastic Interest Rate Modeling
by N. Privault

Vol. 13 Change of Time and Change of Measure
by O. E. Barndorff-Nielsen and A. Shiryaev

Advanced Series on
Statistical Science &
Applied Probability
Vol. 13

Change of Time and Change of Measure

Ole E. Barndorff-Nielsen
Aarhus University, Denmark

Albert Shiryaev
Steklov Mathematical Institute and Moscow State University, Russia

NEW JERSEY • LONDON • SINGAPORE • BEIJING • SHANGHAI • HONG KONG • TAIPEI • CHENNAI

Published by

World Scientific Publishing Co. Pte. Ltd.
5 Toh Tuck Link, Singapore 596224
USA office: 27 Warren Street, Suite 401-402, Hackensack, NJ 07601
UK office: 57 Shelton Street, Covent Garden, London WC2H 9HE

British Library Cataloguing-in-Publication Data
A catalogue record for this book is available from the British Library.

CHANGE OF TIME AND CHANGE OF MEASURE
Advanced Series on Statistical Science and Applied Probability — Vol. 13

ISBN-13 978-981-4324-47-2
ISBN-10 981-4324-47-7

Printed in Singapore.

Foreword

The conception of the book, based on LECTURE COURSES delivered by the authors in the last years (Aarhus, Moscow, Barcelona, Halmstad, *etc.*), is defined in many respects by the desire to state the main ideas and results of the stochastic theory of "change of time and change of measure". These ideas and results have manifold applications, particularly in Mathematical Finance, Financial Economics, Financial Engineering and Actuarial Business, when constructing probabilistic and statistical models adequate to statistical data, when investigating the problems of arbitrage, hedging, rational (fair) pricing of financial and actuarial instruments, when making decisions minimizing the financial and actuarial risks, *etc.* The lecture-based character of the book defined as well the style of presentation—we have not aimed to give all and complete proofs, many of which are rather long. Our purpose was different, namely to specify the main, essential topics and results of "change of time and change of measure", so that the readers could make use of them in their theoretical and applied activity.

Acknowledgments. We express our gratitude to our colleagues, especially Ernst Eberlein and Neil Shephard, for stimulating discussions. We are grateful to the Thiele Centre (Department of Mathematical Sciences, Aarhus University) and the Steklov Mathematical Institute (Moscow) for providing excellent opportunities to work on the monograph. The support of INTAS, RFBR, Manchester University (School of Mathematics), and Moscow State University (Department of Mechanics and Mathematics) is gratefully acknowledged. We thank T. B. Tolozova for her help in preparation of the text for publication.

O. E. B.-N.,
A. N. Sh.

Foreword

Contents

Foreword v

Introduction xi

1. Random Change of Time 1

1.1 Basic Definitions 1
1.2 Some Properties of Change of Time 4
1.3 Representations in the Weak Sense ($X \stackrel{\text{law}}{=} \widehat{X} \circ T$), in the Strong Sense ($X = \widehat{X} \circ T$) and the Semi-strong Sense ($X \stackrel{\text{a.s.}}{=} \widehat{X} \circ T$). I. Constructive Examples 8
1.4 Representations in the Weak Sense ($X \stackrel{\text{law}}{=} \widehat{X} \circ T$), Strong Sense ($X = \widehat{X} \circ T$) and the Semi-strong Sense ($X \stackrel{\text{a.s.}}{=} \widehat{X} \circ T$). II. The Case of Continuous Local Martingales and Processes of Bounded Variation 15

2. Integral Representations and Change of Time in Stochastic Integrals 25

2.1 Integral Representations of Local Martingales in the Strong Sense 25
2.2 Integral Representations of Local Martingales in a Semi-strong Sense 33
2.3 Stochastic Integrals Over the Stable Processes and Integral Representations 35
2.4 Stochastic Integrals with Respect to Stable Processes and Change of Time 38

3. Semimartingales: Basic Notions, Structures, Elements of

Stochastic Analysis 41

3.1 Basic Definitions and Properties 41
3.2 Canonical Representation. Triplets of Predictable Characteristics 52
3.3 Stochastic Integrals with Respect to a Brownian Motion, Square-integrable Martingales, and Semimartingales . . . 56
3.4 Stochastic Differential Equations 73

4. Stochastic Exponential and Stochastic Logarithm. Cumulant Processes 91

4.1 Stochastic Exponential and Stochastic Logarithm 91
4.2 Fourier Cumulant Processes 96
4.3 Laplace Cumulant Processes 99
4.4 Cumulant Processes of Stochastic Integral Transformation $X^{\varphi} = \varphi \cdot X$ 101

5. Processes with Independent Increments. Lévy Processes 105

5.1 Processes with Independent Increments and Semimartingales 105
5.2 Processes with Stationary Independent Increments (Lévy Processes) 108
5.3 Some Properties of Sample Paths of Processes with Independent Increments 113
5.4 Some Properties of Sample Paths of Processes with Stationary Independent Increments (Lévy Processes) 117

6. Change of Measure. General Facts 121

6.1 Basic Definitions. Density Process 121
6.2 Discrete Version of Girsanov's Theorem 123
6.3 Semimartingale Version of Girsanov's Theorem 126
6.4 Esscher's Change of Measure 132

7. Change of Measure in Models Based on Lévy Processes 135

7.1 Linear and Exponential Lévy Models under Change of Measure 135
7.2 On the Criteria of Local Absolute Continuity of Two Measures of Lévy Processes 142

7.3 On the Uniqueness of Locally Equivalent Martingale-type Measures for the Exponential Lévy Models 144
7.4 On the Construction of Martingale Measures with Minimal Entropy in the Exponential Lévy Models 147

8. Change of Time in Semimartingale Models and Models Based on Brownian Motion and Lévy Processes 151

8.1 Some General Facts about Change of Time for Semimartingale Models 151
8.2 Change of Time in Brownian Motion. Different Formulations 154
8.3 Change of Time Given by Subordinators. I. Some Examples 156
8.4 Change of Time Given by Subordinators. II. Structure of the Triplets of Predictable Characteristics 158

9. Conditionally Gaussian Distributions and Stochastic Volatility Models for the Discrete-time Case 163

9.1 Deviation from the Gaussian Property of the Returns of the Prices 163
9.2 Martingale Approach to the Study of the Returns of the Prices 166
9.3 Conditionally Gaussian Models. I. Linear (AR, MA, ARMA) and Nonlinear (ARCH, GARCH) Models for Returns 171
9.4 Conditionally Gaussian Models. II. IG- and $\mathbb{G}$IG-distributions for the Square of Stochastic Volatility and $\mathbb{G}$H-distributions for Returns 175

10. Martingale Measures in the Stochastic Theory of Arbitrage 195

10.1 Basic Notions and Summary of Results of the Theory of Arbitrage. I. Discrete Time Models 195
10.2 Basic Notions and Summary of Results of the Theory of Arbitrage. II. Continuous-Time Models 207
10.3 Arbitrage in a Model of Buying/Selling Assets with Transaction Costs 215
10.4 Asymptotic Arbitrage: Some Problems 216

11. Change of Measure in Option Pricing 225

11.1 Overview of the Pricing Formulae for European Options . 225
11.2 Overview of the Pricing Formulae for American Options . 240
11.3 Duality and Symmetry of the Semimartingale Models . . 243
11.4 Call-Put Duality in Option Pricing. Lévy Models 254

12. Conditionally Brownian and Lévy Processes. Stochastic Volatility Models 259

12.1 From Black–Scholes Theory of Pricing of Derivatives to the Implied Volatility, Smile Effect and Stochastic Volatility Models 259
12.2 Generalized Inverse Gaussian Subordinator and Generalized Hyperbolic Lévy Motion: Two Methods of Construction, Sample Path Properties 270
12.3 Distributional and Sample-path Properties of the Lévy Processes L($\mathbb{G}$IG) and L($\mathbb{G}$H) 275
12.4 On Some Others Models of the Dynamics of Prices. Comparison of the Properties of Different Models 283

Afterword 289

Bibliography 291

Index 301

Introduction

1. One of the topical problems of Probability Theory and the Theory of Stochastic Processes is the following:

> How, for the given stochastic processes (maybe with "complicated" structure), to obtain a relatively simple representation via some "simple" processes of the type of "white noise" in discrete-time case or Brownian motion or Lévy processes in the case of continuous time?

For example, from the theory of stationary sequences we know that every "regular" sequence $X = (X_n)$ admits the "Wold decomposition"

$$X_n = \sum_{k=0}^{\infty} a_k \varepsilon_{n-k}, \qquad n \in \mathbb{Z} = \{\ldots, -1, 0, 1, \ldots\},$$

where $\varepsilon = (\varepsilon_n)$ is a sequence of pairwise orthogonal random variables ("white noise") with $\mathsf{E}\varepsilon_n = 0$, $\mathsf{E}|\varepsilon_n|^2 = 1$.

Another example. If we agree that the Brownian motion $B = (B_t)_{t\geq 0}$ is a process of a "simple" structure, then a solution $X = (X_t)_{t\geq 0}$ to the Itô stochastic differential equation

$$dX_t = a(t, X_t)\, dt + \sigma(t, X_t)\, dB_t$$

can be considered as a version of (a candidate for) the Kolmogorov diffusion process with the local characteristics $a(t, x)$ and $\sigma(t, x)$.

In the present book our main interest will be related to the following two methods for getting "simple" representations:

CHANGE OF TIME and *CHANGE OF MEASURE.*

We also shall consider another method of representation of the processes based on stochastic integrals with respect to some "simple" processes. This method is convenient as an intermediate step for getting the change of time representation for the "complicated" processes.

The *change of time* is based on the idea of representation of a given process $X = (X_t)_{t\geq 0}$ *via* a "simple" process $\widehat{X} = (\widehat{X}_\theta)_{\theta\geq 0}$ and a "change of time" $T = (T(t))_{t\geq 0}$:

$$X = \widehat{X} \circ T$$

or, in detail, $X_t = \widehat{X}_{T(t)}$. In other words, the process X is a time-deformation of the process $\widehat{X}$. This can be considered as a way to change the velocity in moving along the trajectories of $\widehat{X}$.

The technique of *change of measure* does not operate with the transformation of the trajectories. Instead it is based on the construction of a new probability measure $\widetilde{\mathsf{P}}$ equivalent to the given measure P and a process $\widetilde{X} = (\widetilde{X}_t)_{t\geq 0}$ with a "simple" structure such that

$$\mathrm{Law}(X \mid \widetilde{\mathsf{P}}) = \mathrm{Law}(\widetilde{X} \mid \mathsf{P}).$$

From the point of view of applications the general problem of *change of measure* is of central interest in mathematical finance, where so-called "martingale measures" $\widetilde{\mathsf{P}}$ play a key role for criteria of "No-Arbitrage" and for the "Pricing and Hedging" machinery.

The concept of *change of time* is also of substantial interest for understanding the nature of financial time series; witness the common phrase that "Prices on the financial markets are Brownian motions in the operational (or business) time".

2. Let us give a more detailed description of the content of the chapters of the book.

Chapter 1 contains some material about Brownian motion and Lévy processes as main "simple" driving processes used in constructing the change of time. Because these important processes and the processes constructed from them usually belong to the class of semimartingales, we have included also some text (Chap. 3) about semimartingales which become more and more popular in many fields and in mathematical finance in particular.

The general scheme of change of time ("old time" $\to$ "new time" $\to$ "old time" discussed in Chap. 1 is the following.

Assume a stochastic process $X = (X_t)_{t\geq 0}$ to be given (in "old" time t) on a filtered probability space $(\Omega, \mathcal{F}, \mathbb{F} = (\mathcal{F}_t)_{t\geq 0}, \mathsf{P})$, where $\mathbb{F} = (\mathcal{F}_t)_{t\geq 0}$ is the "flow" of information ($\mathcal{F}_s \subseteq \mathcal{F}_t \subseteq \mathcal{F}$ for $s \leq t$). We construct an increasing family $\widehat{T} = (\widehat{T}(\theta))_{\theta\geq 0}$ of stopping times $\widehat{T}(\theta)$ (with respect to $\mathbb{F}$)

and we introduce a "new" process $\widehat{X} = (\widehat{X}_\theta)_{\theta \geq 0}$ (in a "new" time θ) by the formula

$$\widehat{X}_\theta = X_{\widehat{T}(\theta)}$$

which we shall usually write in a short form

$$\widehat{X} = X \circ \widehat{T}.$$

Suppose that the process $\widehat{X}$ has a simple structure, then it is reasonable to try to find a new increasing family of stopping times $T = (T(\theta))_{\theta \geq 0}$ (with respect to $\widehat{\mathbb{F}} = (\widehat{\mathcal{F}}_\theta)_{\theta \geq 0}$, where $\widehat{\mathcal{F}}_\theta = \mathcal{F}_{\widehat{T}(\theta)}$), such that the following representation in the *strong* sense holds:

$$X = \widehat{X} \circ T, \qquad \textit{i.e.}, \qquad X_t = \widehat{X}_{T(t)}, \quad t \geq 0.$$

The given description distinguishes between "old" ("physical", "calendar") time t and a "new" ("operational", "business") time θ, with $\theta = T(t)$ and $t = \widehat{T}(\theta)$ as the formulae which define the transitions: $t \to \theta \to t$.

All previous considerations had the following aim: given an "old" (initial) process X (in time t), to construct a simple "new" process $\widehat{X}$ (in time θ) and to construct two changes of time $\widehat{T}(\theta)$ and $T(t)$ such that X and $\widehat{X}$ can be obtained by the transformations $\widehat{X} = X \circ \widehat{T}$ and $X = \widehat{X} \circ T$.

So far we have assumed that the property $\widehat{X} = X \circ \widehat{T}$ (and $X = \widehat{X} \circ T$) holds identically (for all $\omega \in \Omega$ and all $t \geq 0$). However, sometimes such representations are hard to find but one can find representations of the type $X \overset{\text{a.s.}}{=} \widehat{X} \circ T$ (so-called *semi-strong representations*) or $X \overset{\text{law}}{=} \widehat{X} \circ T$ (so-called *weak representations*).

Chapter 2 is about

STOCHASTIC VOLATILITY REPRESENTATION or
STOCHASTIC INTEGRAL REPRESENTATION

$$\boxed{X = H \cdot \widetilde{X}}$$

given the process X, where $H \cdot \widetilde{X}$ is the *stochastic integral* with respect to some "simple" process $\widetilde{X}$ (usually assumed to be a semimartingale); the integrand H is often called a *stochastic volatility*.

Having the

CHANGE OF TIME REPRESENTATION

$$\boxed{X = \widehat{X} \circ T}$$

of the process X *via* some "simple" process $\widehat{X}$ and a *change of time* T we get very transparent connection between the stochastic volatility models and the change of time models:

$$\boxed{H \cdot \widetilde{X} = \widehat{X} \circ T}\,.$$

We emphasize that this duality of the "volatility" and the "change of time" plays a very important role in the construction of convenient models. Especially it is important for the financial modeling.

For many popular models in finance the processes $\widetilde{X}$ are Brownian motions or Lévy processes. So, stochastic volatility models with semimartingales $\widetilde{X}$ cover the most commonly used models.

It is useful to note that to define the time-changed process $\widehat{X} \circ T$ we need not assume that $\widehat{X}$ is a semimartingale.

In Chap. 2 we also consider more general stochastic integral representations (using measures of jumps). As in Chap. 1, both "strong" and "weak" representations are discussed.

Chapter 3 contains important material about *semimartingales, i.e.*, stochastic processes $X = (X_t)_{t\geq 0}$ representable as sums $X = X_0 + A + M$, where $A = (A_t)_{t\geq 0}$ is a process of bounded variation and $M = (M_t)_{t\geq 0}$ is a local martingale. This class is rather wide, the stochastic calculus for these processes is well developed, and they proved to be useful for the study of problems in mathematical finance, actuarial mathematics, and many other fields.

Without any doubt, the class of semimartingale models, including those of Brownian and Lévy type, will play the increasingly important roles in applications of stochastic calculus, not least in finance.

In *Chap. 4* some fundamental notions, namely, stochastic exponential, stochastic logarithm, and cumulant processes, are introduced. These will be of high importance in the rest of the monograph.

Chapter 5 provides a short survey of processes with independent increments (PII), in particular of *Lévy processes.* In some sense the class of semimartingales is a very natural extension of the Lévy processes. Indeed, for PII the triplet (B, C, ν) of characteristics, involved in the Kolmogorov–Lévy–Khinchin formula for the processes with independent increments, consists of *deterministic* components. In the case of semimartingales there exists also a similar triplet (B, C, ν) whose components have the *predictability* property which can be interpreted as a *stochastic determinancy.*

Change of measure plays a crucial role in both probability theory and its applications, providing a powerful tool for study of the distributional properties of stochastic processes. *Chapter 6* "Change of Measure. General Facts" serves as a quick introduction to this subject.

In *Chap. 7* we focus on problems of change of measure especially for Lévy processes, which constitute now a basis for construction of different models (in finance, actuarial science, *etc.*).

Chapter 8 is devoted to the other (along with change of measure) key topic of the book, namely, change of time in semimartingale, Brownian, and Lévy models.

Chapter 9 plays an important conceptual role in our monograph. Firstly, this chapter reviews the "martingale-predictable" approach (based on Doob's decomposition) to study of sequences $H = (H_n)_{n\geq 0}$ which describe the evolution of financial indexes $S_n = S_0 e^{H_n}$, $n \geq 0$, and "return" sequences $h = (h_n)_{n\geq 0}$, where $h_n = \log(S_n/S_{n-1})$ ($\equiv \Delta H_n = H_n - H_{n-1}$). This "martingale-predictable" scheme naturally comprises both linear (AR, MA, ARMA, *etc.*) and nonlinear (ARCH, GARCH, *etc.*) models.

Secondly, in this chapter we introduce the class of $\mathbb{GIG}$ (Generalized Inverse Gaussian) distributions for σ^2, the square of stochastic volatility, and the class of $\mathbb{GH}$ (Generalized Hyperbolic) distributions for the "returns" $h = \mu + \beta\sigma^2 + \sigma\varepsilon$, where σ and ε are independent, σ^2 has $\mathbb{GIG}$-distribution, and ε has the standard Gaussian distribution $\mathcal{N}(0, 1)$. The most recent econometric investigations show convincingly that $\mathbb{GIG}$ and $\mathbb{GH}$ distributions fit well the empirical distributions of various financial indexes.

Chapter 10 demonstrates, first of all, how ideas of change of measure allow one to transform the economic notion of arbitrage into the martingale property of (normalized) prices with respect to special measures called "martingale measures". We consider both discrete and continuous time cases.

Chapter 11 provides a short overview of basic results in the option pricing theory. We cite some classical formulae (Bachelier, Black–Scholes, Cox–Ross–Rubinstein) and discuss different properties such as call–put parity and call–put duality in the semimartingale and especially Lévy's models.

Chapter 12 is closely related to the material of Chap. 9. Since $\mathbb{GIG}$ and $\mathbb{GH}$ distributions, introduced in Chap. 9, are infinitely divisible, one can construct Lévy's processes $T = T(t)$, $t \geq 0$, and $H = (H_t)_{t\geq 0}$, such that $\mathrm{Law}(T(1)) = \mathbb{GIG}$ and $\mathrm{Law}(H_1) = \mathbb{GH}$. The process $H = (H_t)_{t\geq 0}$, can be chosen as $H_t = \mu + \beta T(t) + B_{T(t)}$, where $B = (B_t)_{t\geq 0}$ is a Brownian motion which does not depend on $T = (T(t))_{t\geq 0}$. Introduction of these processes (Sec. 12.3) is preceded by a review of a number of classical and modern financial models accentuated on stylized features of observed prices. Different types of models having desirable features are listed in Sec. 12.4.

Concluding the Introduction, we notice that the thorough reading of certain chapters demands sometimes a look into subsequent chapters. For example, already in Chap. 1 we mention stochastic integrals with respect to the Brownian motion (Wiener process) and semimartingales, although the careful construction of these integrals is given only in Chap. 2. In the same way, examples of Chap. 1 operate with hyperbolic and generalized hyperbolic distributions, whose detailed discussion is postponed to Chap. 9. We hope that this will not create any difficulty for the reader.

Chapter 1

Random Change of Time

1.1 Basic Definitions

1. Let $(\Omega, \mathcal{F}, \mathsf{P})$ be a Kolmogorov's *probability space*, where Ω is the space of *elementary events* ω, $\mathcal{F}$ is some σ-algebra of subsets of Ω, and P is a *probability measure*. In all our considerations, a crucial role is played by an additional structure $(\mathcal{F}_t)_{t\geq 0}$ (called a filtration) which is a nondecreasing right-continuous family of sub-σ-algebras of $\mathcal{F}$ (in other words, $\mathcal{F}_s \subseteq \mathcal{F}_t$ for all $0 \leq s \leq t$ and $\mathcal{F}_t = \mathcal{F}_{t+}$, where $\mathcal{F}_{t+} = \bigcap_{s>t} \mathcal{F}_s$).

The collection $(\Omega, \mathcal{F}, (\mathcal{F}_t)_{t\geq 0}, \mathsf{P})$ is called a *filtered probability space* or *stochastic basis* [100]. (As a rule we assume a stochastic basis to satisfy the *usual conditions*, namely, the σ-algebra $\mathcal{F}$ is P-complete and every $\mathcal{F}_t$ contains all P-null sets of $\mathcal{F}$; see [100].) Sometimes it is convenient to consider $\mathcal{F}_t$ as the "information" available on the time interval $[0, t]$.

2. As was mentioned in the Introduction, it is convenient, when defining the notion of "change of time", to distinguish between the "old" (physical, calendar) t-time and a "new" (operational, business) θ-time.

The following definition is useful if we need to construct, starting from the initial process $X = (X_t)_{t\geq 0}$ (adapted to the filtration $(\mathcal{F}_t)_{t\geq 0}$), a new process $\widehat{X} = (\widehat{X}_\theta)_{\theta\geq 0}$ evolving in θ-time and having certain desired properties.

Definition 1.1. A family of random variables $\widehat{T} = (\widehat{T}(\theta))_{\theta\geq 0}$ is said to be a *random change of time* (or rather, a change of θ-time into t-time in accordance with the map $\theta \rightsquigarrow t = \widehat{T}(\theta)$), if

(a) $(\widehat{T}(\theta))_{\theta\geq 0}$ is a nondecreasing (in the terminology of stochastic analysis—*increasing*), right-continuous family of $[0, \infty]$-valued random variables $\widehat{T}(\theta)$, $\theta \geq 0$;

(b) for all $\theta \geq 0$ the random variables $\widehat{T}(\theta)$ are stopping times (or Markov times) with respect to the filtration $(\mathcal{F}_t)_{t\geq 0}$, *i.e.*,

$$\{\widehat{T}(\theta) \leq t\} \in \mathcal{F}_t, \qquad \theta \geq 0, \quad t \geq 0.$$

Definition 1.2. The random variable

$$\widehat{\zeta} = \inf\{\theta : \widehat{T}(\theta) = \infty\}$$

is called the *life time* of the process $\widehat{T} = (\widehat{T}(\theta))_{\theta\geq 0}$. The change of time $(\theta \rightsquigarrow t = \widehat{T}(\theta))$ is said to be *finite*, if $\widehat{T}(\theta) < \infty$ (P-a.s.), for all $\theta \in [0, \infty)$, or, equivalently, $\mathsf{P}(\widehat{\zeta} = \infty) = 1$.

Definition 1.3. The change of time $\widehat{T} = (\widehat{T}(\theta))_{\theta\geq 0}$ bears the name of *subordinator*, if this random process $\widehat{T}$ on the interval $[0, \widehat{\zeta}]$ is a Lévy process.

If $\mathsf{P}(\widehat{\zeta} = \infty) = 1$, then the change of time $\widehat{T}$ is said to be a *subordinator in the strong sense.*

The above definitions presumed that the initial stochastic basis is a filtered probability space $(\Omega, \mathcal{F}, (\mathcal{F}_t)_{t\geq 0}, \mathsf{P})$. But if we assume that the initial basis is $(\Omega, \mathcal{F}, (\widehat{\mathcal{F}}_\theta)_{\theta\geq 0}, \mathsf{P})$, then, in the analogous way, we can define a change of time $T = (T(t))_{t\geq 0}$, realizing the map $t \rightsquigarrow \theta = T(t)$, the random variables $T(t)$ being stopping times with respect to the filtration $(\widehat{\mathcal{F}}_\theta)_{\theta\geq 0}$.

3. Let us dwell on some constructions and then on some examples of change of time.

Let $(\Omega, \mathcal{F}, (\mathcal{F}_t)_{t\geq 0}, \mathsf{P})$ be a stochastic basis, and let $A = (A_t)_{t\geq 0}$ be an (adapted to filtration $(\mathcal{F}_t)_{t\geq 0}$) increasing, right-continuous random process with $A_0 = 0$.

Put

$$\widehat{T}(\theta) = \inf\{t : A_t > \theta\}, \qquad \theta \geq 0, \tag{1.1}$$

where, as usual, $\inf \varnothing = \infty$.

Lemma 1.1. *The family of random variables $\widehat{T} = (\widehat{T}(\theta))_{\theta\geq 0}$ constitutes a (random) change of time.*

Proof. We need to verify that the properties (a) and (b) in Definition 1.1 are satisfied for this family.

The property (a) follows directly from the definition (1.1).

Since

$$\{t : A_t > \theta\} = \bigcup_{\varepsilon>0} \{t : A_t > \theta + \varepsilon\},$$

we have $\widehat{T}(\theta) = \lim_{\varepsilon \downarrow 0} \widehat{T}(\theta + \varepsilon)$, *i.e.*, the process $\widehat{T} = (\widehat{T}(\theta))_{\theta \geq 0}$ is right-continuous at each point $\theta \geq 0$.

Finally, the requirement of the random variables $\widehat{T}(\theta)$ to be, for each $\theta \geq 0$, stopping times with respect to the filtration $(\mathcal{F}_t)_{t \geq 0}$ can be established in the following way.

The filtration $(\mathcal{F}_t)_{t \geq 0}$ being assumed *right-continuous* ($\mathcal{F}_t = \mathcal{F}_{t+} \equiv \bigcap_{s>t} \mathcal{F}_s$ for any $t \geq 0$), the property "$\{\widehat{T}(\theta) \leq t\} \in \mathcal{F}_t$, $t \geq 0$" is equivalent to the following: "$\{\widehat{T}(\theta) < t\} \in \mathcal{F}_t$, $t \geq 0$". Indeed, from the definition (1.1) we see that

$$\{\widehat{T}(\theta) < t\} = \bigcup_{s<t,\, s\in\mathbb{Q}} \{A_s > \theta\} \in \mathcal{F}_t,$$

where $\mathbb{Q}$ is the set of the rational numbers on $\mathbb{R}_+ = [0, \infty)$. So, for each $t \geq 0$ the times $\widehat{T}(\theta)$ are Markov times relative to the filtration $(\mathcal{F}_t)_{t \geq 0}$, *i.e.*, the property (b) does hold. □

4. Since $\widehat{T}(\theta)$ is nondecreasing in θ, the limit $\lim_{\tilde{\theta} \uparrow \theta} \widehat{T}(\tilde{\theta})$ $(= \widehat{T}(\theta-))$ exists, and

$$\widehat{T}(\theta-) = \inf\{t : A_t \geq \theta\}. \tag{1.2}$$

Lemma 1.2. *If* $\widehat{T} = (\widehat{T}(\theta))_{\theta \geq 0}$ *is defined by* (1.1), *then*

$$A_t = \inf\{\theta : \widehat{T}(\theta) > t\}, \tag{1.3}$$

and, for all $t \geq 0$, *the random variables* A_t *are* $(\widehat{\mathcal{F}}_\theta)_{\theta \geq 0}$*-stopping times (Markov times), where*

$$\widehat{\mathcal{F}}_\theta = \mathcal{F}_{\widehat{T}(\theta)}. \tag{1.4}$$

Proof. Recall, first of all, that if τ is a Markov time with respect to the filtration $(\mathcal{F}_t)_{t \geq 0}$, then $\mathcal{F}_\tau$ is the σ-algebra of the sets $A \in \mathcal{F}$ for which $A \cap \{\tau \leq t\} \in \mathcal{F}_t$ for all $t \geq 0$.

To prove formula (1.3), observe that if, for given $\theta \in [0, \infty)$, we have $\widehat{T}(\theta) > t$, then, obviously, $A_t \leq \theta$, and so $A_t \leq \inf\{\theta : \widehat{T}(\theta) > t\}$. On the other hand, $\widehat{T}(A_t) \geq t$ for each $t \in [0, \infty)$, and so $\widehat{T}(A_{t+\varepsilon}) \geq t + \varepsilon > t$. Therefore $A_{t+\varepsilon} \geq \inf\{\theta : \widehat{T}(\theta) > t\}$, whence, $A = (A_t)_{t \geq 0}$ being right-continuous, $A_t \geq \inf\{\theta : \widehat{T}(\theta) > t\}$. Together with the already established inverse inequality this proves the required property (1.3).

Further, (1.3) implies that

$$\{A_t < \theta\} = \bigcup_{s<\theta,\, s\in\mathbb{Q}} \{\widehat{T}(s) > t\} \in \widehat{\mathcal{F}}_\theta = \mathcal{F}_{\widehat{T}(\theta)}. \tag{1.5}$$

Now notice that the filtration of σ-algebras $(\widehat{\mathcal{F}}_\theta)_{\theta\geq 0}$ is right-continuous (if $\theta_n \downarrow \theta$, then $\widehat{T}(\theta_n) \downarrow \widehat{T}(\theta)$ and for the right-continuous family $(\mathcal{F}_t)_{t\geq 0}$ we then have $\widehat{\mathcal{F}}_\theta \equiv \mathcal{F}_{\widehat{T}(\theta)} = \bigcap_n \mathcal{F}_{\widehat{T}(\theta_n)} \equiv \bigcap_n \widehat{\mathcal{F}}_{\theta_n} = \widehat{\mathcal{F}}_{\theta+}$). Therefore (1.5) implies $\{A_t \leq \theta\} \in \mathcal{F}_{\widehat{T}(\theta)}$ which proves that the A_t are $(\widehat{\mathcal{F}}_\theta)_{\theta\geq 0}$-stopping times for all $t \geq 0$. □

5. Recall that when defining (in Subsec. 2) the time change $T = (T(t))_{t\geq 0}$ (by analogy with Definition 1.1 for $\widehat{T} = (\widehat{T}(\theta))_{\theta\geq 0}$) we assumed that a certain filtration $(\widehat{\mathcal{F}}_\theta)_{\theta\geq 0}$ of σ-algebras is chosen on $(\Omega, \mathcal{F}, \mathsf{P})$. Thus, we see that if $\widehat{\mathcal{F}}_\theta = \mathcal{F}_{\widehat{T}(\theta)}$, where $\widehat{T}(\theta)$ is constructed according to (1.1) with the use of the chosen process $A = (A_t)_{t\geq 0}$, then the "dual" change of time $T = (T(t))_{t\geq 0}$ just *coincides* with $A = (A_t)_{t\geq 0}$: $T(t) = A_t$, $t \geq 0$. This property explains why the changes of time T and $\widehat{T}$ are often said to be *mutually inverse* (dual): $\widehat{T}$ may be constructed starting from T by $\widehat{T}(\theta) = \inf\{t : T(t) > \theta\}$, and then T may be restored from $\widehat{T}$ as $T(t) = \inf\{\theta : \widehat{T}(\theta) > t\}$.

6. In the sequel when considering random processes $X = (X_t)_{t\geq 0}$, given on a filtered probability space $(\Omega, \mathcal{F}, (\mathcal{F}_t)_{t\geq 0}, \mathsf{P})$, we will always assume them to be *progressively measurable* (see the definition (3.52)). This assumption guarantees that, for the change of time $\widehat{T} = (\widehat{T}(\theta))_{\theta\geq 0}$, the "composition" process $\widehat{X} = X \circ \widehat{T}$, that is the process

$$\widehat{X}_\theta = X_{\widehat{T}(\theta)}, \qquad \theta \geq 0,$$

is $(\widehat{\mathcal{F}}_\theta)$-adapted, in other words, the random variables $\widehat{X}_\theta$ are $\widehat{\mathcal{F}}_\theta$-measurable for each $\theta \geq 0$, where $\widehat{\mathcal{F}}_\theta = \mathcal{F}_{\widehat{T}(\theta)}$. (See Lemma 1.8 in [125].)

1.2 Some Properties of Change of Time

1. Let a change of time $\widehat{T} = (\widehat{T}(\theta))_{\theta\geq 0}$ be determined starting from a generating process $A = (A_t)_{t\geq 0}$ (see Subsec. 3 in Sec. 1.1) by the formula $\widehat{T}(\theta) = \inf\{t : A_t > \theta\}$ and $T(t) = A_t$. The following properties may be verified immediately:

(a) If the process $A = (A_t)_{t\geq 0}$ is *continuous* and σ is a Markov time with respect to $(\mathcal{F}_t)$, then

$$\mathcal{F}_\sigma \subseteq \widehat{\mathcal{F}}_{A_\sigma} = \widehat{\mathcal{F}}_{T(\sigma)} = \mathcal{F}_{\widehat{T}(T(\sigma))};$$

(b) if the process $A = (A_t)_{t\geq 0}$ is *continuous* and *strictly increasing*, then

$$\widehat{T}(T(t)) = t,$$
$$T(\widehat{T}(\theta)) = \theta,$$
$$\widehat{T}(\theta) = T^{-1}(\theta),$$
$$T(t) = \widehat{T}^{-1}(t),$$

and if σ is an $(\mathcal{F}_t)$-Markov time, then

$$\mathcal{F}_\sigma = \widehat{\mathcal{F}}_{A_\sigma} = \widehat{\mathcal{F}}_{T(\sigma)} = \mathcal{F}_{\widehat{T}(T(\sigma))};$$

(c) if the process $A = (A_t)_{t\geq 0}$ is *continuous*, *strictly increasing* and $A_\infty = \infty$ (P-a.s.), then the corresponding process $\widehat{T} = (\widehat{T}(\theta))_{\theta\geq 0}$ is continuous and strictly increasing as well and its life time $\widehat{\zeta} = \infty$ (P-a.s.).

2. From the point of view of mathematical analysis the change of time in (usual, *i.e.*, nonstochastic) Lebesgue–Stieltjes integrals is a well-known technique. For example, let $A = (A(t))_{t\geq 0}$, $A(0) = 0$, be a nonrandom increasing continuous function. Put (for the symmetry of notation) $\widehat{A}(\theta) = \inf\{t : A(t) > \theta\}$. Then, by Lemma 1.2, $A(t) = \inf\{\theta : \widehat{A}(\theta) > t\}$ (so that $\widehat{A}(\theta) = A^{-1}(\theta)$ and $A(t) = \widehat{A}^{-1}(t)$), and if $\varphi = \varphi(t)$ is a nonnegative Borel function on $[0,\infty)$, we have

$$\int_0^{\widehat{A}(b)} \varphi(t)\, dA(t) = \int_0^b \varphi(\widehat{A}(\theta))\, d\theta, \tag{1.6}$$

since, clearly,

$$\int_0^{\widehat{A}(b)} \varphi(t)\, dA(t) = \int_0^b \varphi(\widehat{A}(\theta))\, dA(\widehat{A}(\theta))$$

and $A(\widehat{A}(\theta)) = \theta$. (The identity (1.6) remains valid if $(A(t))_{t\geq 0}$ is only right-continuous.)

Note that (1.6) could also be rewritten (by symmetry) in the form:

$$\int_0^{A(a)} \varphi(\theta)\, d\widehat{A}(\theta) = \int_0^a \varphi(A(t))\, dt. \tag{1.7}$$

3. Let us give a stochastic generalization of (1.6). We shall use the notation of Sec. 1.1 and the following definition: we say that a process G is $\widehat{T}$-*continuous* if G is a constant on each interval $[\widehat{T}(\theta-), \widehat{T}(\theta)]$.

Lemma 1.3. *Let $\varphi = \varphi(t, \omega)$ be a progressively measurable (with respect to the filtration $(\mathcal{F}_t)_{t \geq 0}$) nonnegative random function, and let $G = G(t, \omega)$ be an $\mathcal{F}_t$-measurable (in ω for each $t \geq 0$), right-continuous (in t for each $\omega \in \Omega$) process with bounded variation which is $\widehat{T}$-continuous with respect to the change of time $\widehat{T} = (\widehat{T}(\theta))_{\theta \geq 0}$. Then, identically for all $\omega \in \Omega$,*

$$\int_0^{\widehat{T}(b)} \varphi(t, \omega)\, dG(t, \omega) = \int_0^b \varphi(\widehat{T}(\theta), \omega)\, dG(\widehat{T}(\theta), \omega), \tag{1.8}$$

or, in symbolic form,

$$\widehat{\varphi \cdot G} = \widehat{\varphi} \cdot \widehat{G} \tag{1.9}$$

(with the above accepted notation $\widehat{X} = X_{\widehat{T}}$ and the notation $\varphi \cdot G$ for the integral $\int_0^{\cdot} \varphi\, dG$).

By symmetry with (1.7) the following formula

$$\int_0^{T(a)} \varphi(\theta, \omega)\, dG(\theta, \omega) = \int_0^a \varphi(T(t), \omega)\, dG(T(t), \omega) \tag{1.10}$$

holds.

Proof. All formulae given above can be established in a standard way: first we make sure of their correctness for *simple* functions, and then, using the results from the range of "theorems on monotone classes" (see, for example, [146; Chap. 0, § 2 and § 4]), we prove that they are true under the formulated conditions. Let us dwell only on the proof of Equation (1.6), assuming that $b = \infty$ and $\widehat{A}(b) = \infty$, *i.e.*, that $A(\infty) = \infty$. We must show that, for nonnegative Borel functions $\varphi = \varphi(t)$,

$$\int_0^\infty \varphi(t)\, dA(t) = \int_0^\infty \varphi(\widehat{A}(\theta))\, d\theta. \tag{1.11}$$

Let $\varphi(t) = I_{[0,v]}(t)$. Then, on the one hand,

$$\int_0^\infty \varphi(t)\, dA(t) = A(v)$$

and, on the other hand,

$$\begin{aligned}\int_0^\infty \varphi(\widehat{A}(\theta))\, d\theta &= \int_0^\infty I_{[0,v]}(\widehat{A}(\theta))\, d\theta \\ &= \int_0^\infty I\{\theta : \widehat{A}(\theta) \leq v\}\, d\theta = \mathrm{Leb}\{\theta : \widehat{A}(\theta) \leq v\},\end{aligned}$$

where Leb is the linear Lebesgue measure. But

$$\mathrm{Leb}\{\theta : \widehat{A}(\theta) \leq v\} = \inf\{\theta : \widehat{A}(\theta) > v\} = A(v).$$

Thus, the formula (1.11) is correct for indicator functions $\varphi(t) = I_{[0,v]}(t)$, and, therefore, for the functions $\varphi(t) = I_{(u,v]}(t)$ and for functions (called simple), which are linear combinations of indicators of the form $I_{(u,v]}(t)$.

If a function $\varphi = \varphi(t)$ is measurable and nonnegative, then there exists a sequence of simple functions $\varphi^{(n)} = \varphi^{(n)}(t)$, $n \geq 1$, such that $\varphi^{(n)}(t) \uparrow \varphi(t)$ pointwise. Then, by the theorem on monotone convergence in the Lebesgue–Stieltjes integral, we get that the property (1.11) holds for each measurable nonnegative function $\varphi = \varphi(t)$.

Statements (1.8), (1.10) can be established, as we already mentioned, in an analogous way. □

Remark 1.1. In what follows the argument ω of random functions $\varphi(t,\omega)$, $G(t,\omega)$, *etc.*, will be omitted in accordance with common (in the theory of stochastic processes) conventions.

Corollary 1.1. *Assume that the process $A = (A_t)_{t\geq 0}$, generating the change of time $\widehat{T}$* (see (1.1)), *is continuous and strictly increasing. Let $\varphi(t) = \Phi(T(t))$, where $T(t) = A_t$, $t \geq 0$, and $\Phi = \Phi(s)$, $s \geq 0$, is a Borel function on $[0,\infty) = \mathbb{R}_+$. Then*

$$\int_0^{\widehat{T}(b)} \Phi(T(t))\, dG(t) = \int_0^b \Phi(\theta)\, dG(\widehat{T}(\theta)). \tag{1.12}$$

If, in addition, $G(t) = t$, then

$$\int_0^{\widehat{T}(b)} \Phi(T(t))\, dt = \int_0^b \Phi(\theta)\, d\widehat{T}(\theta), \tag{1.13}$$

and if $G(t) = T(t)$, then

$$\int_0^{\widehat{T}(b)} \Phi(T(t))\, dT(t) = \int_0^b \Phi(\theta)\, d\theta, \tag{1.14}$$

or, equivalently,

$$\int_0^a \Phi(T(t))\, dT(t) = \int_0^{T(a)} \Phi(\theta)\, d\theta. \tag{1.15}$$

Now let $\varphi(\theta) = \Phi(\widehat{T}(\theta))$. Then

$$\int_0^{T(a)} \Phi(\widehat{T}(\theta))\, dG(\theta) = \int_0^a \Phi(t)\, dG(T(t)). \tag{1.16}$$

If, in addition, $G(\theta) = \theta$, then

$$\int_0^{T(a)} \Phi(\widehat{T}(\theta))\, d\theta = \int_0^a \Phi(t)\, dT(t), \tag{1.17}$$

and if $G(\theta) = \widehat{T}(\theta)$, *then*

$$\int_0^{T(a)} \Phi(\widehat{T}(\theta))\, d\widehat{T}(\theta) = \int_0^a \Phi(t)\, dt, \tag{1.18}$$

or, equivalently,

$$\int_0^b \Phi(\widehat{T}(\theta))\, d\widehat{T}(\theta) = \int_0^{\widehat{T}(b)} \Phi(t)\, dt. \tag{1.19}$$

Corollary 1.2. *If* $h = h(\theta)$ *is a nondecreasing continuous (deterministic) function, then* (1.19) *implies that*

$$\int_{h(a)}^{h(b)} \Phi(t)\, dt = \int_a^b \Phi(h(\theta))\, dh(\theta). \tag{1.20}$$

If, in addition, $dh(\theta) = h'(\theta)\, d\theta$, *then*

$$\int_{h(a)}^{h(b)} \Phi(t)\, dt = \int_a^b \Phi(h(\theta))h'(\theta)\, d\theta. \tag{1.21}$$

The formulae (1.20) and (1.21) bear in Analysis the name *formulae of change of variables in the Lebesgue–Stieltjes integral* or *formulae of integration by substitution* (see, for example, [160; Chap. II, § 6]).

1.3 Representations in the Weak Sense ($X \overset{\text{law}}{=} \widehat{X} \circ T$), in the Strong Sense ($X = \widehat{X} \circ T$) and the Semi-strong Sense ($X \overset{\text{a.s.}}{=} \widehat{X} \circ T$). I. Constructive Examples

1. We have said above that one of the primary problems of time change is how to construct, starting with a given process X, a "simple" process $\widehat{X}$ and a change of time T so as to have one (or more) of the following properties:

$$X = \widehat{X} \circ T, \qquad X \overset{\text{a.s.}}{=} \widehat{X} \circ T, \qquad X \overset{\text{law}}{=} \widehat{X} \circ T.$$

From the point of view of the general theory of stochastic processes, we ought to recognize that the natural candidates for "simple" processes are Brownian motion (Wiener process), Poisson process and, more generally, Lévy processes.

The exceptional role of *Brownian motion* can be explained first of all by the following facts:

a) The *Central Limit Theorem* of probability theory which reveals the universal role of the normal (Gaussian) distribution as the limit distribution for sums of independent (or weakly dependent) small random variables;
b) The *Functional Limit Theorem* ("invariance principle") which shows the universal role of Brownian motion and Wiener measure on the functional space $C[0,\infty)$ of continuous functions when we consider weak limits of sequences of continuous processes;
c) The possibility to represent a wide class of processes (in fact, semimartingales) X in the form $X \stackrel{\text{law}}{=} \widehat{X} \circ T$ (perhaps on different probability spaces; see "Monroe's theorem" in Chap. 8) for a certain Brownian motion $\widehat{X}$ and change of time T.

If we turn to Mathematical Finance, it was Brownian motion that underlaid the initial formulation of models for dynamics of financial asset prices $S = (S_t)_{t\geq 0}$:

(i) *Bachelier model* [4], assuming that prices $S = (S_t)_{t\geq 0}$ have the representation

$$S_t = S_0 + \mu t + \sigma B_t, \tag{1.22}$$

where $B = (B_t)_{t\geq 0}$ is a Brownian motion;

(ii) *Samuelson model* [152], used by Black, Merton, and Scholes in the theory of pricing of options, in which not the prices $S = (S_t)_{t\geq 0}$ but their *logarithms* $\log(S_t/S_0)$ have the representation:

$$\log \frac{S_t}{S_0} = \mu t + \sigma B_t. \tag{1.23}$$

If one takes account of "Monroe's theorem" (Chap. 8), assuming that prices $S = (S_t)_{t\geq 0}$ are *semimartingales* (in agreement with the assumption of markets; see Chap. 10), then the suggestion to consider the following two models instead of (1.22) and (1.23) is quite natural: quite natural:

$$S_t = S_0 + \mu T(t) + \sigma B_{T(t)} \tag{1.24}$$

and

$$\log \frac{S_t}{S_0} = \mu T(t) + \sigma B_{T(t)} \tag{1.25}$$

with a Brownian motion B and a "change of time" $T = (T(t))_{t\geq 0}$, which in financial works is interpreted as "operational", or "business", time and is closely related to the well-known concept of "volatility".

2. It is quite clear that in the above-mentioned "Monroe's theorem", which gives the representation $X = B \circ T$ for a semimartingale X, the *change of time* T can turn out to be highly intricate. This is caused by the fact that if X has *jump* components, then, because of the continuity of trajectories of Brownian motion $B = (B_\theta)_{\theta \geq 0}$, the process $T = (T(t))_{t\geq 0}$ must have discontinuous trajectories. Thus, accepting the Brownian motion as a "simple" process, we may find, at the same time, in the representation $X = B \circ T$, a process T with "nonsimple" structure.

Thereupon it would be quite natural to consider also representations $X = \widehat{X} \circ T$ with some $\widehat{X}$ different from the Brownian motion, for example, $\widehat{X}$ could be some "nice" *Lévy process.* Such considerations are rather natural, if we suppose that the initial process X is "close" to a Lévy process. Then one can expect the change of time T in the representation $X = \widehat{X} \circ T$ to be not too sophisticated.

3. The facts, stated above, make it clear that getting representations of the form $X = \widehat{X} \circ T$, for a given process X and selected process $\widehat{X}$, is generally not an easy task. Therefore it is natural to take another route, namely, to try to understand, at which classes of processes X we can arrive starting from given $\widehat{X}$ and T.

Many models of such type are accumulated in the literature. Below we present the main ones by determining constructively the processes $\widehat{X}$ and T.

Example 1.1. Let $(\Omega, \mathcal{F}, \mathsf{P})$ be a probability space, $\widehat{B} = (\widehat{B}_\theta)_{\theta\geq 0}$ a Brownian motion (playing the role of $\widehat{X} = (\widehat{X}_\theta)_{\theta\geq 0}$) and $T = (T(t))_{t\geq 0}$ a *deterministic* nondecreasing right-continuous function, $T(0) = 0$. Then the process $X_t = \widehat{B}_{T(t)}$, $t \geq 0$, is, obviously, a Gaussian right-continuous process with $\mathsf{E}X_t = 0$, $\mathsf{E}X_t^2 = T(t)$ and characteristic function

$$\mathsf{E}e^{iu(X_t - X_s)} = e^{-u^2(T(t)-T(s))/2}. \tag{1.26}$$

Example 1.2. Again we assume that $\widehat{B} = (\widehat{B}_\theta)_{\theta\geq 0}$ is a Brownian motion and $T = (T(t))_{t\geq 0}$ is a deterministic function with the same properties as in the previous example.

Let $f = f(t)$, $g = g(t)$, and $F = F(x)$ be continuous functions ($t \geq 0$, $x \in \mathbb{R}$).

We consider the process $X = (X_t)_{t\geq 0}$ generated by $\widehat{B}$ and T, which has the form:

$$X_t = F\big(f(t) + g(t)\widehat{B}_{T(t)}\big). \tag{1.27}$$

Models of such a type turn out to be useful in relation to representations of solutions to stochastic differential equations by means of "change of time" in (certain) Brownian motions.

For concreteness, let $X = (X_t)_{t \geq 0}$ be a process satisfying the stochastic differential equation (Sec. 3.4) of the Ornstein–Uhlenbeck type:

$$dX_t = (\alpha(t) - \beta(t)X_t)\,dt + \gamma(t)\,dW_t, \qquad X_0 = \text{const}, \tag{1.28}$$

where $W = (W_t)_{t \geq 0}$ is a Wiener process (Brownian motion).

This equation has a unique strong solution (see [125; § 4.4])

$$X_t = g(t)\left[X_0 + \int_0^t \frac{\alpha(s)}{g(s)}\,ds + \int_0^t \frac{\gamma(s)}{g(s)}\,dW_s\right], \tag{1.29}$$

where

$$g(t) = \exp\left\{-\int_0^t \beta(s)\,ds\right\}.$$

Here we assumed that

$$\int_0^t \left|\frac{\alpha(s)}{g(s)}\right| ds < \infty, \qquad \int_0^t \left|\frac{\gamma(s)}{g(s)}\right|^2 ds < \infty, \qquad t \geq 0;$$

these conditions ensure that the right-hand side in (1.29) is well defined.

Put

$$T(t) = \int_0^t \left(\frac{\gamma(s)}{g(s)}\right)^2 ds,$$

and suppose $T(t) \uparrow \infty$ as $t \to \infty$. In the sequel it will be shown (see Sec. 1.4 below) that one can find a "new" Brownian motion $\widehat{B} = (\widehat{B}_\theta)_{\theta \geq 0}$ such that

$$\int_0^t \frac{\gamma(s)}{g(s)}\,dW_s = \widehat{B}_{T(t)}. \tag{1.30}$$

Thus, it results from (1.29) and (1.30) that the process X has the representation of type (1.27):

$$X_t = f(t) + g(t)\widehat{B}_{T(t)} \tag{1.31}$$

with

$$f(t) = g(t)\left[X_0 + \int_0^t \frac{\alpha(s)}{g(s)}\,ds\right]. \tag{1.32}$$

(To look ahead, notice that the Brownian motion $\widehat{B} = (\widehat{B}_\theta)_{\theta \geq 0}$ is constructed as follows:

$$\widehat{B}_\theta = \int_0^{\widehat{T}(\theta)} \frac{\gamma(s)}{g(s)}\,dW_s,$$

where $\widehat{T}(\theta) = \inf\{t : T(t) > \theta\}$; see Sec. 1.4 below.)

Example 1.3. Let a Brownian motion $\widehat{B} = (\widehat{B}_\theta)_{\theta\geq 0}$ and a time change $T = (T(t))_{t\geq 0}$ be both given on $(\Omega, \mathcal{F}, (\widehat{\mathcal{F}}_\theta)_{\theta\geq 0}, \mathsf{P})$. (Each $T(t)$ is a stopping time with respect to $(\widehat{\mathcal{F}}_\theta)_{\theta\geq 0}$.) Assume that $\widehat{B}$ and T are independent (notation: $\widehat{B} \perp\!\!\!\perp T$) and that $T(t) < \infty$ for all $t < \infty$ and $T(\infty) = \lim_{t\to\infty} T(t) = \infty$. Write $\mathcal{F}^T = \sigma(T(t),\, t \geq 0)$, and let $X_t = \widehat{B}_{T(t)} + f(T(t))$, where $f = f(\theta)$ is a continuous deterministic function.

Then

$$\mathsf{E}\Big[e^{iu(X_t - X_s)} \,\Big|\, \mathcal{F}^T\Big] = e^{iu[f(T(t)) - f(T(s))]} \cdot e^{-u^2(T(t)-T(s))/2},$$

which implies that

$$X = \widehat{B} \circ T + f \circ T \tag{1.33}$$

is a process with conditionally Gaussian increments and, consequently,

$$\mathrm{Law}(X_t - X_s) = \mathsf{E}'_{(s,t)} \mathcal{N}(\mu, \sigma^2),$$

where $\mathcal{N}(\mu, \sigma^2)$ is the normal distribution with the parameters μ and σ^2 and $\mathsf{E}'_{(s,t)}$ stands for averaging by distribution of the (random) variables $\mu = f(T(t)) - f(T(s))$ and $\sigma^2 = T(t) - T(s)$.

The following case is of particular importance. Let $f(t) = \mu t$, and let $T = (T(t))_{t\geq 0}$ be a nondecreasing process with *independent increments*, $T(0) = 0$. In this case X is a *Markov* process. Moreover, if the process T is a subordinator, *i.e.*, a *nondecreasing Lévy process*, then X is a *Lévy process* as well. (General statements on the structure of processes of type $X = \widehat{X} \circ T$, where $\widehat{X}$ and T are independent Lévy processes and T is nondecreasing, can be found in Chap. 8.)

Example 1.4. Let us assume that two processes are given on the probability space $(\Omega, \mathcal{F}, (\widehat{\mathcal{F}}_\theta)_{\theta\geq 0}, \mathsf{P})$: a Brownian motion $\widehat{B} = (\widehat{B}_\theta)_{\theta\geq 0}$ (with $\widehat{B}_0 = 0$, $\mathsf{E}\widehat{B}_\theta = 0$, and $\mathsf{E}\widehat{B}^2_\theta = 2\theta$) and a process $T = (T(t))_{t\geq 0}$, which is an $\alpha/2$-stable subordinator, *i.e.*, the nonnegative nondecreasing $\alpha/2$-stable $(0 < \alpha < 2)$ process (Sec. 5.2) with

$$\mathsf{E}e^{-uT(t)} = e^{-tu^{\alpha/2}}, \qquad u \geq 0. \tag{1.34}$$

Under the assumption of *independence* of $\widehat{B}$ and T, we find that

$$\mathsf{E}e^{iu\widehat{B}_{T(t)}} = \mathsf{E}\mathsf{E}\big(e^{iu\widehat{B}_{T(t)}} \,\big|\, T(t)\big) = \mathsf{E}e^{-\frac{u^2}{2}\cdot 2T(t)} = \mathsf{E}e^{-u^2T(t)} = e^{-u^\alpha t}. \tag{1.35}$$

Thus, starting from the independent processes $\widehat{B}$ and T, we get the process X, which is a symmetric α-stable Lévy process. This process underlies the Mandelbrot–Taylor model ([129], 1967), which they proposed for describing the dynamics of (logarithmic) market prices.

Example 1.5. Let again $\widehat{B} = (\widehat{B}_\theta)_{\theta\geq 0}$ be a *standard* Brownian motion (*i.e.*, such that $\mathsf{E}\widehat{B}_\theta = 0$, $\mathsf{E}\widehat{B}_\theta^2 = \theta$). Let τ^1 and τ^2 be nonnegative random variables such that

Law(τ^1) is an *Inverse Gaussian distribution* IG(a,b) with the probability density

$$p_{\tau^1}(t) = \sqrt{\frac{b}{2\pi}}\, e^{\sqrt{ab}}\, t^{-3/2}\, e^{-(at+b/t)/2}, \tag{1.36}$$

where $a \geq 0$, $b > 0$;

Law(τ^2) is a *positive hyperbolic distribution* PH(a,b) (also denoted by $\mathrm{H}^+(a,b)$) with the probability density

$$p_{\tau^2}(t) = \frac{\sqrt{a/b}}{2K_1(\sqrt{ab})} e^{-(at+b/t)/2}, \tag{1.37}$$

where $a > 0$, $b > 0$ and $K_1(x)$ is a modified third-order Bessel function of index 1.

(See Sec. 9.4 for more details about these distributions.) It is shown in [14], [90] that the random variables τ^1 and τ^2 are *infinitely divisible.* Thus it is possible to define two Lévy processes $T^1 = (T^1(t))_{t\geq 0}$ and $T^2 = (T^2(t))_{t\geq 0}$ such that

$$\mathrm{Law}(T^1(1)) = \mathrm{Law}(\tau^1) \quad \text{and} \quad \mathrm{Law}(T^2(1)) = \mathrm{Law}(\tau^2).$$

The Laplace transform of τ^1 has a simple structure:

$$\mathsf{E}e^{-\lambda\tau^1} = \exp\Big\{\sqrt{ab}\,\big(1 - \sqrt{1 + 2\lambda/a}\,\big)\Big\}, \qquad \lambda \geq 0$$

(see (9.59)) and the Lévy measure of τ^1 has density

$$f(y) = \sqrt{\frac{b}{2\pi}}\,\frac{e^{-ay/2}}{y^{3/2}}$$

(see (9.54)).

The corresponding formulae for τ^2 are given in (9.69) and (9.50).

Starting from the processes T^1 and T^2, assumed to be independent of $\widehat{B}$, let us construct the processes $H^1 = (H_t^1)_{t\geq 0}$ and $H^2 = (H_t^2)_{t\geq 0}$:

$$H_t^i = \mu_i t + \beta_i T^i(t) + \widehat{B}_{T^i(t)}, \qquad i = 1, 2.$$

The processes constructed in this way are Lévy processes and have particular names: H^1 is called the *Normal Inverse Gaussian* ($\mathbb{N} \circ \mathrm{IG}$) Lévy process and H^2 the *hyperbolic* ($\mathbb{H}$ or $\mathbb{N} \circ \mathrm{H}^+$) Lévy process. In the financial literature these processes are intensively used in building models of the evolution of (logarithms) of statistical data. (The processes T^1, T^2 and H^1, H^2 will be discussed in Sec. 12.2, where we will show that H^1 and H^2 belong to the class of the *generalized hyperbolic* ($\mathbb{G}$H) Lévy processes.)

Example 1.6. Let $T = (T(t))_{t\geq 0}$ be a *Gamma process*, defined as the nonnegative nondecreasing Lévy process (subordinator) for which the probability density of $T(t)$ is the Gamma probability density function

$$p_{T(1)}(s) = \frac{(a/2)^\nu}{\Gamma(\nu)}\, s^{\nu-1} e^{-as/2}, \qquad a > 0, \quad \nu > 0, \quad s > 0. \tag{1.38}$$

The characteristic function of $T(1)$ has a simple structure:

$$\mathsf{E} e^{iuT(1)} = \Big(1 - i\frac{2\lambda}{a}\Big)^{-\nu}.$$

A straightforward calculation shows that the density $f(y)$ of the Lévy measure of the random variable $T(1)$ is given by

$$f(y) = \frac{\nu e^{-ay/2}}{y}, \qquad y > 0. \tag{1.39}$$

Take a Brownian motion $\widehat{B} = (\widehat{B}_\theta)_{\theta\geq 0}$ and, assuming that $\widehat{B}$ and T are independent, form a new process $H = (H_t)_{t\geq 0}$ with

$$H_t = \mu t + \beta T(t) + \widehat{B}_{T(t)}. \tag{1.40}$$

This process introduced in [128] is a Lévy process and is called the $\mathbb{V}$G ($\mathbb{V}$*ariance Gamma*) process; see also Sec. 12.2. Its characteristic function is

$$\mathsf{E} e^{iuH_t} = e^{i\mu u t}\Big(1 - 2i\frac{\beta u}{a} + \frac{u^2}{a}\Big)^{-\nu t} \tag{1.41}$$

and, for $\mu = 0$, its density is

$$p_{H_t}(x) = \frac{2\exp\{\beta x\}(a/2)^{\nu t}}{\sqrt{2\pi}\,\Gamma(\nu t)}\Big(\frac{x^2}{a+\beta^2}\Big)^{\frac{\nu t}{2}-\frac{1}{4}} K_{\nu t-1/2}\big(|x|\sqrt{a+\beta^2}\,\big), \tag{1.42}$$

where $K_b(x)$ is the modified Bessel function of the third kind with the index b (see (9.46), (9.47) and (9.92)). The Lévy measure $\nu_{H_1}(dx)$ of H_1 for $a = 2\nu$ is given by the formula (see [127])

$$\nu_{H_1}(dx) = \frac{\nu e^{\beta x}}{|x|}\exp\big\{-\sqrt{2\nu+\beta^2}\,|x|\big\}\,dx. \tag{1.43}$$

Notice that from (1.41) it is straightforward that, as $\nu = a/2 \to \infty$,

$$\mathsf{E} e^{iuH_t} \longrightarrow \exp\Big\{iu\widetilde{\mu}t - \frac{u^2 t}{2}\Big\} = \mathsf{E} e^{iu(\widetilde{\mu}t + \widehat{B}_t)},$$

where $\widetilde{\mu} = \mu + \beta$.

Thus for large a the random variables H_t are close (from the standpoint of their distributions) to $\mu t + \widehat{B}_t$. (The $\mathbb{V}$G process is another special case of the $\mathbb{G}$H-processes referred to above; see also Chap. 12.)

Example 1.7. Let $T = (T(t))_{t\geq 0}$ and $\widehat{S} = (\widehat{S}_\theta)_{\theta\geq 0}$ be two independent processes, where $T = (T(t))_{t\geq 0}$ is the Gamma process from the previous example with $a = 2$, $\nu = 1$ and $\widehat{S} = (\widehat{S}_\theta)_{\theta\geq 0}$ is the nonnegative stable process with the Laplace transform

$$\mathsf{E}e^{-u\widehat{S}_\theta} = e^{-\theta u^\alpha} \tag{1.44}$$

with $0 < \alpha < 1$.

Let $X_t = \widehat{S}_{T(t)}$, $t \geq 0$. We find directly that

$$\begin{aligned}\mathsf{E}e^{-uX_t} &= \mathsf{E}e^{-u\widehat{S}_{T(t)}} = \mathsf{E}\mathsf{E}\big(e^{-u\widehat{S}_{T(t)}} \,\big|\, T(t)\big) = \mathsf{E}e^{-T(t)u^\alpha} \\ &= \frac{1}{\Gamma(t)}\int_0^\infty e^{-su^\alpha} s^{t-1}e^{-s}\,ds = (1+u^\alpha)^{-t}.\end{aligned} \tag{1.45}$$

Such a process $X = (X_t)_{t\geq 0}$ is called a *Mittag-Leffler process*, as is justified by the fact (see [140], 1990) that for $0 < \alpha < 1$ the well-known Mittag-Leffler function

$$F_\alpha(x) = \sum_{k=0}^\infty (-1)^k \frac{x^{\alpha(k+1)}}{\Gamma(1+\alpha(k+1))}, \qquad x \geq 0, \tag{1.46}$$

is a distribution function (on $[0,\infty)$) with the Laplace transform $(1+u^\alpha)^{-1}$ (cf. (1.45)).

1.4 Representations in the Weak Sense ($X \overset{\text{law}}{=} \widehat{X} \circ T$), Strong Sense ($X = \widehat{X} \circ T$) and the Semi-strong Sense ($X \overset{\text{a.s.}}{=} \widehat{X} \circ T$). II. The Case of Continuous Local Martingales and Processes of Bounded Variation

1. In the previous section while considering the representation $X = \widehat{X} \circ T$ we proceeded in the following direction: from $\widehat{X}$ and T to X. However, the inverse situation is, of course, more interesting: how, given the process X, can one construct the corresponding "simple" process $\widehat{X}$ and (which is also desirable) "simple" time-scale T such that the following representation $X = \widehat{X} \circ T$ holds (identically or almost everywhere).

Below we consider two "classical" representations—for the case of *continuous* local martingales and for that of local martingales obtained by compensation of "one-point" point processes. (For the "one-point" terminology see [100; Chap. II, § 3c].)

2. Let $(\Omega, \mathcal{F}, (\mathcal{F}_t)_{t\geq 0}, \mathsf{P})$ be a stochastic basis, and let $M = (M_t, \mathcal{F}_t)_{t\geq 0}$ be a *continuous* local martingale, $M_0 = 0$. Denote by $\langle M\rangle = (\langle M\rangle_t, \mathcal{F}_t)_{t\geq 0}$

its *quadratic characteristic* (*i.e.*, the nondecreasing continuous process with the property that the process $M^2 - \langle M \rangle$ is a local martingale; the existence of such a process follows from the *Doob–Meyer decomposition* for local submartingales; see [100; Chap. I, § 4a] and below).

In accordance with Sec. 1.1 we let $A_t = \langle M \rangle_t$, $t \geq 0$, and we assume that $\langle M \rangle_\infty$ $(= \lim_{t\to\infty} \langle M \rangle_t) = \infty$. Starting from such a process $A = (A_t, \mathcal{F}_t)_{t\geq 0}$, let us construct a (random) change of time $\widehat{T} = (\widehat{T}(\theta))_{\theta \geq 0}$:

$$\widehat{T}(\theta) = \inf\{t : A_t > \theta\} \tag{1.47}$$

(cf. (1.1)). Also let

$$\widehat{B}_\theta = M_{\widehat{T}(\theta)} \tag{1.48}$$

and

$$\widehat{\mathcal{F}}_\theta = \mathcal{F}_{\widehat{T}(\theta)}. \tag{1.49}$$

Since $\langle M \rangle_\infty = \infty$, the times $\widehat{T}(\theta)$ are finite for all $\theta \geq 0$. We assert that the process $\widehat{B} = (\widehat{B}_\theta, \widehat{\mathcal{F}}_\theta)_{\theta\geq 0}$ is a *Brownian motion.*

From the famous *Doob optional sampling theorem* (or, what is the same, the theorem on preservation of the martingale property under a change of time, [146]) we find that $\widehat{B}$ is a $(\widehat{\mathcal{F}}_\theta)_{\theta\geq 0}$-local martingale. Trivially, $\widehat{B}_0 = 0$.

The process $\widehat{B}$ is continuous as well. This results from the following considerations.

By the construction itself, $\widehat{B}_\theta = M_{\widehat{T}(\theta)}$, and by continuity of M, it is clear that if at the point θ the process $\widehat{T}(\theta)$ is continuous, then the same is true at this point for the process $\widehat{B}$.

Some difficulties arise, however, when the point θ is a discontinuity point of $\widehat{T}$, *i.e.*, $\widehat{T}(\theta-) = \lim_{s\uparrow\theta} \widehat{T}(s) < \widehat{T}(\theta)$. Let us consider the interval $[\widehat{T}(\theta-), \widehat{T}(\theta)]$. For all t within this interval the process A_t $(= \langle M \rangle_t)$ does not change its values, *i.e.*, this interval is an interval of constancy of the function $\langle M \rangle$. It is well known that for continuous local martingales M their intervals of constancy *coincide* with that of the process $\langle M \rangle$. (See, for example, [146; Chap. IV, (1.13)].) This implies that at the point θ of discontinuity of the function $\widehat{T}(\theta)$ the function $\widehat{B}_\theta$ is continuous.

Now let us show that the process $\widehat{B} = (\widehat{B}_\theta)_{\theta\geq 0}$, being, as is mentioned above, a continuous local martingale, is, moreover, a Brownian motion. By the well-known characterization theorem of P. Lévy (see [146; Chap. IV, (3.6)]), to this end it suffices just to verify that the quadratic characteristic $\langle \widehat{B} \rangle_\theta = \theta$.

We have

$$\widehat{B}_\theta = M_{\widehat{T}(\theta)} = \int_0^\infty I(s \le \widehat{T}(\theta))\, dM_s \tag{1.50}$$

whence, by the properties of stochastic integrals,

$$\langle \widehat{B} \rangle_\theta = \int_0^\infty I(s \le \widehat{T}(\theta))\, d\langle M \rangle_s = \langle M \rangle_{\widehat{T}(\theta)}. \tag{1.51}$$

Now notice that whenever $\langle M \rangle_t$ is strictly increasing, the property (b) ($T(\widehat{T}(\theta)) = \theta$) in Subsec. 1 of Sec. 1.2 implies that $\langle M \rangle_{\widehat{T}(\theta)} = \theta$. As to the intervals of constancy of $\langle M \rangle$, on them this equality is straightforward from the definition of the time $\widehat{T}(\theta)$.

Thus, starting from the quadratic characteristic $\langle M \rangle$ we have constructed a "simply" organized process $\widehat{B}$, which is a Brownian motion. Now let us show that in fact the initial process M can be represented by means of the process $\widehat{B}$ in the following way:

$$M_t = \widehat{B}_{\langle M \rangle_t}, \qquad t \ge 0, \tag{1.52}$$

i.e., $M = \widehat{B} \circ \langle M \rangle$, or, in more usual notation,

$$M = \widehat{B} \circ T,$$

where $T = (T(t))_{t \ge 0}$ with $T(t) = \langle M \rangle_t$.

This results from the equalities

$$\widehat{B}_{T(t)} = \widehat{B}_{\langle M \rangle_t} = M_{\widehat{T}(\langle M \rangle)_t} = M_{\widehat{T}(T(t))}.$$

If we now assume that the process $T(t)$ ($= \langle M \rangle_t$) is strictly increasing, then $\widehat{T}(T(t)) = t$ by the first property (b) in Subsec. 1 in Sec. 1.2. Consequently, $\widehat{B}_{T(t)} = M_t$. But if $T(t)$ is not a strictly increasing process, then it may turn out that $\widehat{T}(T(t)) > t$. However, once again $M_{\widehat{T}(T(t))} = M_t$, because the values of M do not change on the intervals of constancy of the quadratic characteristic $\langle M \rangle$, *i.e.*, of the process $T = (T(t))_{t \ge 0}$.

Thus, we have proved the following theorem.

Theorem 1.1 (Dambis, Dubins and Schwarz; see [146]). *Let $M = (M_t, \mathcal{F}_t)_{t \ge 0}$ be a continuous local martingale with $M_0 = 0$ and $\langle M \rangle_\infty = \infty$. Then there exists a Brownian motion $\widehat{B} = (\widehat{B}_\theta)_{\theta \ge 0}$ such that for the change of time $T = \langle M \rangle$,* i.e., *$T(t) = \langle M \rangle_t$, $t \ge 0$, we have the strong representation $M = \widehat{B} \circ T$.*

Remark 1.2. If we assume from the beginning that the process $\langle M \rangle$ is *strictly* increasing, then the proof becomes somewhat simpler, because in this case we need not consider the problem of coincidence of intervals of constancy for M and $\langle M \rangle$.

Remark 1.3. In the above theorem we have assumed $\langle M\rangle_\infty = \infty$, that guaranteed the finiteness of the times $\widehat{T}(\theta)$ for each $\theta \geq 0$. But if $\langle M\rangle_\infty < \infty$ with a positive probability, the theorem on the representability of continuous local martingales remains true, although in a slightly different form.

To give the corresponding formulation we need the following definition.

Definition 1.4. Let $(\Omega, \mathcal{F}, (\mathcal{F}_s)_{s\geq 0}, \mathsf{P})$ be a filtered probability space. We define its *extension* as a filtered probability space $(\widetilde{\Omega}, \widetilde{\mathcal{F}}, (\widetilde{\mathcal{F}}_s)_{s\geq 0}, \widetilde{\mathsf{P}})$, constructed from the auxiliary filtered space $(\Omega', \mathcal{F}', (\mathcal{F}'_s)_{s\geq 0}, \mathsf{P}')$ in the pattern of *direct product* of spaces:

$$\widetilde{\Omega} = \Omega \times \Omega', \quad \widetilde{\mathcal{F}} = \mathcal{F} \otimes \mathcal{F}', \quad \widetilde{\mathcal{F}}_s = \bigcap_{\varepsilon>0} (\mathcal{F}_{s+\varepsilon} \otimes \mathcal{F}'_{s+\varepsilon}), \quad \widetilde{\mathsf{P}} = \mathsf{P} \otimes \mathsf{P}'.$$

If $X = (X_s(\omega))_{s\geq 0}$, $\omega \in \Omega$, is a process defined on $(\Omega, \mathcal{F}, (\mathcal{F}_s)_{s\geq 0}, \mathsf{P})$, it can also be considered as a process $\widetilde{X}$ defined on $(\widetilde{\Omega}, \widetilde{\mathcal{F}}, (\widetilde{\mathcal{F}}_s)_{s\geq 0}, \widetilde{\mathsf{P}})$ by means of the following natural redefinition:

$$\widetilde{X} = (\widetilde{X}_s(\widetilde{\omega}))_{s\geq 0}, \qquad \widetilde{\omega} = (\omega, \omega') \in \widetilde{\Omega}, \tag{1.53}$$

where $\widetilde{X}_s(\widetilde{\omega}) = \widetilde{X}_s(\omega, \omega') = X_s(\omega)$.

It is clear that

$$\mathrm{Law}(\widetilde{X} \mid \widetilde{\mathsf{P}}) = \mathrm{Law}(X \mid \mathsf{P}).$$

Taking into account these definitions we can reformulate the Dambis–Dubins–Schwarz theorem in the following way.

Let $(\Omega', \mathcal{F}', (\mathcal{F}'_\theta)_{\theta\geq 0}, \mathsf{P}')$ be a ("rich" enough) filtered probability space with a Brownian motion $\beta' = (\beta'_\theta(\omega'))_{\theta\geq 0}$ defined on it. Consider also the probability space $(\Omega, \bigvee \widehat{\mathcal{F}}_\theta, (\widehat{\mathcal{F}}_\theta)_{\theta\geq 0}, \mathsf{P})$, where $\widehat{\mathcal{F}}_\theta = \mathcal{F}_{\widehat{T}(\theta)}$, $\widehat{T}(\theta) = \inf\{t : A_t > \theta\}$, $A_t = \langle M\rangle_t$, $\bigvee \widehat{\mathcal{F}}_\theta = \sigma\big(\bigcup_{\theta\geq 0} \widehat{\mathcal{F}}_\theta\big)$, and construct the space $(\widetilde{\Omega}, \widetilde{\mathcal{F}}, (\widetilde{\mathcal{F}}_\theta)_{\theta\geq 0}, \widetilde{\mathsf{P}})$ according to the above definition.

Finally, taking account of (1.53), let

$$\widetilde{B}_\theta = \begin{cases} M_{\widehat{T}(\theta)}, & \theta < A_\infty = \langle M\rangle_\infty, \\ M_\infty + \beta'_\theta - \beta'_{\langle M\rangle_\infty}, & \theta \geq A_\infty = \langle M\rangle_\infty, \end{cases}$$

or, what is the same,

$$\widetilde{B}_\theta = M_{\widehat{T}(\theta)} + \int_0^\theta I(s > \langle M\rangle_\infty)\, d\beta'_s = M_{\widehat{T}(\theta)} + \beta'_\theta - \beta'_{\theta\wedge\langle M\rangle_\infty}.$$

This process $\widetilde{B} = (\widetilde{B}_\theta, \widetilde{\mathcal{F}}_\theta)_{\theta \geq 0}$ is a Brownian motion and the initial process M can be represented (on the extended space) as $M = \widetilde{B} \circ T$, where $T(t) = \langle M \rangle_t$, $t \geq 0$. (See details in [146; Chap. V, § 1], [110; Theorem 16.4].)

Remark 1.4. The Dambis–Dubins–Schwarz theorem admits a multidimensional generalization (F. Knight): *if $M = (M^1, \ldots, M^d)$ is a continuous d-dimensional local martingale with the property that the quadratic covariation $\langle M^i, M^j \rangle = 0$ for $i \neq j$* (see definition of $\langle M^i, M^j \rangle$ in Sec. 3.1) *and $\langle M^i \rangle_\infty = \infty$ for all $i = 1, \ldots, d$, then there exist mutually independent Brownian motions $\widehat{B}^1, \ldots, \widehat{B}^d$ such that*

$$M^i = \widehat{B}^i \circ T^i, \qquad i = 1, \ldots, d,$$

with $T^i = \langle M^i \rangle$.

Notice that in general one does not manage to find a *common* change of time T (so that $M^i = \widehat{B}^i \circ T$ *for all* i). The above statement asserts nothing but the fact that for each component there is *its own* change of time. (See [108], [151; Chap. V, § 1].)

Remark 1.5. Suppose now that the components of the given d-dimensional martingale $M = (M^1, \ldots, M^d)$ are not orthogonal (*i.e.*, generally speaking, $\langle M^i, M^j \rangle \neq 0$ if $i \neq j$). What one can then say about an analogue of the Knight result?

The very natural idea is to orthogonalize first the martingale M to a local martingale $X = (X^1, \ldots, X^d)$ with the orthogonal components (*i.e.*, $\langle X^i, X^j \rangle = 0$ for $i \neq j$). Having such a process, one can then construct a d-dimensional Brownian motion $\widetilde{B}_\theta = (\widetilde{B}^1_\theta, \ldots, \widetilde{B}^d_\theta)$ in a similar way as (before Remark 1.4) the process $\widetilde{B}_\theta$ was constructed using the process M. Further one can check that the column-vector $M = (M^1, \ldots, M^d)$ admits the following representation:

$$M_t = \int_0^t U(s)\, d\widehat{B} \circ \langle M, M \rangle_s,$$

where the column-vector $\widehat{B} \circ \langle M, M \rangle_s$ has the form

$$\widehat{B} \circ \langle M, M \rangle_s = \left(\widetilde{B}^1_{\langle M^1, M^1 \rangle_s}, \ldots, \widetilde{B}^d_{\langle M^d, M^d \rangle_s} \right)$$

and the matrix $U(s)$ is determined as follows. Introduce the notation

$$C(t) = \left(\frac{d\langle M, M \rangle}{da}(t) \right)_{1 \leq i,j \leq n},$$

where $a(t) = \operatorname{tr}\langle M \rangle_t = \sum_{i=1}^n \langle M^i, M^i \rangle_t$. The matrix C is predictable, non-negative and symmetric. So there exist a predictable orthogonal matrix U

(which is just the required matrix) and a predictable diagonal matrix D such that

$$U'CU = D = (d_i)_{1\le i\le n},$$

where all d_i are nonnegative $\mathbb{Q}$-a.s. ($\mathbb{Q}$ is a measure on the predictable σ-field defined by $d\mathbb{Q} = da \times d\mathsf{P}$) and U' stands for the transpose of U. For details of this construction and for the proofs see [78].

3. The following proposition, resulting from Theorem 1.1, has a number of applications related to models of stochastic volatility.

Corollary 1.3. *Let a continuous local martingale be defined as*

$$M_t = \int_0^t \sigma_s\, dB_s, \qquad t \ge 0, \tag{1.54}$$

where $B = (B_s, \mathcal{F}_s)_{s\ge 0}$ is a Brownian motion, $\sigma = (\sigma_s, \mathcal{F}_s)_{s\ge 0}$ is a positive process, $\int_0^t \sigma_s^2\, ds < \infty$ and $\int_0^\infty \sigma_s^2\, ds = \infty$. If

$$\widehat{T}(\theta) = \inf\left\{t : \int_0^t \sigma_s^2\, ds \ge \theta\right\}, \tag{1.55}$$

then the process $\widehat{B}_\theta = M_{\widehat{T}(\theta)}$ is a Brownian motion (with respect to the filtration $(\widehat{\mathcal{F}}_\theta)_{\theta\ge 0}$) and $M = \widehat{B}\circ T$ with $T = (T(t))_{t\ge 0}$, $T(t) = \int_0^t \sigma_s^2\, ds$, i.e.,

$$M_t = \widehat{B}_{\int_0^t \sigma_s^2\, ds}, \qquad t \ge 0. \tag{1.56}$$

4. In the previous subsections 1–3 we considered a problem on representability in the strong sense for the case of *continuous* local martingales. There it was Brownian motion that played the role of a "simple" process. It is interesting, of course, to consider not only *continuous*, but also *discontinuous* processes, without restricting oneself to the assumption of their martingale structure. We begin with considering the case of "one-point" point processes X (in other words, *counting processes*; [125; Chap. 3, § 4], [100; Chap. I, § 4a]). Among them, surely, the Poisson process stands out for its "simplicity", and it is quite natural to expect that, for the problems of representations considered, the Poisson process will play the role analogous to that of the Brownian motion in the case discussed in Subsecs. 2 and 3.

Subsequently we shall proceed to consider the more general case of processes of *bounded variation*.

5. So, let us assume that the process $X = (X_t)_{t\ge 0}$, which is of interest for us, is a *counting process* $N = (N_t, \mathcal{F}_t)_{t\ge 0}$ with the compensator

$A = (A_t, \mathcal{F}_t)_{t\geq 0}$ and the Doob–Meyer decomposition $N = A + M$, where $M = (M_t, \mathcal{F}_t)_{t\geq 0}$ is a local martingale. (The process N is assumed to be "nonexploding", *i.e.*, $N_t < \infty$ (a.s.) for each $t \geq 0$.)

Notice that if M is a square-integrable martingale, then (see [124; Chap. 18, Lemma 18.12])

$$\langle M\rangle_t = \int_0^t (1 - \Delta A_s)\, dA_s = A_t - \sum_{s\leq t} (\Delta A_s)^2. \tag{1.57}$$

Thus, if the compensator A is *continuous*, then $\langle M\rangle = A$. By analogy with the definition (1.47), for this case of *continuous* compensator $A = \langle M\rangle$ let

$$\widehat{T}(\theta) = \inf\{t : \langle M\rangle_t > \theta\}.$$

Suppose that $\langle M\rangle_\infty = \infty$. Let us show that the process $\widehat{N} = (\widehat{N}_\theta)_{\theta\geq 0}$ with

$$\widehat{N}_\theta = N_{\widehat{T}(\theta)} \tag{1.58}$$

(*i.e.*, $\widehat{N} = N \circ \widehat{T}$) is a *standard Poisson process* (with the parameter $\lambda = 1$) and, on the other hand, the initial counting process $N = (N_t)_{t\geq 0}$ admits the following representation (cf. (1.52)):

$$N_t = \widehat{N}_{T(t)}, \qquad t \geq 0, \tag{1.59}$$

where $T(t) = \langle M\rangle_t$.

To prove that the process $\widehat{N}$ is Poissonian (with parameter $\lambda = 1$), it suffices to verify that, for each pair (θ, θ') such that $0 \leq \theta \leq \theta'$, we have

$$\mathsf{E}\left[e^{ia(\widehat{N}_{\theta'} - \widehat{N}_\theta)} \,\middle|\, \widehat{\mathcal{F}}_\theta\right] = \exp\left\{(e^{ia} - 1)(\theta' - \theta)\right\} \qquad (a \in \mathbb{R}). \tag{1.60}$$

(Indeed, if this property is fulfilled, then the process $(\widehat{N}_\theta)_{\theta\geq 0}$ is a process with independent increments whose characteristic function is

$$\mathsf{E} e^{ia(\widehat{N}_{\theta'} - \widehat{N}_\theta)} = \exp\left\{(e^{ia} - 1)(\theta' - \theta)\right\}, \tag{1.61}$$

which is well known to correspond only to the Poisson distribution with the parameter $\lambda = 1$. (See [125; § 18.5].)

We have

$$\begin{aligned} e^{ia\widehat{N}_{\theta'}} - e^{ia\widehat{N}_\theta} &= \sum_{\theta < u \leq \theta'} \left(e^{ia\widehat{N}_u} - e^{ia\widehat{N}_{u-}}\right) \\ &= \sum_{\theta < u \leq \theta'} e^{ia\widehat{N}_{u-}} (e^{ia} - 1)(\widehat{N}_u - \widehat{N}_{u-}) \\ &= (e^{ia} - 1) \int_\theta^{\theta'} e^{ia\widehat{N}_{u-}}\, d\widehat{N}_u. \end{aligned} \tag{1.62}$$

By virtue of (1.8),

$$\int_{\theta}^{\theta'} e^{ia\widehat{N}_{u-}}\, d\widehat{N}_u = \int_{\widehat{T}(\theta)}^{\widehat{T}(\theta')} e^{iaN_{u-}}\, dN_u. \tag{1.63}$$

Therefore

$$\begin{aligned} e^{ia\widehat{N}_{\theta'}} &= e^{ia\widehat{N}_\theta} + (e^{ia}-1)\int_{\widehat{T}(\theta)}^{\widehat{T}(\theta')} e^{iaN_{u-}}\, dN_u \\ &= e^{ia\widehat{N}_\theta} + (e^{ia}-1)\int_{\widehat{T}(\theta)}^{\widehat{T}(\theta')} e^{iaN_{u-}}\, d(N_u - A_u) \\ &\qquad + (e^{ia}-1)\int_{\widehat{T}(\theta)}^{\widehat{T}(\theta')} e^{iaN_{u-}}\, dA_u. \end{aligned} \tag{1.64}$$

The process $(M_u = N_u - A_u, \mathcal{F}_u)_{u\geq 0}$ is a local martingale, and so is the process

$$\left(\int_0^t e^{iaN_{u-}}\, d(N_u - A_u),\ \mathcal{F}_t\right)_{t\geq 0}, \tag{1.65}$$

because the integrand is bounded ($|e^{iaN_{u-}}| = 1$). Now let us assume that this process is not only a local martingale but in fact a *martingale*. Then, by the Doob optional sampling theorem, we find from (1.64) that

$$\mathsf{E}\big(e^{ia\widehat{N}_{\theta'}} \,|\, \widehat{\mathcal{F}}_\theta\big) = e^{ia\widehat{N}_\theta} + (e^{ia}-1)\mathsf{E}\left(\int_{\widehat{T}(\theta)}^{\widehat{T}(\theta')} e^{iaN_{u-}}\, dA_u \,\Big|\, \widehat{\mathcal{F}}_\theta\right). \tag{1.66}$$

Here, in view of the continuity of the process $(A_u)_{u\geq 0}$ and the property (1.8)

$$\begin{aligned} \int_{\widehat{T}(\theta)}^{\widehat{T}(\theta')} e^{ia\widehat{N}_{u-}}\, d(A_u) &= \int_{\widehat{T}(\theta)}^{\widehat{T}(\theta')} e^{ia\widehat{N}_{u}}\, d(A_u) \\ &= \int_{\theta}^{\theta'} e^{iaN_{\widehat{T}(v)}}\, dA_{\widehat{T}(v)} = \int_{\theta}^{\theta'} e^{iaN_{\widehat{T}(v)}}\, dv, \end{aligned} \tag{1.67}$$

since $A_{\widehat{T}(v)} = v$.

From (1.66) and (1.67) we find that

$$\mathsf{E}\big(e^{ia\widehat{N}_{\theta'}} \,|\, \widehat{\mathcal{F}}_\theta\big) = e^{ia\widehat{N}_\theta} + (e^{ia}-1)\int_{\theta}^{\theta'} \mathsf{E}\big(e^{ia\widehat{N}_{v}} \,|\, \widehat{\mathcal{F}}_\theta\big)\, dv. \tag{1.68}$$

If we denote, for $\theta' \geq 0$,

$$V_\theta(\theta') = \mathsf{E}\big(e^{ia\widehat{N}_{\theta'}} \,|\, \widehat{\mathcal{F}}_\theta\big),$$

then (1.68) can be rewritten in the form

$$V_\theta(\theta') = V_\theta(\theta) + (e^{ia}-1)\int_{\theta}^{\theta'} V_\theta(v)\, dv. \tag{1.69}$$

The bounded solution to this equation is

$$V_\theta(\theta') = V_\theta(\theta)\exp\{(e^{ia} - 1)(\theta' - \theta)\},$$

which proves the required formula (1.60), under the assumption that $M = N - A$ is a martingale.

The general case of *local* martingales reduces to the one discussed by passing from M to the localized martingales $M^n = (M_t^n, \mathcal{F}_t)$ for which $|M_t^n| \le n$. (For the details see [125; Theorem 18.10].)

Summarizing the above considerations, we come to the following result (see, for example, [125; Chap. 19]).

Theorem 1.2. *Let $N = (N_t, \mathcal{F}_t)_{t\ge 0}$ be a "one-point" point process with continuous compensator $A = (A_t, \mathcal{F}_t)_{t\ge0}$, $N_0 = 0$, $A_0 = 0$. Let $M = (M_t, \mathcal{F}_t)_{t\ge0}$, where $M_t = N_t - A_t$. If $A_\infty = \infty$, then there exists a standard Poisson process $\widehat{N} = (\widehat{N}_\theta, \widehat{\mathcal{F}}_\theta)_{\theta\ge0}$ (with $\lambda = 1$) such that $N = \widehat{N} \circ T$, $T = A$ and $M = \widehat{M} \circ T$, where $\widehat{M}_t = \widehat{N}_t - t$ is a "Poisson martingale".*

6. Assume now that there are two processes defined on the probability space $(\Omega, \mathcal{F}, \mathsf{P})$:

1) a Lévy process $X = (X_t)_{t\ge0}$, $X_0 = 0$,

and

2) a Brownian motion $\widehat{B} = (\widehat{B}_t)_{t\ge0}$ which does not depend on X.

Let us study the problem of how to construct a nondecreasing process $T = (T(t))_{t\ge0}$, $T(0) = 0$, so that, at least for each fixed $t > 0$, the representation

$$X_t = \widehat{B}_{T(t)} \tag{1.70}$$

holds with probability one.

Remark 1.6. We do not introduce any filtration (*i.e.*, flow of σ-algebras) on the considered probability space. The matter is that the questions of measurability will not be of interest for us.

Assume that the initial *Lévy process* X is *nondecreasing*. Then the natural candidate for the desired process $T = (T(t))_{t\ge0}$ is the process

$$T(t) = \inf\{\theta \ge 0 : \widehat{B}_\theta = X_t\}. \tag{1.71}$$

Indeed, firstly, it is evident that the process $T = (T(t))_{t\ge0}$ is nondecreasing. Secondly, in virtue of continuity of the process $\widehat{B}$ and finiteness P-a.s. of the time $T(t)$ for each $t > 0$, the property (1.70) holds.

The considered setting of the representability problem when we are given not only the process X, but also the process $\widehat{B}$ which does not depend on X, is contained in the paper by B. W. Huff ([93], 1969). In this paper it is shown (Theorem 2) that the process $T = (T(t))_{t \geq 0}$ is a Lévy process as well, and, consequently, by virtue of its nondecreasingness, it is (according to Bochner's terminology) a *subordinator*.

Denote by $\nu_X = \nu_X(dx)$ the Lévy measure of $X = (X_t)_{t \geq 0}$. Then

$$\mathsf{E}e^{iuX_t} = \exp\left\{t\left(iub + \int_0^\infty (e^{iux} - 1)\,\nu_X(dx)\right)\right\}. \tag{1.72}$$

Now let $\nu_T = \nu_T(dx)$ be the Lévy measure of the process $T = (T(t))_{t \geq 0}$. It is shown in [93] that

$$\mathsf{E}e^{iuT(t)} = \exp\left\{t\left(iub + \int_0^\infty (e^{iux} - 1)\,\nu_T(dx)\right)\right\}, \tag{1.73}$$

where

$$\nu_T(dx) = \left[\frac{b}{\sqrt{2\pi}}\,x^{-3/2} + \int_0^\infty \frac{|y|}{\sqrt{2\pi x^3}}\,e^{-y^2/(2x)}\,\nu_X(dy)\right] dx. \tag{1.74}$$

If, for example, $X_t \equiv t$, then the corresponding Lévy measure in the representation $t = \widehat{B}_{T(t)}$ (P-a.s.) is given (according to (1.74)) by the formula

$$\nu_T(dx) = \frac{1}{\sqrt{2\pi}}\,x^{-3/2}\,dx. \tag{1.75}$$

In the above-mentioned paper [93] there are a number of results related to the representation (1.70) for the general case of Lévy processes with the trajectories of bounded variation (not only for subordinators).

Chapter 2

Integral Representations and Change of Time in Stochastic Integrals

2.1 Integral Representations of Local Martingales in the Strong Sense

1. In the previous chapter we considered the change of time as one of the methods which allows one to represent processes X with "complicated" structure in the form $X = \widehat{X} \circ T$, where $\widehat{X}$ is a process with "simple" structure (for example, a Brownian motion $\widehat{B}$). In this chapter we will study another, rather effective, method of "integral representation" of stochastic functionals and stochastic processes X in the form of stochastic integrals: $X = H \cdot \overline{X}$, where $\overline{X}$ is, *e.g.*, a Brownian motion B or another process with "simple" structure, say, a stable process or, more generally, a Lévy process. The stochastic integral representations of type $X = W * (p-q)$, where $p-q$ is a martingale stochastic measure and $W = W(\omega, s, x)$ (see Subsec. 5), will be considered as well.

Stochastic integral representations $X = H \cdot \overline{X}$ are closely related to representations of the type of random change of time $X = \widehat{X} \circ T$, where T is a change of time and $\widehat{X}$ is a process with "simple" structure.

2. Let us begin with some classical results related to representations with respect to a Brownian motion ($\overline{X} = B$).

Let $(\Omega, \mathcal{F}, (\mathcal{F}_t)_{t\geq 0}, \mathsf{P})$ be a stochastic basis, $B = (B_t, \mathcal{F}_t)_{t\geq 0}$ a Brownian motion, $\mathcal{F}^B = (\mathcal{F}_t^B)_{t\geq 0}$ the Brownian filtration (*i.e.*, $\mathcal{F}_t^B = \mathcal{F}_t^+ \vee \mathcal{N}$, where $\mathcal{F}_t^+ = \bigcap_{s>t} \sigma(B_u, u \leq s)$ and $\mathcal{N}$ is the σ-algebra of P-null subsets of $\mathcal{F}$).

One of the first general results on stochastic integral representations for Brownian functionals is given by the following theorem.

Theorem 2.1. *Let $X = X(\omega)$ be an $\mathcal{F}_T^B$-measurable random variable ("Brownian functional").*

1. *If* $\mathsf{E}X^2 < \infty$, *then there exists a stochastic process* $H = (H_t, \mathcal{F}_t^B)_{t \le T}$ *with* $\mathsf{E}\int_0^T H_t^2\,dt < \infty$ *such that* (P-*a.s.*)

$$X = \mathsf{E}X + \int_0^T H_t\,dB_t, \tag{2.1}$$

or, in shorthand form,

$$X = \mathsf{E}X + (H \cdot B)_T. \tag{2.2}$$

The process H *is unique up to an evanescent set* (*see* Sec. 3.1.

2. *If* $\mathsf{E}|X| < \infty$, *then the representation* (2.1) *holds, with a certain process* $H = (H_t, \mathcal{F}_t^B)_{t \le T}$ *such that*

$$\mathsf{P}\left(\int_0^T H_t^2\,dt < \infty\right) = 1. \tag{2.3}$$

3. *If* X *is an* $\mathcal{F}_T^B$-*measurable positive random variable with* $\mathsf{E}X < \infty$, *then there exist a process* $h = (h_t, \mathcal{F}_t^B)_{t \le T}$ *with* $\mathsf{P}(\int_0^T h_t^2\,dt < \infty) = 1$ *and a positive constant* c *such that the* (*multiplicative*) *representation*

$$X = c\exp\left(\int_0^T h_t\,dB_t - \frac{1}{2}\int_0^T h_t^2\,dt\right) \qquad (\mathsf{P}\text{-}a.s.)$$

holds.

4. *In the general case* (i.e., *when nothing but* $|X| < \infty$ (P-*a.s.*) *is assumed*) *one can find a constant* c *and a process* $H = (H_t, \mathcal{F}_t^B)$ *such that*

$$X = c + \int_0^T H_t\,dB_t. \tag{2.4}$$

For a proof see [146], [124], [125].

The following result (on integral representation of local martingales) can be deduced directly from Theorem 2.1 (and *vice versa*).

Theorem 2.2. 1. *Let* $X = (X_t, \mathcal{F}_t^B)_{t \le T}$ *be a square-integrable martingale. Then there exists a process* $H = (H_t, \mathcal{F}_t^B)_{t \le T}$ *with* $\mathsf{E}\int_0^T H_t^2\,dt < \infty$ *such that*

$$X_t = X_0 + \int_0^t H_s\,dB_s, \qquad t \le T, \tag{2.5a}$$

or, in more compact coordinate-free form,

$$X = X_0 + H \cdot B. \tag{2.5b}$$

2. *Let* $X = (X_t, \mathcal{F}_t^B)_{t \le T}$ *be a local martingale. Then representation* (2.5a) *remains true, with a process* $H = (H_t, \mathcal{F}_t^B)_{t \le T}$, *satisfying* (2.3).

3. *Let* $X = (X_t, \mathcal{F}_t^B)_{t \le T}$ *be a positive local martingale,* $X_0 = \text{const} > 0$. *Then there exists a process* $h = (h_t, \mathcal{F}_t^B)_{t \le T}$ *with* $\mathsf{P}\big(\int_0^T h_t^2\,dt < \infty\big) = 1$, *such that* ($\mathsf{P}$-*a.s.*)

$$X_t = X_0 \exp\left\{ \int_0^t h_s\,dB_s - \frac{1}{2}\int_0^t h_s^2\,ds \right\}, \qquad t \le T. \tag{2.6}$$

Remark 2.1. The processes H and h in Theorem 2.2 are unique up to evanescent sets.

Remark 2.2. In the theorems above we assumed that $T < \infty$. In fact the assertion of Theorem 2.1 remains true if $X = X(\omega)$ is an $\mathcal{F}_\infty^B$-measurable random variable ($\mathcal{F}_\infty^B = \sigma\big(\bigcup_{t \ge 0} \mathcal{F}_t^B\big)$). In this case it suffices to put $T = \infty$ in (2.1)–(2.3). (For different definitions and readings of stochastic integrals $(H \cdot B)_\infty$ see the paper [46] devoted to this subject. In our context $(H \cdot B)_\infty$ is treated as stochastic integral of H with respect to B over $[0, \infty)$. See also Sec. 3.3.) Just the same, Theorem 2.2 remains valid for the processes $X = (X_t, \mathcal{F}_t^B)$ defined on $t \in [0, \infty)$. In that case formulae (2.5a) and (2.6) are considered for $t < \infty$.

Remark 2.3. The problem of finding the explicit form of the processes H and h in the above representations is far from easy. Indeed, consider representation (2.5a). The properties of stochastic integrals over Brownian motion (see [146], [100]) imply that the predictable quadratic covariation (Sec. 3.1)

$$\langle X - X_0, B \rangle_t = \int_0^t H_s\,ds.$$

Thus, the process $H = (H_t, \mathcal{F}_t^B)$ can be chosen to be equal to

$$\frac{d\langle X - X_0, B\rangle_t}{dt}. \tag{2.7}$$

But in practice this representation is difficult to use because of the problem of finding $\langle X - X_0, B\rangle_t$.

Remark 2.4. Let us consider the representation (2.1) with assumption that the variables X and $B = (B_t)_{t \le T}$ form a *Gaussian* system. Then (using the theorem on normal correlation; [125; Chap. 11]) one can establish that H_t can be chosen as

$$H_t = \frac{d}{dt}\,\mathsf{E}\,[(X - X_0)B_t].$$

(See more detail in [125; Chap. 5].)

3. Let us outline the main steps of the proof of representation (2.5a). (For the details of this proof as well as for the proofs of other representations in Theorems 2.1 and 2.2 the reader is referred to [124], [145], [123].)

Assume $X = (X_t, \mathcal{F}_t)_{t \le T}$ is a square-integrable martingale given on a filtered probability space $(\Omega, \mathcal{F}, (\mathcal{F}_t)_{t \le T}, \mathsf{P})$. Let $X = (X_t, \mathcal{F}_t)_{t \le T}$ be a Brownian motion given on this stochastic basis. Denote by $\langle X, B \rangle$ the predictable quadratic covariation (see Subsec. 9 in Sec. 3.1) of the processes X and B. The proof of representation (2.5a) is divided into three assertions (A), (B) and (C).

(A) There is a process $f = (f_t, \mathcal{F}_t)_{t \le T}$ with $\int_0^T \mathsf{E} f_t^2 \, dt < \infty$ such that $\langle X, B \rangle = \int_0^t \mathsf{E} f_s^2 \, ds$ (P-a.s.), $t \le T$.

(B) Let f be the process from assertion (A). Introduce $Y_t = X_0 + \int_0^t \mathsf{E} f_s^2 \, dB_s$, $t \le T$. This process is a square-integrable martingale (Subsec. 3 in Sec. 3.3) with $\langle Y \rangle_t = \int_0^t \mathsf{E} f_s^2 \, ds$. Let $Z_t = X_t - Y_t$, $t \le T$. This process is a square-integrable martingale. Since $\langle X, Y \rangle_t = \left\langle X, \int_0^{\cdot} f_s \, dB_s \right\rangle_t = \int_0^t f_s \, d\langle X, B \rangle_s = \int_0^t f_s^2 \, ds$ we deduce that $\langle Z, Y \rangle_t = \langle X - Y, Y \rangle_t = \langle X, Y \rangle_t - \langle Y \rangle_t = 0$. Thus, Z and Y are orthogonal martingales.

(C) Assume now that the σ-algebras $\mathcal{F}_t$ coincide with the σ-algebras $\mathcal{F}_t^B$ generated by the Brownian motion and completed with all P-null sets from $\mathcal{F}$. The property $\langle Z, Y \rangle_t = 0$ established in (B) implies that $\mathsf{E} Z_t \int_0^t f_s^2 \, dB_s = 0$, $T \ge t$. In Theorem 5.5 in [124], it is shown that the latter identity implies in turn that $\mathsf{E} Z_t \prod_{j=1}^n F_j(B_{t_j}) = 0$ for all bounded Borel functions $F_j(x)$ and $0 \le t_1 \le \cdots \le t_n \le t$. Since the Z_t are $\mathcal{F}_t^B$-measurable, the latter property leads to $Z_t = 0$ (P-a.s.), $t \le T$. Consequently, $X_t = Y_t$, which proves representation (2.5a).

4. Here are some examples to illustrate Theorems 2.1 and 2.2.

Example 2.1. Let $S_T = \max_{t \le T} B_t$, and let $M^{(T)} = (M_t^{(T)}, \mathcal{F}_t^B)_{t \le T}$ be the martingale having the representation

$$M_t^{(T)} = \mathsf{E}(S_T \,|\, \mathcal{F}_t^B)$$

("Lévy martingale").

The random variable S_T admits for every $T > 0$ the following representation:

$$S_T = \mathsf{E} S_T + \int_0^T H_t^{(T)} \, dB_t, \tag{2.8}$$

where

$$H_t^{(T)} = 2\left[1 - \Phi\left(\frac{S_t - B_t}{\sqrt{T - t}}\right)\right], \qquad \mathsf{E} S_T = \mathsf{E}|B_T| = \sqrt{\frac{2T}{\pi}}, \tag{2.9}$$

and $S_t = \max_{u \le t} B_u$, $t \le T$ (Φ being the standard normal distribution function).

For $M_t^{(T)}$, we find from (2.8) that

$$M_t^{(T)} = \sqrt{\frac{2T}{\pi}} + \int_0^t H_u^{(T)}\, dB_u.$$

We sketch, following [163], the essential ideas of finding the explicit form of the integrand $H_t^{(T)}$ in (2.8).

We have

$$S_T = \max_{t \le T} B_t = \int_0^\infty I\Big(a < \max_{t \le T} B_t\Big)\, da = \int_0^\infty I(T_a < T)\, da, \qquad (2.10)$$

where $T_a = \inf\{t \ge 0 \colon B_t = a\}$, $a \ge 0$.

Since the stochastic exponential $\mathcal{E}(\lambda)_t = \exp\{\lambda B_t - \lambda^2 t/2\}$ solves (by Itô's formula; see [163]) the equation

$$\mathcal{E}(\lambda)_t = 1 + \lambda \int_0^t \mathcal{E}(\lambda)_s\, dB_s$$

and $\mathcal{E}(\lambda)_{T_a} = \exp\{\lambda a - \lambda^2 T_a/2\}$ (P-a.s.), we find that for all $t > 0$

$$\mathsf{E}\big(e^{-\lambda^2 T_a/2} \,\big|\, \mathcal{F}_t^B\big) = e^{-\lambda a} + \lambda \int_0^{T_a \wedge t} e^{-\lambda(a-B_s)-\lambda^2 s/2}\, dB_s.$$

Letting $t \to \infty$ leads to the representation:

$$e^{-\lambda^2 T_a/2} = e^{-\lambda a} + \lambda \int_0^{T_a} e^{-\lambda(a-B_s)-\lambda^2 s/2}\, dB_s.$$

which implies (see Lemma 1 in [174]) that for any $a > 0$

$$I(T_a < T) = \mathsf{P}(T_a < T) + 2 \int_0^{T_a \wedge T} \varphi_{T-s}(B_s - a)\, dB_s, \qquad (2.11)$$

where $\varphi_t(x) = (2\pi t)^{-1/2}\, e^{-x^2/(2t)}$.

Substituting (2.11) into (2.10), after easy transformations we find that

$$\begin{aligned}
S_T &= \int_0^\infty \mathsf{P}(T_a < T)\, da + 2 \int_0^\infty \left[\int_0^{T_a \wedge T} \varphi_{T-s}(B_s - a)\, dB_s\right] da \\
&= \mathsf{E} S_T + 2 \int_0^T \left[\int_0^\infty \varphi_{T-u}(B_u - a) I(S_u < a)\, da\right] dB_u \\
&= \mathsf{E} S_T + 2 \int_0^T \left[\int_{S_u}^\infty \varphi_{T-u}(B_u - a)\, da\right] dB_u \\
&= \mathsf{E} S_T + 2 \int_0^T \left[1 - \Phi\left(\frac{S_u - B_u}{\sqrt{T-u}}\right)\right] dB_u,
\end{aligned}$$

which provides the required representation (2.8).

Example 2.2. For $a > 0$, let

$$T_{-a} = \inf\{t > 0 : B_t \leq -a\}.$$

Then (see [163])

$$\max_{t \leq T_{-a}} B_t = -\int_0^{T_{-a}} \log\left(1 + \frac{S_t}{a}\right) dB_t. \tag{2.12}$$

Notice that whereas in the first example $\mathsf{E} \max_{t \leq T} B_t < \infty$, in the second one $\mathsf{E} \max_{t \leq T_{-a}} B_t = +\infty$. Indeed, assume that in the latter case $\mathsf{E} \max_{t \leq T_{-a}} B_t < \infty$. Then the process $(B_{t \wedge T_{-a}})_{t \geq 0}$ should be a uniformly integrable martingale (see [146], [100]). So, by the optional sampling theorem (see again [100]), $\mathsf{E} B_{T_{-a}} = 0$ but at the same time, since $B_{T_{-a}} < \infty$ (P-a.s.) we have $\mathsf{E} B_{T_{-a}} = -a$, where $a > 0$. This contradiction shows that $\mathsf{E} \max_{t \leq T_{-a}} B_t = +\infty$.

Example 2.3. Let the process $X = (X_t, \mathcal{F}_t^B)_{t \geq 0}$ be defined by (2.6). Then the Itô formula gives

$$dX_t = X_t h_t \, dB_t \tag{2.13}$$

and so

$$X_t = X_0 + \int_0^t X_s h_s \, dB_s. \tag{2.14}$$

Note that X_0 is constant (P-a.s.) since $\mathcal{F}_0^B = \{\varnothing, \Omega\}$.

5. Before going to the general results on integral representation (in a strong sense) for martingales let us turn to the analogs of Theorems 2.1 and 2.2 for the case when the "generating" process is not a Brownian motion B but a ("nonexploding") point process $N = (N_t, \mathcal{F}_t)_{t \geq 0}$.

Let $\mathcal{F}_t^N = \left[\bigcap_{\varepsilon > 0} \sigma\{N_s, s \leq t + \varepsilon\}\right] \vee \mathcal{N}$, where $\mathcal{N}$ is the sub-σ-algebra which consists of P-null sets from $\mathcal{F}$. With respect to the filtration $\mathcal{F}^N = (\mathcal{F}_t^N)_{t \geq 0}$ the process $N = (N_t)_{t \geq 0}$ is again a point process. For such a process $N = (N_t, \mathcal{F}_t^N)_{t \geq 0}$ consider its Doob–Meyer decomposition (Sec. 3.1)

$$N_t = \Pi_t + A_t,$$

where $\Pi = (\Pi_t, \mathcal{F}_t^N)_{t \geq 0}$ is a local martingale and $A = (A_t, \mathcal{F}_t^N)_{t \geq 0}$ is an increasing (more precisely, nondecreasing) predictable process named the *compensator* (of the process N).

Let $X = (X_t, \mathcal{F}_t^N)_{t \geq 0}$ be a local martingale. The following result is well known (see, for example, [125]).

Theorem 2.3. *One can find a predictable process* $H = (H_t, \mathcal{F}_t^N)_{t\geq 0}$ *with*

$$\mathsf{P}\left(\int_0^t |H_s|\, dA_s < \infty\right) = 1, \qquad t \geq 0,$$

such that for all $t > 0$

$$X_t = X_0 + \int_0^t H_s\, d\Pi_s, \tag{2.15}$$

or, in more compact form,

$$X = X_0 + H \cdot \Pi, \tag{2.16}$$

where the integral is interpreted as a stochastic integral with respect to the local martingale $\Pi = (\Pi_t, \mathcal{F}_t^N)_{t\geq 0}$.

It is interesting to note that the stochastic integral $H \cdot \Pi$, where $\Pi = N - A$, is in fact equal to the difference of two Stieltjes integrals $H \cdot N$ and $H \cdot A$, *i.e.*,

$$H \cdot \Pi = H \cdot N - H \cdot A. \tag{2.17}$$

(See [100; Chap. III, Theorem 4.37].)

For a better understanding of the result of Theorem 2.4 below it is useful to write representations (2.16) and (2.17) in somewhat different form.

To this end let μ be the "time-space" measure of jumps of the process N:

$$\mu(\omega, dt, dx) = \sum_{s>0} \delta_{(s, \Delta N_s(\omega)=1)}(dt, dx), \tag{2.18}$$

where $\delta_{(a)}$ is the Dirac measure, "sitting" at a point a. The compensator of the measure μ will be denoted by ν (see Sec. 3.2). In terms of these notions the processes N and Π can be rewritten in the form

$$N_t = \int_0^t \int_{\mathbb{R}} x\, d\mu \tag{2.19}$$

and

$$\Pi_t = \int_0^t \int_{\mathbb{R}} x\, d(\mu - \nu). \tag{2.20}$$

The integral in (2.19) is interpreted as a pathwise Stieltjes integral, and the integral in (2.20) as a stochastic integral over the "martingale" measure $\mu - \nu$.

If we write (2.20) formally in the form

$$d\Pi_t = \int_{\mathbb{R}} x\, d(\mu - \nu), \tag{2.21}$$

then (2.15) can be represented as

$$X_t = X_0 + \int_0^t \int_{\mathbb{R}} W(\omega, s, x)\, d(\mu - \nu), \tag{2.22a}$$

or, in more compact coordinate-free form,

$$X = X_0 + W * (\mu - \nu), \tag{2.22b}$$

where $W(\omega, s, x) = xH_s(\omega)$.

6. Representations (2.5a) and (2.22a) (or their "compact" forms (2.5b) and (2.22b)) suggest that they might be particular cases of more general representations which in their turn are combinations of expressions like (2.5a) and (2.22a).

In [100; Chap. III, § 4d] the "Fundamental theorem on representation" of local martingales generated be semimartingales is given. Consider a particular case of this theorem.

Let us assume that we are given a filtered probability space $(\Omega, \mathcal{F}, (\mathcal{F}_t)_{t\geq 0}, \mathsf{P})$, which satisfies the "usual" conditions (see [100; Chap. I, § 1a] and Sec. 3.1 below). Let $Y = (Y_t, \mathcal{F}_t)_{t\geq 0}$ be a semimartingale with triplet (B, C, ν) (see Sec. 3.2). Generally speaking, besides the measure P there can exist another measure Q, with respect to which the process $Y = (Y_t, \mathcal{F}_t)_{t\geq 0}$ is a semimartingale as well with the same triplet (B, C, ν). In what follows our main assumption will consist in *uniqueness*—in that sense—of the measure P. (In fact it suffices to assume that the measure P is extremal in the set of all measures which have the property that under each of them the process Y is a semimartingale with the triplet (B, C, ν).)

Let (in the context under consideration) $X = (X_t, \mathcal{F}_t^Y)_{t\geq 0}$ be a local martingale.

Theorem 2.4. *Under the above-described conditions each local martingale $X = (X_t, \mathcal{F}_t^Y)_{t\geq 0}$ admits the following stochastic integral representation:*

$$X = X_0 + H \cdot Y^c + W * (\mu^Y - \nu^Y), \tag{2.23}$$

where Y^c is the continuous martingale part of the semimartingale Y, μ^Y is the measure of the jumps of Y, and ν^Y is the compensator of the measure μ^Y. The predictable functions $H = (H_t(\omega))$ and $W = (W(\omega, t, x,))$ are such that $H \in L^2_{\text{loc}}(\langle Y^c \rangle)$ and $W \in G_{\text{loc}}(\mu)$ (see [100; Chap. II, § 1d]).

Remark 2.5. If we abandon the assumption about extremality of measure P then representation (2.23) is in general invalid. Instead of it the following representation will take place:

$$X = X_0 + H \cdot Y^c + W * (\mu^Y - \nu^Y) + N, \tag{2.24}$$

where N is a local martingale, which meets certain conditions of "orthogonality" to Y^c and to μ^Y (see Lemma 4.24 in Chap. III of [100]). Unfortunately, this representation (2.24) does not give much, because we are not able to describe effectively the process N.

7. Let us dwell on an important particular case of Theorem 2.4. Assume that the considered probability space is canonical and the semimartingale Y is a (continuous in probability) process with independent increments. The triplet (B, C, ν) of such a process is deterministic, so, by the Kolmogorov–Lévy–Khinchin formula, the measure P is determined uniquely by the collection (B, C, ν) (see Chap. 5). The measure μ, usually denoted by $p = p(\omega, dt, dx)$, is a Poisson random measure; its compensator is commonly denoted by $q = q(dt, dx)$. The process Y^c is Gaussian, so its quadratic characteristic $\langle Y^c \rangle_t$ coincides with the variance $\mathsf{D}Y_t^c$. The canonical representation (Sec. 3.2) of the process $Y = (Y_t)_{t \geq 0}$ itself has the form

$$Y_t = B_t + Y_t^c + \int_0^t \int_{|x| \leq 1} x \, d(p - q) + \int_0^t \int_{|x| \geq 1} x \, dp, \tag{2.25}$$

which can be rewritten also in the form

$$Y_t = B_t + Y_t^c + \int_{|x| \leq 1} x \, (p - q)((0, t], dx) + \int_{|x| > 1} x \, p((0, t], dx). \tag{2.26}$$

(Recall that in the considered case $q = \mathsf{E}p$.)

Note also that Theorem 2.4 implies that every local martingale $X = (X_t, \mathcal{F}_t^Y)_{t \geq 0}$ admits the following representation:

$$X_t = X_0 + \int_0^t H_s \, dY_t^c + \int_0^t \int W(\omega, s, x) \, d(p - q), \tag{2.27}$$

where the integrability conditions imposed on H and W are as follows: for any $t > 0$ (P-a.s.)

$$\int_0^t H_s^2 \, d\langle Y^c \rangle_s < \infty, \qquad \int_0^t \int W^2(\omega, s, x) \, dq < \infty. \tag{2.28}$$

2.2 Integral Representations of Local Martingales in a Semi-strong Sense

In Sec. 2.1, when considering the representations of local martingales X, we assumed that they are generated by a certain semimartingale Y given *a priori*. Now we will take a somewhat different point of view and pose the question as follows.

Let the filtered probability space $(\Omega, \mathcal{F}, (\mathcal{F}_t)_{t\geq 0}, \mathsf{P})$ be endowed with a local martingale $X = (X_t, \mathcal{F}_t)_{t\geq 0}$. The question is as to whether it is possible to obtain, in some sense, for this local martingale a representation, say of type (2.27) with certain Y^c, p, q, and functions H and W, which might be given on an extension of the initial probability space.

Let $X = X^c + X^d$ be a decomposition of the local martingale X into its continuous, X^c, and purely discontinuous, X^d, components.

If X^c has the representation

$$X_t^c = \int_0^t H_s\, dB_s, \tag{2.29}$$

where B is a Brownian motion, then the quadratic characteristic

$$\langle X^c \rangle_t = \int_0^t H_s^2\, ds \quad (= A_t). \tag{2.30}$$

Thus it becomes clear that to guarantee the representation (2.29), we must demand X^c to meet condition (2.30).

Now consider the purely discontinuous martingale part X^d. By means of the measure of jumps μ^X and its compensator ν^X the variables X_t^d can be represented (see details in [100]) in the form

$$X_t^d = \int_0^t \int x\, d(\mu^X - \nu^X). \tag{2.31}$$

As for our aim, it consists in finding a process like the last term in formula (2.26) from Sec. 2.1, which, at least in distribution, coincides with the process X_t^d, $t \geq 0$.

In addition to (2.30) we will assume that the compensator ν^X of the measure μ^X is such that there exist a σ-finite measure $q = q(dt, dx)$ (on $(\mathbb{R}_+ \times \mathbb{R}, \mathcal{B}(\mathbb{R}_+) \otimes \mathcal{B}(\mathbb{R} \setminus \{0\})))$ with $q(\{t\} \times \mathbb{R}) = 0$ for all $t > 0$, and a predictable function $W(\omega, s, x)$ such that for each nonnegative measurable function $g = g(s, x)$

$$\int_0^\infty \int g(s,x)\, \nu(\omega, ds, dx) = \int_0^\infty \int g(s, W(\omega, s, x))\, q(ds, dx). \tag{2.32}$$

Theorem 2.5 (see [107]). *Under assumptions* (2.30) *and* (2.32) *one can construct on a filtered probability space* $(\widetilde{\Omega}, \widetilde{\mathcal{F}}, (\widetilde{\mathcal{F}}_t)_{t\geq 0}, \widetilde{\mathsf{P}})$, *which is an extension* (see Definition 1.4) *of the filtered probability space* $(\Omega, \mathcal{F}, (\mathcal{F}_t)_{t\geq 0}, \mathsf{P})$, *a continuous Gaussian martingale* $m = (m_t)_{t\geq 0}$ *with* $\langle m \rangle_t = A_t$, *and an integer-valued random measure* $p = p(\widetilde{\omega}, dt, dx)$ *with deterministic compensator* $q = q(dt, dx)$, *such that*

$$X_t(\widetilde{\omega}) = \int_0^t H_s(\widetilde{\omega})\, dm_s + \int_0^t \int W(\widetilde{\omega}, s, x)\, d(p - q) \qquad (\widetilde{\mathsf{P}}\text{-a.s.}), \tag{2.33}$$

where for $\widetilde{\omega} = (\omega, \omega')$

$$X_t(\widetilde{\omega}) = X_t(\omega), \quad H_t(\widetilde{\omega}) = H_t(\omega), \quad \dots .$$

In the next section we will show that under certain special assumptions on the structure of the compensator $q(dt, dx)$ the double integration in (2.33) can be reduced to a simpler single-fold integration.

2.3 Stochastic Integrals Over the Stable Processes and Integral Representations

1. Assume that local martingale X is purely discontinuous and has a representation

$$X_t = \int_0^t \int W(\omega, s, x)\, d(\mu - \nu), \tag{2.34}$$

where μ is an integer-valued random measure, ν its compensator and the integral in (2.34) is well defined. Our prime interest now is the following question: Is there any way to rewrite the representation (2.34) (in which participates a *two-fold* integral) as a *single* integral of the form

$$X_t = \int_0^t H_s\, dZ_s, \tag{2.35}$$

where the integration is carried over a certain process $Z = (Z_t)_{t \geq 0}$ with "simple" structure? It turns out that the considered question admits an affirmative answer (with a stable process Z), if the compensator ν in (2.34) is assumed to have special structure defined by the compensator of the process Z, over which the integration in (2.35) is taken.

In Sec. 5.4 we will show that if $Z = (Z_t)_{t \geq 0}$ is a stable process with the parameter $1 < \alpha < 2$, then it is a purely discontinuous martingale and Z admits the representation

$$Z_t = \int_0^t \int x\, d(p - q), \tag{2.36}$$

where p is the Poisson measure of the jumps of the process Z and its compensator q has the following structure:

$$q(dt, dx) = q_\alpha(x)\, dt\, dx$$

with

$$q_\alpha(x) = \begin{cases} c_+ x^{-(\alpha+1)}, & x > 0, \\ c_- |x|^{-(\alpha+1)}, & x < 0. \end{cases} \tag{2.37}$$

Assume, for the sake of simplicity, that $c_+ = c_- = c$ ("symmetric case"), *i.e.*, let

$$q_\alpha(x) = \frac{c}{|x|^{\alpha+1}}. \tag{2.38}$$

It follows from the results of Sec. 2.4 below that if $H = (H_t(\omega))_{t\geq 0}$ is a predictable process (adapted to $(\mathcal{F}_t^Z)_{t\geq 0}$) such that

$$\mathsf{P}\Big(\int_0^t |H_s(\omega)|^\alpha \, ds < \infty\Big) = 1, \qquad t > 0, \tag{2.39}$$

then the stochastic integral

$$(H \cdot Z)_t = \int_0^t H_s(\omega) \, dZ_s \tag{2.40}$$

is well defined and the process $H \cdot Z$ is a local martingale.

2. Assume for the moment that the process X admits the representation (2.34), in which the compensator ν has the ("special") form:

$$\nu(\omega, dt, dx) = |H_t(\omega)|^\alpha q_\alpha(x) \, dt \, dx, \tag{2.41}$$

where $q_\alpha(x)$ is given by (2.38) and $H_t(\omega) \neq 0$.

From the previous section we know that if (for nonnegative functions $g = g(s,x)$) the property

$$\int_0^\infty \int g(s,x) \, \nu(\omega, ds, dx) = \int_0^\infty \int g(s, W(\omega, s, x)) q_\alpha(x) \, dx \, ds \tag{2.42}$$

is satisfied, with a predictable function $W = W(\omega, s, x)$, then X admits (on an extended probability space) the representation

$$X_t = \int_0^t \int W \, d(p - q) \tag{2.43}$$

with $dq = q_\alpha(x) \, dx \, ds$.

From (2.41) it follows that (2.42) takes the form

$$\int_0^\infty \int g(s,x) |H_s(\omega)|^\alpha q_\alpha(x) \, dx \, ds = \int_0^\infty \int g(s, W(\omega, s, x)) q_\alpha(x) \, dx \, ds. \tag{2.44}$$

In this representation we can choose

$$W(\omega, x, s) = |H_s(\omega)| x. \tag{2.45}$$

Indeed, if (2.45) holds, then

$$\begin{aligned}\int_0^\infty \int g(s, W(\omega, s, x))q_\alpha(x)\,dx\,ds &= \int_0^\infty \int g(s, |H_s(\omega)|x)\,\frac{c}{|x|^{\alpha+1}}\,dx\,ds\\ &= c\int_0^\infty \int g(s,y)\,\frac{1}{(y/|H|)^{1+\alpha}}\,\frac{dy\,ds}{|H|}\\ &= c\int_0^\infty \int g(s,y)|H|^\alpha\,\frac{dy\,ds}{|y|^{1+\alpha}},\end{aligned}$$

which proves the required statement.

Thus, under assumption (2.41), representation (2.43) takes the form

$$X_t = \int_0^t \int x|H_s(\widetilde{\omega})|\,d(p-q) \tag{2.46}$$

with $dq = q_\alpha(x)\,dx\,dt$.

We have already noted that Z has representation (2.36), which in "differential" form can be rewritten as

$$dZ = \int x\,d(p-q). \tag{2.47}$$

From (2.46) and (2.47) it becomes clear that for X_t the representation

$$X_t = \int_0^t |H_s|\,dZ_s \tag{2.48}$$

holds.

Remark 2.6. The exact formulation of result (2.48) consists in the following: under assumption (2.38), on an extended probability space the right-hand sides in (2.46) and (2.48) are indistinguishable processes. For details of the proof see [175] and [125; Chap. 3, § 5, Problem 5].

3. Let us address the case $0 < \alpha < 1$. According to Sec. 5.4, in this case the stable processes Z are "purely discontinuous" processes of bounded variation with representations

$$Z_t = \int_0^t \int x\,dp, \tag{2.49}$$

where p is a Poisson measure with the compensator $q(dt, dx) = q_\alpha(x)\,dt\,dx$ (see (2.37)). It is important to note the difference between the cases $1 < \alpha < 2$ and $0 < \alpha < 1$. Whereas in the first one Z is a local martingale, in the second Z is only a semimartingale, which is not special (Sec. 3.2), since here

$$\int (x^2 \wedge |x|)I(|x| > 1)\,dq = \infty$$

(see (3.35) on page 55). Thus the process Z cannot be a local martingale, and so the expressions of type (2.35) cannot lead in the case $0 < \alpha < 1$ to local martingales.

Taking this in mind, it is expedient (in the considered case $0 < \alpha < 1$) to seek for a possibility of integral representations of type (2.35) for "purely discontinuous" semimartingales with the trajectories of bounded variation and triplet $(0, 0, \nu)$, where (by analogy with the case $1 < \alpha < 2$)

$$\nu(\omega, dt, dx) = |H_t(\omega)|^\alpha q_\alpha(x)\, dt\, dx$$

with $q_\alpha(x) = c|x|^{-(1+\alpha)}$ (in the symmetrical case).

We assert that again (on an extension of the initial probability space) one can find an α-stable process $Z = (Z_t)_{t\geq 0}$ such that representation (2.48) is valid.

The method of proof is just the same as in the case $1 < \alpha < 2$.

4. Finally, let us address the case $\alpha = 1$. Assume the initial process $X = (X_t)_{t\geq 0}$ has the canonical representation

$$X_t = \int_0^t \int_{|x|\leq 1} x\, d(\mu - \nu) + \int_0^t \int_{|x|>1} x\, d\mu \tag{2.50}$$

with

$$\nu(\omega, dx, dt) = |H_t(\omega)| \frac{c\, dx}{|x|^2}\, dt. \tag{2.51}$$

In the case $\alpha = 1$ the process $Z = (Z_t)_{t\geq 0}$ is a Cauchy process having the representation

$$Z_t = \int_0^t \int_{|x|\leq 1} x\, d(p - q) + \int_0^t \int_{|x|>1} x\, dp. \tag{2.52}$$

Making use of the same arguments as above and comparing (2.50) with (2.52), we arrive at representation (2.48). (The details of the proof can be found in [175].)

2.4 Stochastic Integrals with Respect to Stable Processes and Change of Time

1. Let $X_t = \int_0^t H_s\, dB_s$, where $B = (B_t)_{t\geq 0}$ is a Brownian motion and the adapted process $H = (H_t)_{t\geq 0}$ is such that

$$\int_0^t H_s^2\, ds < \infty \quad \text{and} \quad \int_0^\infty H_s^2\, ds = \infty.$$

It is easy to see that $X = (X_t)_{t\geq 0}$ is a local martingale with the quadratic variation $\langle X\rangle_t = \int_0^t H_s^2\, ds$.

From Sec. 1.4 of Chap. 1 it follows that the process $\widehat{B}_\theta = X_{\widehat{T}(\theta)}$ with $\widehat{T}(\theta) = \inf\{t \geq 0 : \langle X\rangle_t > \theta\}$ is a Brownian motion and the initial process $X = (X_t)_{t\geq 0}$ itself admits the representation

$$X_t = \widehat{B}_{\langle X\rangle_t} = \widehat{B}_{T(t)} \tag{2.53}$$

with $T(t) = \langle X\rangle_t = \int_0^t H_s^2\, ds$.

Thus, in the considered case

$$X = H \cdot B = \widehat{B} \circ T,$$

In other words, the stochastic integral over a Brownian motion B coincides with the process which is a time-changed process of a new Brownian motion $\widehat{B}$.

Theorem 2.6. *Assume the processes H and B to be independent. Then the processes $\langle X\rangle$ $(= T = \int_0^\cdot H_s^2\, ds)$ and $\widehat{B}$ are independent as well.*

Proof. Let $\mathcal{F}^H = \sigma(H_s, s \geq 0)$. Then for any deterministic function $g = g(s)$ with $\int_0^\infty g^2(s)\, ds < \infty$ we find that

$$\begin{aligned}
\mathsf{E}\Big(e^{\int g(\theta)\, d\widehat{B}_\theta}\,\Big|\,\mathcal{F}^H\Big) &= \mathsf{E}\Big(e^{\int g(\theta)\, dX_{\widehat{T}(\theta)}}\,\Big|\,\mathcal{F}^H\Big) \\
&= \mathsf{E}\Big(e^{\int g(\langle X\rangle_s)\, dX_s}\,\Big|\,\mathcal{F}^H\Big) = \mathsf{E}\Big(e^{\int g(\langle X\rangle_s) H_s\, dB_s}\,\Big|\,\mathcal{F}^H\Big) \\
&= e^{\frac{1}{2}\int g^2(\langle X\rangle_s) H_s^2\, ds} = e^{\frac{1}{2}\int g^2(\langle X\rangle_s)\, d\langle X\rangle_s} \\
&= e^{\frac{1}{2}\int g^2(s)\, ds} = \mathsf{E}\, e^{\int g(\theta)\, d\widehat{B}_\theta},
\end{aligned} \tag{2.54}$$

which clearly implies the independence of $\widehat{B}$ and H. □

2. Consider the analogous questions for the stable processes.

Let

$$X_t = \int_0^t H_s\, dZ_s^{(\alpha)}, \qquad t \geq 0, \tag{2.55}$$

where $Z^{(\alpha)} = (Z_t^{(\alpha)})_{t\geq 0}$ is an α stable process, $T(t) = \int_0^t |H_s|^\alpha\, ds < \infty$ for $t < \infty$ and $\int_0^\infty |H_s|^\alpha\, ds = \infty$.

Set

$$\widehat{T}(\theta) = \inf\{t \geq 0 : T(t) > \theta\}. \tag{2.56}$$

We assert that

$$\widehat{Z}_\theta^{(\alpha)} = X_{\widehat{T}(\theta)}, \qquad \theta \geq 0, \tag{2.57}$$

is an α-stable process (cf. $\widehat{B}_\theta = X_{\widehat{T}(\theta)}$ with a Brownian motion $\widehat{B}$ in Subsec. 1) and that for the process X itself, given by (2.55), we have the representation of "change of time" type:

$$X_t = \widehat{Z}^{(\alpha)}_{T(t)}, \qquad t \geq 0, \tag{2.58}$$

i.e., $X = \widehat{Z}^{(\alpha)} \circ T$ (cf. the representation $X = \widehat{B} \circ T$ in Subsec. 1).

To prove that $\widehat{Z}^{(\alpha)} = (\widehat{Z}^{(\alpha)}_\theta)_{\theta \geq 0}$ is α-stable (and so $\widehat{Z}^{(\alpha)} \overset{\text{law}}{=} Z^{(\alpha)}$), consider the process

$$M_t = \frac{e^{i\lambda X_t}}{e^{-|\lambda|^\alpha \int_0^t |H_s|^\alpha \, ds}}. \tag{2.59}$$

According to Sec. 4.2, this process is a local (complex-valued) martingale. From the Doob optional sampling theorem [100; Chap. I, Theorem 1.39] we deduce that the process $(M_{\widehat{T}(\theta)})_{\theta \geq 0}$ is a local martingale as well. So,

$$M_{\widehat{T}(\theta)} = \exp\left\{ i\lambda X_{\widehat{T}(\theta)} + |\lambda|^\alpha \int_0^{\widehat{T}(\theta)} |H_s|^\alpha \, ds \right\} = \exp\{i\lambda \widehat{Z}^{(\alpha)}_\theta + |\lambda|^\alpha \theta\}.$$

From this, together with the characteristic property of semimartingales (see Sec. 4.4 and [100; Chap. II, Corollary 2.48]), it follows that the process $\widehat{Z}^{(\alpha)} = (\widehat{Z}^{(\alpha)}_\theta)_{\theta \geq 0}$ is an α-stable process.

Finally, to prove (2.58) it is enough to note that

$$\widehat{Z}^{(\alpha)}_{T(t)} = X_{\widehat{T}(T(t))} = X_t. \tag{2.60}$$

Remark 2.7. If one assume that $H \neq 0$, then (2.60) follows immediately from the fact that in this case the process $T(t) = \int_0^t |H_s|^\alpha \, ds$ is both continuous and increasing (and so, by virtue of property (b) in Sec. 1.2, $\widehat{T}(T(t)) = t$). As for the case when H may vanish, the process $T = (T(t))_{t \geq 0}$ has intervals of constancy. But on such intervals of constancy the process X does not change its values either, and property (2.60) remains in force. (Cf. [146; 2nd edn., p. 174].)

In conclusion of this section, note that if the processes $Z^{(\alpha)}$ and H in (2.55) are independent, then so are the processes $\widehat{Z}^{(\alpha)}$ and T in (2.60).

Chapter 3

Semimartingales: Basic Notions, Structures, Elements of Stochastic Analysis

3.1 Basic Definitions and Properties

1. Above we said a lot about the fact that in constructing models with complicated structure (in particular, in mathematical finance) the basic role is played by

Brownian motion

first introduced as a mathematical object in the work by L. Bachelier ([4], 1900), in connection with the analysis of dynamics of prices of financial assets and in the work by A. Einstein ([65], 1905), who used it when analyzing the chaotic motion of particles in a liquid.

At nearly the same time, namely in 1903, F. Lundberg [126] took the

Poisson process

as basis for constructing mathematical models of the processes describing the dynamics of the capitals of insurance companies.

Brownian motion—by definition—is a process with continuous trajectories. The Poisson process is a typical instance of processes with discrete intervention of chance. Both these processes play a fundamental role in forming a general class of homogeneous processes with independent increments—the so-called

Lévy processes,

whose application in mathematical finance and actuarial science is becoming part of common practice. This can find its explanation in the fact that, firstly, processes of this type have been found to provide good descriptions of a variety of statistical data (see Chap. 12) and, secondly, for these we have available well-developed analytical tools which allow us to analyze stochastic phenomena in insurance and finance—and, indeed, in many subject fields.

However, in many cases one has to address more complicated classes of processes, *e.g.*, processes with independent increments (not necessarily homogeneous as in the case of Lévy processes), processes built from Lévy processes and having dependent increments, diffusion processes, *etc.* It turns out that a great many of these processes belong to the rather large class Sem$\mathcal{M}$ of

semimartingales,

which have the remarkable properties that, firstly, they are well-adapted to describe the dynamics of processes evolving in time (as is typical in both financial and actuarial business), and, secondly, for them there exist well-developed mathematical tools of stochastic analysis (calculus).

These facts explain why, in this book, we will pay much attention to semimartingales. Moreover, from our point of view, the role of such processes in mathematical finance and insurance is bound to increase.

An extensive literature is devoted to the theory of semimartingales. We will refer essentially to the monograph [100], where the reader can find the proofs of many of the propositions formulated here (and many others).

2. The concept underlying the theory of semimartingales is that of *stochastic basis*, *i.e.*, a filtered probability space

$$(\Omega, \mathcal{F}, (\mathcal{F}_t)_{t\geq 0}, \mathsf{P}),$$

which (by the definition given in Sec. 1.1 is a probability space $(\Omega, \mathcal{F}, \mathsf{P})$, equipped with a flow (filtration) $(\mathcal{F}_t)_{t\geq 0}$ of sub-σ-algebras $\mathcal{F}_t$, $t \geq 0$, satisfying to the so-called *usual* conditions

$$\mathcal{F}_s \subseteq \mathcal{F}_t \subseteq \mathcal{F}, \qquad s \leq t,$$

$$\mathcal{F}_t = \bigcap_{s>t} \mathcal{F}_s,$$

$$\mathcal{F}_t = \mathcal{F}_t^{\mathsf{P}}, \qquad t \geq 0$$

($\mathcal{F}_t^{\mathsf{P}}$ stands for completion of the σ-algebra $\mathcal{F}_t$ by the P-null sets from $\mathcal{F}$).

From the point of view of mathematical finance and insurance, it is natural to conceive of σ-algebras $\mathcal{F}_t$ as information (on prices, indices, exchange rates, *etc.*) accumulated on the time interval $[0, t]$.

Remark 3.1. From the beginning we assume given the probability measure P. However, in many cases—when we have to operate with several measures—it proves to be convenient to start with a filtered measurable space $(\Omega, \mathcal{F}, (\mathcal{F}_t)_{t\geq 0})$ (where $(\mathcal{F}_t)_{t\geq 0}$ is a certain nondecreasing family of sub-σ-algebras of $\mathcal{F}$), without specifying one or another probability measure.

3. All stochastic processes $X = (X_t)_{t\geq 0}$ considered on a stochastic basis $(\Omega, \mathcal{F}, (\mathcal{F}_t)_{t\geq 0}, \mathsf{P})$ will be assumed such that the random variables $X_t = X_t(\omega)$ are $\mathcal{F}_t$-measurable. It is common to say that such processes are adapted with respect to the flow $(\mathcal{F}_t)_{t\geq 0}$, or $(\mathcal{F}_t)_{t\geq 0}$-adapted. Instead of $X = (X_t)_{t\geq 0}$ one often writes $X = (X_t, \mathcal{F}_t)_{t\geq 0}$ or just $X = (X_t, \mathcal{F}_t)$. Without additional comment, all stochastic processes $X = (X_t)_{t\geq 0}$ considered in the sequel will be assumed such that their trajectories, for every $\omega \in \Omega$, are right-continuous (for $t \geq 0$) and have limits from the left (for $t > 0$). The space of such trajectories is commonly denoted by D (or $D([0,\infty))$). In French abbreviation, they appear as *càdlàg* processes (*c*ontinuité *à* *d*roite avec des *l*imites *à* *g*auche).

We call two processes X, Y *indistinguishable* if the set $\{X \neq Y\} = \{(\omega, t) : X_t(\omega) \neq Y_t(\omega)\}$ is *evanescent*, *i.e.*, the set $\{\omega$: there exists a $t \in \mathbb{R}_+$ such that $(\omega, t) \in \{X \neq Y\}\}$ is P-null (see also [100; Chap. I, § 1a]).

In all subsequent exposition the important role will be played by the following two classes of (adapted) processes:

$$\mathcal{V} = \{A : A = (A_t, \mathcal{F}_t)\text{—processes of bounded variation, } i.e., \text{ with the property } \textstyle\int_0^t |dA_s(\omega)| < \infty,\ t \geq 0,\ \omega \in \Omega\}$$

and

$$\mathcal{M}_{\text{loc}} = \{M : M = (M_t, \mathcal{F}_t)\text{—processes which are local martingales}\}.$$

To give the definition of a local martingale, let us first recall the notion of martingale and Markov time.

A stochastic process $M = (M_t, \mathcal{F}_t)_{t\geq 0}$ is said to be a *martingale* (*submartingale*), if

$$\mathsf{E}|M_t| < \infty, \qquad t \geq 0,$$

$$\mathsf{E}(M_t \,|\, \mathcal{F}_s) \underset{(\geq)}{=} M_s \quad (\mathsf{P}\text{-a.s.}), \qquad s \leq t.$$

The classes of martingales and submartingales will be denoted by $\mathcal{M}$ and Sub$\mathcal{M}$, respectively.

A random variable $\tau = \tau(\omega)$ taking values in $[0, \infty]$ is called a *Markov time* (notation: $\tau \in \overline{\mathfrak{M}}$) if

$$\{\omega : \tau(\omega) \leq t\} \in \mathcal{F}_t, \quad t \geq 0.$$

The Markov times which are finite ($\tau(\omega) < \infty$, $\omega \in \Omega$, or $\mathsf{P}(\tau(\omega) < \infty) = 1$) are commonly named *stopping times* (notation: $\tau \in \mathfrak{M}$).

A process $M = (M_t, \mathcal{F}_t)_{t\geq 0}$ is called a *local martingale* (*local submartingale*), if there exists a sequence $(\tau_n)_{n\geq 0}$ of stopping times such that $\tau_n(\omega) \leq \tau_{n+1}(\omega)$, $\tau_n(\omega) \uparrow \infty$ (P-a.s.) as $n \to \infty$ and the "stopped" processes $M^{\tau_n} = (M_{\tau_n \wedge t}, \mathcal{F}_t)$ are martingales (submartingales). (Such a sequence $(\tau_n)_{n\geq 0}$ is said to form a *localizing* sequence.)

4. A stochastic process $X = (X_t, \mathcal{F}_t)_{t\geq 0}$ is called a *semimartingale*, if it admits the following decomposition:

$$X_t = X_0 + A_t + M_t, \quad t \geq 0, \tag{3.1}$$

where $A \in \mathcal{V}$ and $M \in \mathcal{M}_{\text{loc}}$.

Of particular importance for us will be the so-called *special semimartingales*, *i.e.*, semimartingales for which there exists a decomposition of the form (3.1) with *predictable* process $A = (A_t, \mathcal{F}_t)_{t\geq 0}$. Here "predictable" is understood in the sense of the following definition.

Let $\mathcal{P}$ be the σ-algebra of subsets in $\Omega \times \mathbb{R}_+$ generated by all adapted (to $(\mathcal{F}_t)_{t\geq 0}$) processes with continuous (or just left-continuous) trajectories. The σ-algebra $\mathcal{P}$ is called the σ-algebra of *predictable* sets, and any process which is measurable with respect to $\mathcal{P}$ is called *predictable*. Note that $\mathcal{P}$ coincides with the σ-algebra generated by the stochastic intervals $[\![0, \tau]\!] = \{(\omega, t) : 0 \leq t \leq \tau(\omega)\}$, $\tau \in \overline{\mathfrak{M}}$.

Another important σ-algebra in $\Omega \times \mathbb{R}_+$ is the so-called *optional* σ-algebra $\mathcal{O}$, which is generated by all adapted càdlàg processes (equivalently, by stochastic intervals $[\![0, \tau[\![= \{(\omega, t) : 0 \leq t < \tau(\omega)\}$). Processes which are measurable with respect to the σ-algebra $\mathcal{O}$ are generally called *optional*. (Thus, the semimartingales under consideration are optional processes.)

It is clear that $\mathcal{P} \subseteq \mathcal{O}$.

Remark 3.2. In the discrete-time case the stochastic basis is $(\Omega, \mathcal{F}, (\mathcal{F}_n)_{n\geq 0}, \mathsf{P})$; the optionality property of a process $X = (X_n, \mathcal{F}_n)$ means that the X_n are $\mathcal{F}_n$-measurable, and the predictability means that the X_n are $\mathcal{F}_{n-1}$-measurable ($\mathcal{F}_{-1} = \mathcal{F}_0$).

Notice also that for the discrete-time case the process $X = (X_n, \mathcal{F}_n)$ is a *local martingale* if and only if $\mathsf{E}|X_0| < \infty$, the conditional expectations $\mathsf{E}(X_n \,|\, \mathcal{F}_{n-1})$, $n \geq 1$, are well defined, and $\mathsf{E}(X_n \,|\, \mathcal{F}_{n-1}) = X_{n-1}$ (P-a.s.). On the other hand, the process $X = (X_n, \mathcal{F}_n)$ is a *local martingale* if and only if X is a *martingale transform*, *i.e.*, there exist a martingale $M = (M_n, \mathcal{F}_n)$ and a predictable process $\gamma = (\gamma_n, \mathcal{F}_{n-1})$ such that $X_n = X_0 +$

$\sum_{k=1}^{n} \gamma_k \Delta M_k$, $n \geq 1$. For the proof of equivalence of these definitions see [161; Chap. II, § 1c].

So, if there is a decomposition (3.1), with a predictable process A of the class $\mathcal{V}$, then the semimartingale X is *special*. This decomposition, if it exists, is always unique. In this case the decomposition (3.1) is said to be *canonical*. If the semimartingale X has bounded jumps ($|\Delta X_t| \leq C$, where $\Delta X_t = X_t - X_{t-}$, $t > 0$), then it is special. (See [100; Chap. I, § 4c].)

Remark 3.3. It is not *a priori* clear from the above definition, why the class of semimartingales deserves particular interest. As an initial motivation we note the following. Firstly, for many physical processes, we can neatly distinguish the two components—low-frequency and high-frequency. The first component can often be well modeled by the processes of bounded variation, and the second by the martingales (whose trajectories in typical cases are as for Brownian motion). Secondly, the class of semimartingales proves to be stable under many transformations, namely, change of time, change of measure, change of filtration, *etc.* Finally, it turns out that semimartingales form, in a certain sense, the maximal class with respect to which one can define a stochastic integral with natural properties (see [100]).

5. Let $B = (B_t, \mathcal{F}_t)$ be a Brownian motion, and let $N = (N_t, \mathcal{F}_t)$ be a Poisson process independent of B with the parameter λ ($\mathsf{E} N_t = \lambda t$). It is clear that the process $M = (M_t, \mathcal{F}_t)$ with

$$M_t = \Pi_t + B_t, \tag{3.2}$$

where $\Pi_t = N_t - \lambda t$, is a martingale. Here the process Π is, so to say, purely discontinuous and the process B has continuous trajectories. It turns out that the analogous result is true in the general case. Namely, any local martingale $M = (M_t, \mathcal{F}_t)_{t \geq 0}$ admits the decomposition

$$M = M_0 + M^c + M^d, \tag{3.3}$$

where $M^c = (M_t^c, \mathcal{F}_t)_{t \geq 0}$ is a continuous local martingale and $M^d = (M_t^d, \mathcal{F}_t)_{t \geq 0}$ is a *purely discontinuous* local martingale, *i.e.*, such that M^d is *orthogonal* to every continuous local martingale Y (in the sense that their product $M^d Y$ is a local martingale). See [100; Chap. I, § 4b].

Thus, if X is a semimartingale with the representation (3.1), then this representation (which is generally not unique) can be detailed:

$$X_t = X_0 + A_t + M_t^c + M_t^d \tag{3.4}$$

(with $A_0 = M_0^c = M_0^d = 0$).

6. As was mentioned above, for semimartingales we dispose of a well-developed mathematical apparatus of stochastic calculus. This is based on the fact that for semimartingales and a wide class of predictable processes H (which includes all locally bounded predictable processes) one can define the *stochastic integral* $H \cdot X = (H \cdot X)_t$, where (in more explicit form)

$$(H \cdot X)_t = \int_0^t H_s \, dX_s \tag{3.5}$$

is the stochastic integral on the interval $(0, t]$.

In the case $X = B$ is a Brownian motion, the stochastic integral $(H \cdot B)_t$ is well-defined if the predictable (or—in this case—even only adapted) process H is such that $\mathsf{P}(\int_0^t H_s^2 \, ds < \infty) = 1$. If $X = X^{(\alpha)}$ is an α-stable process, then the stochastic integral $(H \cdot X^{(\alpha)})_t$ is defined for predictable functions H such that $\mathsf{P}(\int_0^t |H_s|^\alpha \, ds < \infty) = 1$. In more detail the questions of defining the stochastic integrals $(H \cdot X)_t$ will be considered in Sec. 3.4, where we shall describe the corresponding constructions as well as give conditions (in predictable terms) on processes H, which guarantee their integrability with respect to X.

7. Just as in the probability theory an important role is assigned to the notions of variance and covariance, in stochastic calculus it is the notions of variation and quadratic covariation of semimartingales that are of particular importance.

Definition 3.1. For two semimartingales X and Y the *quadratic covariation* $[X, Y]$ is defined as

$$[X, Y] = XY - X_0 Y_0 - X_- \cdot Y - Y_- \cdot X, \tag{3.6}$$

where $X_- \cdot Y$ and $Y_- \cdot X$ are stochastic integrals.

Notice that the integrals $X_- \cdot Y$ and $Y_- \cdot X$ are defined, because the processes $X_- = (X_{t-})_{t>0}$ and $Y_- = (Y_{t-})_{t>0}$, being left-continuous, are locally bounded.

If one puts $Y = X$ in (3.6), then one obtains the definition of the *quadratic variation* $[X]$ (or $[X, X]$):

$$[X] = X^2 - X_0^2 - 2X_- \cdot X. \tag{3.7}$$

The above-introduced names (quadratic variation and quadratic covariation) are justified by the fact that the expressions of type $\sum \Delta X \Delta Y$ and $\sum (\Delta X)^2$ (in a proper understanding; see [100; Chap. I, § 4]) converge in probability to $[X, Y]$ and $[X]$.

8. Let $X = X_0 + A + M$ be a representation (of the form (3.1)) of a semimartingale X. We already know that any local martingale M admits a representation $M = M_0 + M^c + M^d$, where M^c is the continuous martingale component of M. If $X = X_0 + A' + M'$ is another representation of X, it turns out that the continuous martingale component $(M')^c$ coincides (up to stochastic equivalence) with M^c. This gave grounds to call M^c in the representation $X = X_0 + A + M^c + M^d$ the continuous martingale component of the semimartingale X and to denote it by X^c (see [100; Chap. I, § 4]).

By means of the process X^c one can give the following representation for the quadratic variation $[X]$:

$$[X] = \langle X^c \rangle + \sum_{s \le \cdot} (\Delta X_s)^2, \tag{3.8}$$

where $\langle X^c \rangle$ is an increasing predictable process such that

$$(X^c)^2 - \langle X^c \rangle \in \mathcal{M}_{\text{loc}}. \tag{3.9}$$

The existence of such a (unique) process $\langle X^c \rangle$ follows from the following *Doob–Meyer decomposition*:

> If X is a local submartingale, then there exists a (unique) increasing predictable process A such that $X - A \in \mathcal{M}_{\text{loc}}$.

(The process A is called a *compensator* of the process X, for it compensates X to a local martingale.) This is a difficult result, although for the discrete-time case the corresponding representation (*Doob decomposition*)

$$X_n = X_0 + A_n + M_n \tag{3.10}$$

can be obtained quite easily. Indeed, if $X = (X_n, \mathcal{F}_n)$ is an $(\mathcal{F}_n)_{n \ge 0}$-adapted integrable ($\mathsf{E}|X_n| < \infty$, $n \ge 0$) sequence, then, evidently,

$$X_n = X_0 + \sum_{k=1}^{n} \mathsf{E}(\Delta X_k \,|\, \mathcal{F}_{k-1}) + \sum_{k=1}^{n} \big[\Delta X_k - \mathsf{E}(\Delta X_k \,|\, \mathcal{F}_{k-1})\big]. \tag{3.11}$$

Therefore (3.10) is valid with

$$\begin{aligned} A_n &= \sum_{k=1}^{n} \mathsf{E}(\Delta X_k \,|\, \mathcal{F}_{k-1}), \\ M_n &= \sum_{k=1}^{n} [\Delta X_k - \mathsf{E}(\Delta X_k \,|\, \mathcal{F}_{k-1})]. \end{aligned} \tag{3.12}$$

In many problems the following particular case of the above-formulated Doob-Meyer decomposition proves to be useful.

Let $X = (X_t)_{t\geq 0}$ be an $(\mathcal{F}_t)_{t\geq 0}$-adapted increasing (to be more exact nondecreasing) process which belongs to the class $\mathcal{A}^+_{\mathrm{loc}}$ (*i.e.*, such that $\mathsf{E}X_{\tau_n} < \infty$ for a certain localizing sequence $(\tau_n)_{n\geq 0}$). Then there exists a predictable process $A \in \mathcal{A}^+_{\mathrm{loc}}$ such that $X - A \in \mathcal{M}_{\mathrm{loc}}$. For the proof it is enough to note that the processes $X^{\tau_n} = (X_{t\wedge\tau_n})_{t\geq 0}$ are submartingales and, therefore, the process X itself is a local submartingale.

Because the variable $[X]_t$ (for each $t > 0$) is finite (P-a.s.), we have

$$\sum_{s\leq t}(\Delta X_s)^2 < \infty \quad (\mathsf{P}\text{-a.s.}), \quad t > 0. \tag{3.13}$$

9. The local martingale X^c in (3.9) is continuous. Actually, a more general result is also true: if $M = (M_t)_{t\geq 0}$ is a square-integrable ($\mathsf{E}M_t^2 < \infty$, $t \geq 0$) martingale, then there exists an increasing predictable process $\langle M\rangle = (\langle M\rangle_t)_{t\geq 0}$ such that $M^2 - \langle M\rangle$ is a local martingale.

This result admits a further generalization: to each pair (M, N) of square-integrable martingales corresponds a (unique) predictable process $\langle M, N\rangle$ such that $MN - \langle M, N\rangle \in \mathcal{M}_{\mathrm{loc}}$. (The proof follows immediately from the previous result, if only we use the "polarizing" formula

$$MN = \frac{1}{4}\big[(M+N)^2 - (M-N)^2\big].) \tag{3.14}$$

If X, Y are semimartingales and X^c, Y^c are their continuous martingale components, then the above-formulated result implies the existence of a predictable process $\langle X^c, Y^c\rangle$ such that $X^cY^c - \langle X^c, Y^c\rangle \in \mathcal{M}_{\mathrm{loc}}$. This process $\langle X^c, Y^c\rangle$ is called the *predictable quadratic covariation.* Often the process $\langle X^c\rangle$ is also referred to as the *quadratic characteristic* or just the *angular-bracket process.*

Between $[X]$ and $\langle X^c\rangle$ as well as between $[X, Y]$ and $\langle X^c, Y^c\rangle$ there exists a certain connection, already mentioned above:

$$[X] = \langle X^c\rangle + \sum_{s\leq\cdot}(\Delta X_s)^2 \tag{3.15}$$

and

$$[X, Y] = \langle X^c, Y^c\rangle + \sum_{s\leq\cdot}\Delta X_s\Delta Y_s. \tag{3.16}$$

10. In the theory of martingales and semimartingales we often have to deal with the jump-like compensators of these processes. Important tools for investigation of such components are jump measures and their compensators which we are now going to introduce.

Let $X = (X_t, \mathcal{F}_t)_{t\geq 0}$ be a semimartingale. The *jump measure*, $\mu = \mu(\omega; dt, dx)$, is defined by

$$\mu(\omega; dt, dx) = \sum_{s>0} I(\Delta X_s(\omega) \neq 0)\delta(s, \Delta X_s(\omega))(dt, dx), \tag{3.17}$$

where $\Delta X_s(\omega) = X_s(\omega) - X_{s-}(\omega)$ and $\delta(a)$ is the Dirac measure, "sitting" at a point a $(= (s, \Delta X_s(\omega)))$. (Sometimes, to emphasize the dependence of μ on X, we shall write μ^X.) Often one has to integrate with respect to a (random) measure of jumps. In this connection we give the definition of the integral $W * \mu = ((W * \mu)_t(\omega))_{t\geq 0}$.

Let $\mathbb{R}_+ = \{t : t \in [0, \infty)\}$, $\mathbb{R} = \{x : x \in (-\infty, \infty)\}$. Introduce the space $\widetilde{\Omega} = \Omega \times \mathbb{R}_+ \times \mathbb{R}$ and define on it two σ-algebras

$$\widetilde{\mathcal{O}} = \mathcal{O} \otimes \mathcal{B}(\mathbb{R}) \quad \text{and} \quad \widetilde{\mathcal{P}} = \mathcal{P} \otimes \mathcal{B}(\mathbb{R})$$

of subsets in $\widetilde{\Omega}$. These algebras are also called the optional and predictable σ-algebras. Similarly, any $\widetilde{\mathcal{O}}$- or $\widetilde{\mathcal{P}}$-measurable function $W = W(\omega; t, x)$ will be called optional ($\widetilde{\mathcal{O}}$-optional) or predictable ($\widetilde{\mathcal{P}}$-predictable).

Let W be an optional function. If for $\omega \in \Omega$ the Lebesgue–Stieltjes integral

$$\int_{[0,t]\times\mathbb{R}} |W(\omega; s, x)| \, \mu(\omega; ds, dx) < \infty, \tag{3.18}$$

then for such an $\omega \in \Omega$ we define

$$W * \mu_t(\omega) = \int_{[0,t]\times\mathbb{R}} W(\omega; s, x) \, \mu(\omega; ds, dx). \tag{3.19}$$

But if $\omega \in \Omega$ is such that the left-hand side in (3.18) is equal to $+\infty$, then we put $W * \mu_t(\omega) = \infty$.

Consider a set $A \in \mathcal{B}(\mathbb{R})$. Then it is clear that the process

$$X_t^A(\omega) = \mu(\omega; (0, t] \times A), \qquad t > 0,$$

is nondecreasing in t (note that if $A = \{0\}$ then $X^A = 0$). If we assume that there exists a localizing sequence $(\tau_n^A)_{n\geq 0}$ such that $\mathsf{E}X^A_{\tau_n^A} < \infty$, $n \geq 0$ (*i.e.*, $X^A \in \mathcal{A}^+_{\text{loc}}$), then the corollary to the Doob–Meyer decomposition implies the existence of the compensator (*i.e.*, of the predictable increasing process)—which is denoted by $\widetilde{X}_t^A(\omega) = \nu^A(\omega; (0, t])$, $t > 0$,—such that

$$X^A - \widetilde{X}^A \in \mathcal{M}_{\text{loc}}, \tag{3.20}$$

that is,

$$\mu(\omega; (0, t] \times A) - \nu^A(\omega; (0, t]), \quad t > 0,$$

is a local martingale.

The above considerations deal only with a "restriction" of the initial measure μ onto a set A and say nothing about whether there exists a "predictable"—in the sense to be specified below—so to say "collective" measure $\nu = \nu(\omega; dt, dx)$ such that for any set A

$$\nu^A(\omega; (0,t]) = \nu(\omega; (0,t] \times A),$$

i.e., such that its "restriction" onto the set A coincides with ν^A. It turns out that such a measure does exist.

To formulate the corresponding result, it is useful to introduce the notion of a random measure, which generalizes the notion of the measure of jumps of a semimartingale.

We shall say that a family $\{m(\omega; dt, dx), \omega \in \Omega\}$ of nonnegative measures $m(\omega; dt, dx)$, $\omega \in \Omega$, on $(\mathbb{R}_+ \times \mathbb{R}, \mathcal{B}(\mathbb{R}_+) \otimes \mathcal{B}(\mathbb{R}))$ is a random measure, if $m(\omega; \{0\} \times \mathbb{R}) = 0$ for each $\omega \in \Omega$.

A random measure m is called *optional* (*predictable*), if for each optional (predictable) function $W = W(\omega; t, x)$ the process $W * m$ is optional (predictable).

We shall say that a measure $m = m(\omega; dt, dx)$ is $\widetilde{\mathcal{P}}$-*σ-finite*, if there exists a $\widetilde{\mathcal{P}}$-measurable partitioning $(A_n)_{n\geq 1}$ of the space $\widetilde{\Omega}$ such that for any $n \geq 1$

$$\mathsf{E}(I_{A_n} * m)_\infty < \infty.$$

This is the case, for example, for the jump measure of a semimartingale.

The following result is central for the theory of representations of random measures m in the form similar to one in the Doob–Meyer decomposition:

$$m = \widetilde{m} + (m - \widetilde{m}),$$

where $\widetilde{m}$ is a predictable random measure and the measure $m - \widetilde{m}$ has the "local martingality" property.

Theorem 3.1 ([100; Chap. II, Theorem 1.8]). *If m is a $\widetilde{\mathcal{P}}$-σ-finite measure, then there exists a predictable random measure $\widetilde{m}$ (called a compensator of the measure m), such that the following equivalent properties hold:*

(a) $\mathsf{E}(W * \widetilde{m}_\infty) = \mathsf{E}(W * m_\infty)$ *for every nonnegative $\widetilde{\mathcal{P}}$-measurable function W;*

or

(b) *for every $\widetilde{\mathcal{P}}$-measurable function W on $\widetilde{\Omega}$, such that $|W|*m \in \mathcal{A}^+_{\text{loc}}$, the process*

$$W * m - W * \widetilde{m} \in \mathcal{M}_{\text{loc}}.$$

The case when a random measure m is *integer-valued* and equal to the jump measure μ of some semimartingale is of particular interest for us.

In this case the compensator ν of the measure μ can be chosen in such a way that $\nu(\omega, \{t\} \times \mathbb{R}) \leq 1$ identically, *i.e.*, the jump sizes of ν do not exceed 1.

11. The Poisson measures provide a classical example of integer-valued random measures:

Definition 3.2. Let $(\Omega, \mathcal{F}, (\mathcal{F}_t)_{t\geq 0}, \mathsf{P})$ be a probability space. The *Poisson measure* on $\mathbb{R}_+ \times \mathbb{R}$ is an integer-valued random measure $\mu = \mu(\omega; dt, dx)$ such that

(a) for $s > 0$ and $A \in \mathcal{B}(\mathbb{R}_+) \otimes \mathcal{B}(\mathbb{R})$ such that $A \subseteq (s, \infty) \times \mathbb{R}$, the variables $\mu(\,\cdot\,; A)$ are independent of the σ-algebra $\mathcal{F}_s$;
(b) the "intensity" $a(A) = \mathsf{E}\mu(\omega; A)$ is a σ-finite measure;
(c) $\mu(\omega; \{t\} \times \mathbb{R}) = 0$, $t \geq 0$.

If the intensity $a = a(dt, dx)$ is of the form $a(dt, dx) = dt\, F(dx)$, where $F = F(dx)$ is a positive σ-finite measure on $(\mathbb{R}, \mathcal{B}(\mathbb{R}))$, then the measure μ is called a *homogeneous Poisson measure.*

The easiest way to get an example of such a measure is to consider the Poisson process. To this purpose let us consider adapted processes $N = (N_t, \mathcal{F}_t)$, $N_0 = 0$, taking values in the set $\mathbb{N} = \{0, 1, 2, \dots\}$ and changing by jumps of size $+1$. If $T_n = \inf\{t : N_t = n\}$, then

$$N_t = \sum_{n\geq 1} I(T_n \leq t).$$

In the class of such processes—called *simple point processes*—the Poisson processes are specified by the claim that

(i) $\mathsf{E}N_t < \infty$, $t > 0$;
(ii) $(N_t - N_s)_{t>s}$ does not depend on the σ-algebra $\mathcal{F}_s$;
(iii) the function $a(t) = \mathsf{E}N_t$ is *continuous.*

The remarkable fact is that these assumptions imply that the variables $N_t - N_s$ have the Poisson distribution with the mean $a(t) - a(s)$, where $a(t) = \mathsf{E}N_t$.

Consider the measure μ of jumps of a Poisson process N with $\mathsf{E}N_t = \lambda t$:

$$\mu(\omega; dt, dx) = \sum \delta_{\{1\}}(\Delta N_s)(dt, dx) I(\Delta N_s = 1).$$

Since the jumps of the process N are all equal to 1, we find that

$$\mu(\omega; (0,t] \times \{1\}) = N_t.$$

And the compensator ν of this jump measure is

$$\nu(\omega; (0,t] \times \{1\}) = \lambda t,$$

where $\lambda = F(\{1\})$.

3.2 Canonical Representation. Triplets of Predictable Characteristics

1. Let $X = (X_t, \mathcal{F}_t)$ be a semimartingale, *i.e.*, a process which admits the decomposition

$$X = X_0 + A + M, \tag{3.21}$$

where $A = (A_t, \mathcal{F}_t)$ is a process of bounded variation and $M = (M_t, \mathcal{F}_t)$ is a local martingale. Moreover (see [100; Chap. I, Lemma 4.24]), if the jumps of X are bounded, then there exists a decomposition (3.21) with a predictable process A.

This implies that if $h = h(x)$ is a function with bounded support and $h(x) = x$ in a certain neighborhood of zero (a classical example of such "truncation function" is provided by the function $h(x) = xI(|x| \leq 1)$), then the process $X(h) = (X_t(h), \mathcal{F}_t)$, where

$$X_t(h) = X_t - \sum_{s \leq t} (\Delta X_s - h(\Delta X_s)), \tag{3.22}$$

is a special semimartingale and the following decomposition hold:

$$X(h) = X_0 + B(h) + M(h), \tag{3.23}$$

where $B(h)$ is a predictable process of bounded variation and $M(h)$ is a local martingale.

The process $M(h)$ admits, as was said in Sec. 3.1, the representation

$$M(h) = M^c(h) + M^d(h), \tag{3.24}$$

where $M^c(h)$ is a continuous local martingale and $M^d(h)$ is a purely discontinuous local martingale.

Therefore from (3.22)–(3.24) we have

$$X = X_0 + B(h) + M^c(h) + M^d(h) + \sum_{s\leq \cdot}(\Delta X_s - h(\Delta X_s)), \tag{3.25}$$

or, taking account of the definition of the measure of jumps $\mu = \mu^X$,

$$X = X_0 + B(h) + M^c(h) + M^d(h) + \int_0^t \int (x - h(x))\, d\mu. \tag{3.26}$$

We see that if $h' = h'(x)$ is another truncation function, then

$$\left[B(h) + \int_0^{\cdot} \int (x - h(x))\, d\mu\right] - \left[B(h') + \int_0^{\cdot} \int (x - h'(x))\, d\mu\right]$$
$$= [M^c(h') - M^c(h)] - [M^d(h) - M^d(h')]. \tag{3.27}$$

The right-hand side is a local martingale. Therefore so is the left-hand side in (3.27), which is at the same time a process of bounded variation and, consequently (see [100; Chap. I, Lemma 4.14b]), a purely discontinuous martingale. This in conjunction with (3.27) yields that the continuous local martingale $M^c(h') - M^c(h)$ is simultaneously a purely discontinuous local martingale. According to [100; Chap. I, Lemma 4.14b] such a process is identically null, *i.e.*, $M^c(h') = M^c(h)$. From this we deduce that the continuous martingale component of a semimartingale X does not depend on h, which accords with its notation as X^c.

Thus, we can rewrite the representation (3.26) in the form:

$$X = X_0 + B(h) + X^c + M^d(h) + \int_0^{\cdot} \int (x - h(x))\, d\mu. \tag{3.28}$$

The process X^c is a locally square-integrable martingale (Definition 1.45 in [100]), and its quadratic characteristic $\langle X^c \rangle$ (*i.e.*, a predictable process such that $(X^c)^2 - \langle X^c \rangle \in \mathcal{M}_{\text{loc}}$) is well defined.

In the sequel we shall denote this process $\langle X^c \rangle$ by $C = (C_t, \mathcal{F}_t)$.

Now address the purely discontinuous local martingale $M^d(h)$. The general theory of stochastic integration with respect to random measures (see [100; Chap. II, § 1d]) yields that the process $M^d(h)$ can be represented in the form

$$M_t^d(h) = \int_0^t \int h(x)\, d(\mu - \nu), \tag{3.29}$$

where the integral is interpreted as a stochastic integral with respect to the martingale measure $\mu - \nu$.

Thus, the representation (3.28) takes the following form:

$$X = X_0 + B(h) + X^c + \int_0^{\cdot} \int h(x)\, d(\mu - \nu) + \int_0^{\cdot} \int (x - h(x))\, d\mu, \tag{3.30}$$

called the *canonical representation* of the semimartingale X.

2. It is useful to note that the predictable characteristics C and ν do not depend on h (*i.e.*, they are "intrinsic" characteristics). The process $B(h)$ is predictable as well but it depends on h. It is not difficult to see that if h' is another truncation function, then (up to an evanescent set)

$$B(h) - B(h') = (h - h') * \nu. \tag{3.31}$$

Usually when operating with the canonical representation (3.30), one takes the truncation function as fixed and does not indicate the explicit dependence of $B(h)$ on h—one simply writes B instead of $B(h)$.

The collection

$$\mathbb{T} = (B, C, \nu)$$

is commonly named a *triplet of predictable characteristics* of the semimartingale X. (Sometimes instead of (B, C, ν) we write $(B, C, \nu)_h$ if we want to emphasize that the considered truncation function is h.)

Let us list several properties of these characteristics.

As was mentioned, for semimartingales, the process $\sum_{s \le \cdot} |\Delta X_s|^2 \in \mathcal{V}$, and therefore the process $(|X|^2 \wedge 1) * \mu^X = \sum_{s \le \cdot} (|X_s|^2 \wedge 1)$, is locally integrable (belongs to the class $\mathcal{A}^+_{\mathrm{loc}}$) as a process with bounded jumps. Thus, the definition of the compensator ν implies that

$$(|x|^2 \wedge 1) * \nu \in \mathcal{A}^+_{\mathrm{loc}}. \tag{3.32}$$

This property of the compensator is rather important, because it determines the characteristic features of behavior of the measure $\nu = \nu(\omega; dt, dx)$ for "small" and "large" values of $|x|$.

The following remark on the connection between the jumps of the function $B = B^h$ and those of the measure ν can prove to be useful in many questions:

$$\Delta B_t = \int h(x)\, \nu(\omega; \{t\} \times dx), \tag{3.33}$$

where h is a truncation function and

$$\nu(\omega; \{t\} \times dx) = \nu(\omega; (0, t], dx) - \nu(\omega; (0, t), dx).$$

(See [100; Chap. II, Proposition 2.9].)

To make further references more convenient, we formulate several properties of the compensator ν under different assumptions on the underlying semimartingales. See the proofs in [100; Chap. II, § 2b].

Theorem 3.2. *Let $X = (X_t, \mathcal{F}_t)$ be a semimartingale with compensator ν of the jump measure $\mu = \mu^X$. Then*

(a)

$$(x^2 \wedge 1) * \nu \in \mathcal{A}^+_{\text{loc}} \quad \textit{and} \quad \nu(\omega; \{t\} \times \mathbb{R}) \leq 1; \tag{3.34}$$

(b) *the process X is a special semimartingale if and only if*

$$(x^2 \wedge |x|) * \nu \in \mathcal{A}^+_{\text{loc}}; \tag{3.35}$$

(c) *the process X is locally square-integrable if and only if*

$$x^2 * \nu \in \mathcal{A}^+_{\text{loc}}. \tag{3.36}$$

Remark 3.4. One says that X is a locally square-integrable semimartingale, if X is a special semimartingale and, in its canonical decomposition $X = X_0 + A + M$ (with the predictable process $A \in \mathcal{V}$), the local martingale M is locally square-integrable, *i.e.*, such that $\sup_t \mathsf{E} M^2_{t\wedge\tau_n} < \infty$ for a certain localizing system $(\tau_n)_{n\geq 0}$.

3. How can we express in terms of predictable characteristics the property of a semimartingale X to be a local martingale?

Every local martingale is a special semimartingale and thus $(x^2 \wedge |x|) * \nu \in \mathcal{A}^+_{\text{loc}}$.

Let X be a special semimartingale with $X_0 = 0$. Let $X = A + N$ be its canonical decomposition with the process A of bounded variation and a local martingale N. Consider now also a canonical representation of X with $X_0 = 0$ and $B = B(h)$:

$$X = B + X^c + h * (\mu - \nu) + (x - h(x)) * \mu.$$

Because of property $(x^2 \wedge |x|) * \nu \in \mathcal{A}^+_{\text{loc}}$ we can rewrite X in the form

$$X = [X^c + h * (\mu - \nu) + (x - h) * (\mu - \nu)] + [B + (x - h) * \nu].$$

From here we see that

> *a special semimartingale X with triplet of predictable characteristics (B, C, ν) is a local martingale if and only if (up to an evanescent set)*
>
> $$B + (x - h) * \nu = 0.$$

4. The components of the triplet $\mathbb{T} = (B, C, \nu)$ are determined only up to stochastic equivalence and so not uniquely. In many problems of stochastic analysis it is useful to have "good" versions of these predictable characteristics. One of such versions can be described as follows. Starting from the triplet $\mathbb{T} = (B, C, \nu)$, define a predictable process $A = (A_t, \mathcal{F}_t)$ from the class $\mathcal{A}^+_{\text{loc}}$:

$$A_t = \int_0^t |dB_s| + C_t + \int_0^t \int (x^2 \wedge 1)\, d\nu, \tag{3.37}$$

where $\int_0^t |dB_s|$ is the variation of B on $[0, t]$. It is clear that $dB \ll dA$, $dC \ll dA$. Therefore, by the Radon–Nikodým theorem, for B and C there are representations

$$B_t = \int_0^t b_s\, dA_s, \qquad C_t = \int_0^t c_s\, dA_s, \tag{3.38}$$

where $b = (b_s(\omega))_{s\geq 0}$ and $c = (c_s(\omega))_{s\geq 0}$ are predictable processes and $c_s(\omega) \geq 0$.

In [100; Chap. II, Theorem 1.8] it is shown that the measure $\nu = \nu(\omega; dt, dx)$ can be "disintegrated" (in t), using the above-introduced process A, in the following sense: there is a kernel $K = K(\omega, t; dx)$ such that

$$\nu(\omega; dt, dx) = dA_t(\omega)\, K(\omega, t; dx). \tag{3.39}$$

Moreover, the kernel K can be chosen in such a way that (with the notation $K_{\omega,t}(dx)$ for $K(\omega, t; dx)$)

$$K_{\omega,t}(\{0\}) = 0, \qquad \int K_{\omega,t}(dx)\, (|x|^2 \wedge 1) \leq 1,$$

$$\Delta A_t(\omega) > 0 \implies b_t(\omega) = \int K_{\omega,t}(dx)\, h(x)$$

(cf. (3.33)) and $\Delta A_t(\omega)\, K_{\omega,t}(\mathbb{R}) \leq 1$ (cf. (3.34)).

3.3 Stochastic Integrals with Respect to a Brownian Motion, Square-integrable Martingales, and Semimartingales

1. In this section we shall give the essential elements of the theory of stochastic integration with respect to a Brownian motion and square-integrable martingales.

Norbert Wiener seems to be one of the first to define the stochastic integral $(H \cdot W)_t = \int_0^t H_s\, dW_s$ for a Wiener process (Brownian motion) $W = (W_s)_{s\geq 0}$ and *deterministic* functions $H = (H_s)_{s\geq 0}$. In his construction the integral was defined by means of *integration by parts, i.e.*, he put

$$\int_0^t H_s\, dW_s = H_t W_t - \int_0^t W_s H'_s\, ds. \tag{3.40}$$

Remark 3.5. The integral $\int_0^t H_s\,dW_s$ cannot be understood as a trajectory-wise Lebesgue–Stieltjes integral, because the Wiener process has (P-a.s.) unbounded variation ($\int_0^t |dW_s(\omega)| = \infty$).

The next step should consist in construction (not based on ideas of addressing the integration by part formula) of a stochastic integral $(H \cdot Z)_t = \int_0^t H_s\,dZ_s$ for square-integrable processes Z with orthogonal increments ($\mathsf{E}(Z_t - Z_s)(Z_v - Z_u) = 0,\ u < v < s < t$). The importance and necessity to consider such integrals is illustrated by the *spectral* representation of stationary in the wide sense, mean-square-continuous processes: if $X = (X_t)_{t\in\mathbb{R}}$ is such a process, then

$$X_t = \int_{-\infty}^{\infty} e^{it\lambda}\,dZ_\lambda, \tag{3.41}$$

where $Z = (Z_\lambda)_{\lambda\in\mathbb{R}}$ is a process with orthogonal increments.

2. However many questions of stochastic analysis requires the addressing to stochastic integrals of type $H \cdot X$ also for the case that, firstly, H is a random function and, secondly, X is not necessarily a Wiener process or process with orthogonal increments.

Assume, for concreteness and idealizing somewhat, that the prices of a certain financial asset (say, of a stock) are described by a Wiener process $W = (W_t)_{t\geq 0}$. Then if $H = (H_t(\omega))_{t\geq 0}$ is a (*very*) *simple* (piecewise constant, *left*-continuous) function of the form

$$H_t(\omega) = h_0(\omega)I_{\{0\}}(t) + \sum_{i\geq 0} h_i(\omega)I_{(t_i,t_{i+1}]}(t), \tag{3.42}$$

where $0 = t_0 < t_1 < \cdots$ and the variables $h_i(\omega)$ are $\mathcal{F}_{t_i}$-measurable, then the natural definition of the integral $(H \cdot W)_t = \int_0^t H_s\,dW_s$ is to just put

$$(H \cdot W)_t = \sum_{i\geq 0} h_i(\omega)[W_{t_{i+1}} - W_{t_i}]. \tag{3.43}$$

If one interprets the variable $h_i(\omega)$ as an "amount" of stocks of a "buyer" on the time interval $(t_i, t_{i+1}]$, then the variable $h_i(\omega)[W_{t_{i+1}} - W_{t_i}]$ will characterize the change of the capital on $(t_i, t_{i+1}]$. And the total capital, yielded by the considered "strategy" H, at the time t is exactly $(H \cdot W)_t$.

It is clear that considering the functions H specified by the simple functions of the form (3.43) restricts the class of possible "strategies", and a question arises of how to generalize the above definition to a larger collection of "strategies" specified by H. Why can one hope that the stochastic integral $H \cdot W$ and, more generally, the stochastic integral $H \cdot M$, for, say, the class $\mathcal{M}^2$ of square-integrable martingales M, can be defined with the natural properties preserved?

It can be explained as follows. Every square-integrable martingale belongs to the class of L^2-processes Z with orthogonal increments, for which, as already noted, the stochastic integral $H \cdot Z$ is defined for the deterministic functions H. But it is clear that narrowing this class to $\mathcal{M}^2$, we can, in principle, expect that the class of the functions H can be enlarged. This is actually the case—the stochastic integral $H \cdot M$ can be defined for a comparatively large class of *random* functions H.

3. The general idea of definition of the stochastic integral $H \cdot M$, $M \in \mathcal{M}^2$ is that we must consider the functions H which can be approximated (in a certain sense) by the simple functions H^n, $n \geq 0$, for which the stochastic integral $H^n \cdot M$ is defined by the formulae of type (3.43). And then one takes $H \cdot M$ to be equal to the limit (in a certain sense) $\lim H^n \cdot M$.

Thus, first of all we must single out the collection of functions H which can be approximated with simple functions.

If $M = W$ is a Wiener process (= a Brownian motion) and a $\mathcal{B}(\mathbb{R}_+) \otimes \mathcal{F}$-measurable function $H = (H_t(\omega), t \in \mathbb{R}_+, \omega \in \Omega)$ is such that for each $t \in \mathbb{R}_+$

$$H_t(\omega) \text{ is } \mathcal{F}_t\text{-measurable} \tag{3.44}$$

and

$$\mathsf{E} \int_0^t H_s^2(\omega)\, ds < \infty, \qquad t > 0, \tag{3.45}$$

then one succeeds to construct the simple functions H^n, $n \geq 0$, such that

$$\mathsf{E} \int_0^t (H_s - H_s^n)^2\, ds \longrightarrow 0, \qquad n \to \infty, \ t > 0, \tag{3.46}$$

and therefore

$$\mathsf{E} \int_0^t (H_s^n - H_s^m)^2\, ds \longrightarrow 0, \qquad n, m \to \infty, \ t > 0. \tag{3.47}$$

For the simple functions H^n the following property is easy to establish:

$$\mathsf{E}\Big(\int_0^t H_s^n\, dW_s\Big)^2 = \mathsf{E} \int_0^t (H_s^n)^2\, ds, \qquad t > 0. \tag{3.48}$$

Consequently,

$$\mathsf{E}\Big(\int_0^t H_s^n\, dW_s - \int_0^t H_s^m\, dW_s\Big)^2 = \mathsf{E} \int_0^t (H_s^n - H_s^m)^2\, ds \longrightarrow 0, \qquad t > 0.$$

Thus the sequence

$$\left(\int_0^t H_s^n \, dW_s\right)_{n\geq 0}, \qquad t > 0,$$

is *fundamental* in the space L^2 of random variables. This space is complete, and therefore there exists the limit L^2-lim $\int_0^t H_s^n \, dW_s$ which is denoted by $\int_0^t H_s \, dW_s$ (or $(H \cdot W)_t$) and is taken for the definition of the stochastic integral on $[0, t]$ of the function H, satisfying (3.44) and (3.45), with respect to the Wiener process W.

Remark 3.6. The condition (3.45) for the stochastic integral $(H \cdot W)_t$, $t > 0$, to be defined, can be weakened. Namely, assume that, instead of (3.45), the condition

$$\mathsf{P}\left(\int_0^t H_s^2 \, ds < \infty\right) = 1, \qquad t > 0,$$

is fulfilled. For such functions $H = (H_s)_{s\geq 0}$ the stochastic integral can be defined as follows.

Put

$$T^{(n)} = \inf\left\{t : \int_0^t H_s^2 \, ds \geq n\right\},$$

and let $\inf(\varnothing) = \infty$. If $H_s^{(n)} = H_s I(s \geq T^{(n)})$, then, on the one hand, we see that $\mathsf{E} \int_0^t (H_s^{(n)})^2 \, ds < \infty$, $t > 0$, *i.e.*, the condition (3.45) is fulfilled and therefore the integral $H^{(n)} \cdot W$ is defined; on the other hand,

$$\int_0^t (H_s - H_s^{(n)})^2 \, ds \xrightarrow{\mathsf{P}} 0, \qquad t > 0.$$

This condition implies that the sequence $(H^{(n)} \cdot W)_t$ is fundamental in probability, and thus there exists a random variable (denoted by $(H \cdot W)_t$) which is the limit in probability of the variables $(H^{(n)} \cdot W)_t$. This variable $(H \cdot W)_t$ is again denoted by $\int_0^t H_s \, dW_s$ and called the stochastic integral (over $(0, t]$) with respect to the Wiener process of a function H satisfying $\mathsf{P}(\int_0^t H_s^2 \, ds < \infty) = 1$.

4. In essence, the same construction works if instead of a Wiener process W one takes a square-integrable martingale M.

For the simple functions H the property (3.48) is replaced by

$$\mathsf{E}\left(\int_0^t H_s \, dM_s\right)^2 = \mathsf{E} \int_0^t H_s^2 \, d\langle M\rangle_s, \qquad t > 0, \tag{3.49}$$

where $\langle M \rangle$ is the quadratic characteristic of the martingale $M \in \mathcal{M}^2$. Accordingly, the condition (3.45) has to be replaced by

$$\mathsf{E} \int_0^t H_s^2 \, d\langle M \rangle_s < \infty, \qquad t > 0. \tag{3.50}$$

In order to follow the way described above for a Wiener process, we should investigate the following problem of "measurability": for which functions $H = (H_t)_{t \geq 0}$ is there a sequence of simple functions H^n, $n \geq 0$, such that

$$\mathsf{E} \int_0^t (H_s - H_s^n)^2 \, d\langle M \rangle_s \longrightarrow 0, \qquad n \to \infty, \ t > 0\,? \tag{3.51}$$

In [125] this problem got an exhaustive treatment which explains, in particular, the role of the notion 'predictability' (of the functions H) in the construction of the stochastic integrals $H \cdot M$.

In the Wiener case the "measurability" claim imposed on the process $H = (H_t)_{t \geq 0}$ was in some sense quite weak: we assumed only that H is $\mathcal{F}(\mathbb{R}_+) \otimes \mathcal{F}$-measurable and for every $t \geq 0$ the variable $H_t(\omega)$ is $\mathcal{F}_t$-measurable. In the general case the following results hold (Lemmas 5.3, 5.4 and 5.5 in [125]):

A. In the case of square-integrable martingales M such that $\langle M \rangle$ is *absolutely continuous* with respect to Lebesgue's measure Leb ($d\langle M \rangle \ll d\mathrm{Leb}$), any $\mathcal{F}(\mathbb{R}_+) \otimes \mathcal{F}$-measurable function $H = (H_t)_{t \geq 0}$ such that the H_t are $\mathcal{F}_t$-measurable and satisfy (3.50), can be approximated (in the sense of (3.51)) by the simple functions H^n, $n \geq 0$.

B. If the quadratic characteristic $\langle M \rangle$ is a *continuous* process, the mere $\mathcal{F}_t$-measurability of the variables H_t, generally speaking, does no longer suffice for the approximation by the simple functions.
In this case we must claim that for any finite Markov time τ the variables H_τ are $\mathcal{F}_\tau$-measurable. This property is certainly satisfied if the process $H = (H_t)_{t \geq 0}$ is progressively measurable in the following sense.
We shall say that a collection of events $A \in \mathbb{R}_+ \times \Omega$ is a *system of progressively measurable sets*, if for any $t \in \mathbb{R}_+$

$$\{(s, \omega) : s \in [0, t], \omega \in \Omega \text{ such that } I_A(s, \omega) \in B\} \in \mathcal{B}([0, t]) \otimes \mathcal{F}_t \tag{3.52}$$

for each Borel set $B \in \mathcal{B}(\mathbb{R})$.
It is not difficult to make certain that this system (denoted by $\mathcal{P}\mathrm{rog}$) is a σ-algebra and

$$\mathcal{P} \subseteq \mathcal{O} \subseteq \mathcal{P}\mathrm{rog} \subseteq \mathcal{B}(\mathbb{R}_+) \otimes \mathcal{F}_\infty. \tag{3.53}$$

The process $H = (H_t)_{t \geq 0}$ is called *progressively measurable* if the mapping $(t, \omega) \rightsquigarrow H_t(\omega)$ is $\mathcal{P}$rog-measurable.

C. Finally, in the general case of *arbitrary* quadratic characteristics $\langle M \rangle$, for the possibility of approximation of H in the sense of (3.51) it suffices to demand the *predictability* property for H (*i.e.*, the mapping $(t, \omega) \rightsquigarrow H_t(\omega)$ must be $\mathcal{P}$-measurable). Thus, if, for a square-integrable martingale, we require nothing of its quadratic characteristic, the collection of functions H must be restricted to those that are predictable.

5. We considered above the stochastic integrals $(H \cdot M)_t$ for any *fixed* $t \in (0, \infty)$. Often one proceeds in a somewhat different way, defining first the integral $(H \cdot M)_\infty \equiv \int_0^\infty H_s \, dM_s$, and putting next

$$(H \cdot M)_t = \int_0^\infty I(s \leq t) H_s \, dM_s. \tag{3.54}$$

In essence, the construction of the integral $(H \cdot M)_\infty$ is no different from that of the integrals $(H \cdot M)_t$ for finite $t \in (0, \infty)$. One should only write ∞ instead of t in all the formulae (as, for example, in (3.50), (3.51)).

As to (3.54), its validity can be inferred from the constructions of the integrals $(H \cdot M)_t$ and $(I(\cdot \leq t) H \cdot M)_\infty$.

6. In the case of simple functions H the integrals $(H \cdot M)_t$, $t \geq 0$, considered as processes in t, are (P-a.s.) continuous provided M is a continuous and square-integrable martingale. It is natural to expect that the same is true for the functions $H = (H_t)_{t \geq 0}$ of the class for which the conditions (3.50) are fulfilled and, therefore, the integrals $(H \cdot M)_t$, $t \geq 0$, are well defined.

Let H^n, $n \geq 0$, be simple functions satisfying (3.51). Then, by the *Kolmogorov–Doob inequality* (called also *Doob inequality*) for square-integrable martingales [100; Chap. I, Theorem 1.43], we find that for any $T < \infty$

$$\mathsf{P}\left\{ \sup_{t \leq T} \left| \int_0^t H_s^n \, dM_s - \int_0^t H_s^m \, dM_s \right| \geq \lambda \right\} \leq \frac{1}{\lambda^2} \mathsf{E} \left| \int_0^T (H_s^n - H_s^m) \, dM_s \right|^2$$
$$= \frac{1}{\lambda^2} \mathsf{E} \int_0^T (H_s^n - H_s^m)^2 \, d\langle M \rangle_s. \tag{3.55}$$

Choose a sequence of simple functions H^n, $n \geq 0$, such that

$$\mathsf{E} \int_0^T (H_s^{n+1} - H_s^n)^2 \, d\langle M \rangle_s \leq 2^{-n}, \qquad n \geq 1. \tag{3.56}$$

It is clear by (3.49) and (3.56), that for every $t \le T$ the series

$$\int_0^t H_s^1\, dM_s + \sum_{n=1}^{\infty} \int_0^t (H_s^{n+1} - H_s^n)\, dM_s \tag{3.57}$$

converges in mean square, and, by (3.55) and (3.56), assuming $H^0 = 0$, we find that

$$\sum_{n=0}^{\infty} \left\{ \sup_{t \le T} \left| \int_0^t (H_s^{n+1} - H_s^n)^2\, d\langle M \rangle_s \right| \ge \frac{1}{n^2} \right\} \le \sum_{n=0}^{\infty} \frac{n^4}{2^n} < \infty.$$

Whence, by the Borel–Cantelli lemma [160; Chap. II, § 10], there exists an $N(\omega)$ such that for all $n \ge N(\omega)$ (P-a.s.)

$$\sup_{t \le T} \left| \int_0^t \big(H_s^{n+1}(\omega) - H_s^n(\omega)\big)^2\, dM_s(\omega) \right| \le \frac{1}{n^2}.$$

Consequently, the series (3.57), which consists of continuous functions, is uniformly convergent (P-a.s.) on $[0, T]$. It follows that the process $\int_0^t H_s\, dM_s$, $t \le T$, is continuous on $[0, T]$ for any $T < \infty$.

7. Now we address the construction of stochastic integrals $H \cdot X$ for *semimartingales* $X = X_0 + A + M$ and certain integrands H. We shall realize this construction in three steps:

(A) $H \cdot A$ for an arbitrary process A of bounded variation;

(B) $H \cdot M$ for an arbitrary martingale M;

and, finally,

(C) $H \cdot X$ for an arbitrary semimartingale X,

basing ourselves on the following considerations. If X is a semimartingale which admits the decomposition

$$X = X_0 + A + M, \tag{3.58}$$

and the integrals $H \cdot A$ and $H \cdot M$ are already defined, then it is natural to understand $H \cdot X$ as the sum of two integrals $H \cdot A$ and $H \cdot M$. Certainly, because of possible nonuniqueness of the representation (3.58) it will be necessary to show that the so-defined integral $H \cdot X$ *does not depend* on the form of representation (3.58).

Case (A). Let $A = (A_t, \mathcal{F}_t)$ be a process of bounded variation and let

$$L^{\circ}_{\mathrm{var}}(A) = \{H : H = (H_t, \mathcal{F}_t) \text{ be an optional process with } \textstyle\int_0^t |H_s(\omega)|\, d\mathrm{Var}\,(A)_s(\omega) < \infty,\ \omega \in \Omega,\ t > 0\}.$$

It is natural to define the integral $H \cdot A$, or, more exactly, the process-integral $(H \cdot A_t)_{t \ge 0}$, as a trajectory-wise Lebesgue–Stieltjes integral

$$H \cdot A_t = \int_0^t H_s(\omega)\, dA_s(\omega), \qquad \omega \in \Omega,\ t > 0.$$

For all t the variables $H \cdot A_t$ are $\mathcal{F}_t$-measurable.

If one considers a slightly more narrow class of integrands H:

$$L_{\text{var}}(A) = \{H : H = (H_t, \mathcal{F}_t) \text{ is a predictable process} \\ \text{and } \textstyle\int_0^t |H_s(\omega)|\, d\text{Var}\,(A)_s(\omega) < \infty,\ \omega \in \Omega,\ t > 0\,\} \tag{3.59}$$

and assumes that the process A is predictable, then for $H \in L_{\text{var}}(A)$ the process $H \cdot A$ will be predictable as well [100; Chap. I, Proposition 3.5].

The following properties can be verified without great difficulty:

(a′) if $H \in L_{\text{var}}(A)$, then $H \cdot A \in \mathcal{V}$;

(b′) if $c_i \in \mathbb{R}$, $H_i \in L_{\text{var}}(A)$, $i = 1, 2$, then $c_1 H_1 + c_2 H_2 \in L_{\text{var}}(A)$ and $(c_1 H_1 + c_2 H_2) \cdot A = c_1 (H_1 \cdot A) + c_2 (H_2 \cdot A)$;

(c′) if $H \in L_{\text{var}}(A_1) \cap L_{\text{var}}(A_2)$ and $c_1, c_2 \in \mathbb{R}$, then $H \in L_{\text{var}}(c_1 A_1 + c_2 A_2)$ and

$$H \cdot (c_1 A_1 + c_2 A_2) = c_1 (H \cdot A_1) + c_2 (H \cdot A_2);$$

(d′) if $H \in L_{\text{var}}(A)$, then

$$\text{Var}\,(H \cdot A)_\infty = \int_0^\infty |H_s|\, d\text{Var}\, A_s;$$

(e′) if $H \in L_{\text{var}}(A)$ and a set $D \in \mathcal{P}$, then

$$I_D \cdot (H \cdot A) = (H I_D) \cdot A;$$

(f′) if $H \in L_{\text{var}}(A)$, then

$$\Delta(H \cdot M) = H \Delta M;$$

(g′) if $H \in L_{\text{var}}(A)$ and $H^n = H I(|H| \le n)$, then (uniformly in probability)

$$\sup_{s \le t} |(H^n \cdot A)_s - (H \cdot A)_s| \xrightarrow{\mathsf{P}} 0.$$

Case (B). Let $M \in \mathcal{M}_{\text{loc}}$. Introduce the following class of integrands:

$$L^1(M) = \big\{H : H = (H_t, \mathcal{F}_t) \text{ is a predictable process} \\ \text{and } \mathsf{E}\big(\textstyle\int_0^\infty H_t^2\, d[M]_t\big)^{1/2} < \infty\big\}. \tag{3.60}$$

Let $L^1(M)$ stand for the space of equivalence classes of elements from $L^1(M)$ determined by the following equivalence property:

$$H \sim K \iff \int_0^\infty (H_t - K_t)^2\, d[M]_t = 0 \quad (\mathsf{P}\text{-a.s.}).$$

Definition 3.3. An integrand $H = (H_t)_{t \ge 0}$ will be called (*stochastically*) *simple* (or a *simple* random function), if H is of the form

$$H_t(\omega) = h_0(\omega) I(t = 0) + \sum_{k=0}^{n} h_k(\omega) I(\tau_k(\omega) < t \le \tau_{k+1}(\omega)), \tag{3.61}$$

where $0 = \tau_0 \le \tau_1 \le \cdots \le \tau_{n+1}$ are stopping times, $n \ge 0$, and $h_k(\omega)$ are bounded $\mathcal{F}_{\tau_k}$-measurable random variables. (Cf. (3.42).)

For every (stochastically) simple integrand H the stochastic integral is defined in the following natural way:

$$(H \cdot M)_t = \sum_{k=0}^{n} h_k (M_{t\wedge\tau_{k+1}} - M_{t\wedge\tau_k}). \tag{3.62}$$

It is not difficult to verify that in the case under consideration—the case of simple functions—the following properties are fulfilled:

$$H \cdot M \in \mathcal{M}_{\mathrm{loc}}, \tag{3.63}$$

$$[H \cdot M]_t = \int_0^t H_s^2 \, d[M]_s. \tag{3.64}$$

Let $H \in L^1(M)$, and let

$$\|H\|_{L^1(M)} = \mathsf{E}\Big(\int_0^\infty H_t^2 \, d[M]_t\Big)^{1/2}. \tag{3.65}$$

One verifies directly that the function $\|\cdot\|_{L^1(M)}$ is a *norm* (in $L^1(M)$).

Lemma 3.1. *The set of (stochastically) simple functions which belong to $L^1(M)$ is tight in $L^1(M)$ in the norm $\|\cdot\|_{L^1(M)}$.*

Proof. Let $H \in L^1(M)$. Since $[M]^{1/2} \in \mathcal{A}^+_{\mathrm{loc}}$, there exists a sequence $(\sigma_n)_{n\geq 0}$ of stopping times such that

$$\mathsf{E}\Big(\int_0^{\sigma_n} d[M]_t\Big)^{1/2} < \infty. \tag{3.66}$$

Put $H_t^n = H_t I(t \leq \sigma_n)$. Then, since $H \in L^1(M)$, we have $H^n \overset{L^1(M)}{\longrightarrow} H$, and hence we can assume from the beginning that

$$\mathsf{E}[M]_\infty^{1/2} = \mathsf{E}\Big(\int_0^\infty d[M]_t\Big)^{1/2} < \infty. \tag{3.67}$$

Fix $\lambda \in \mathbb{R}$. By (3.67) every process of the form λI_D, where $D \in \mathcal{P}$, belongs to $L^1(M)$.

Let

$$\mathcal{D}_\lambda = \{D \in \mathcal{P} : \lambda I_D \text{ can be approximated in } L^1(M)\text{-norm by the (stochastically) simple functions}\}.$$

The set $\mathcal{D}_\lambda$ is a monotone class and contains both the sets $\{0\} \times B$ with $B \in \mathcal{F}_0$ and the sets $[\![0,\tau]\!] = \{(t,\omega) : 0 \leq t \leq \tau(\omega)\}$ (and also the sets $(s,t] \times B$ with $B \in \mathcal{F}_s$).

Thus, by the theorems on monotone classes ([160] or Lemma 3.2 below), we obtain $\mathcal{D}_\lambda = \mathcal{P}$.

Further, every bounded predictable function H can be *uniformly* approximated (with the help of the common Lebesgue construction) by the finite sums of the form $\sum_k \lambda_k I_{D_k}$ with $\lambda_k \in \mathbb{R}$, $D_k \in \mathcal{P}$.

Thereby every *bounded* predictable function can be approximated in $L^1(M)$-norm by simple (predictable) functions. Finally, for an arbitrary function $H \in L^1(M)$ the sequence of bounded functions

$$H^n = HI(\|H\| \le n) \overset{L^1(M)}{\longrightarrow} H$$

and, therefore, H can be approximated (in $L^1(M)$) by simple functions. □

For the simple integrands $H \in L^1(M)$ the stochastic integral $H \cdot M$ was defined according to (3.62). Let us generalize this definition to the case of arbitrary functions H from the class $L^1(M)$.

To this end introduce the notation:

$$\mathcal{H}^1 = \Big\{ M \in \mathcal{M}_{\text{loc}} : \mathsf{E} \sup_{t \ge 0} |M_t| < \infty \Big\}.$$

If one specifies in $\mathcal{H}^1$ the norm

$$\|M\|_{\mathcal{H}^1} = \mathsf{E} \sup_{t \ge 0} |M_t|,$$

then this space turns into a Banach space.

The next (difficult!) result will be used when defining integrals $H \cdot M$ for $H \in L^1(M)$ (see [146; Chap. IV]).

Theorem 3.3 (Davis' Inequalities). *There exist universal constants $c > 0$ and $C > 0$ such that for any $M \in \mathcal{H}^1$, $M_0 = 0$, the inequalities*

$$c\mathsf{E}[M, M]_\infty^{1/2} \le \mathsf{E} \sup_{t \ge 0} |M_t| \le C\mathsf{E}[M, M]_\infty^{1/2} \tag{3.68}$$

hold.

The proof can be found, for example, in [125].

We have already seen that for every function $H \in L^1(M)$ there exists a sequence $(H_n)_{n \ge 0}$ of (stochastically) simple functions such that $H^n \overset{L^1(M)}{\longrightarrow} H$. The processes $H^n \cdot M$ are local martingales, and the Davis inequalities together with

$$\|H^n - H\|_{L^1(M)} = \mathsf{E}\Big(\int_0^\infty (H_t^n - H_t)^2 \, d[M]_t \Big)^{1/2} \longrightarrow 0, \tag{3.69}$$

imply that the sequence $(H^n \cdot M)_{n\geq 1}$ is fundamental in the Banach space $\mathcal{H}^1$. Therefore there exists the $\mathcal{H}^1$-limit of this sequence.

Definition 3.4. Let $H \in L^1(M)$, where $M \in \mathcal{M}_{\text{loc}}$, $M_0 = 0$. The *stochastic integral* of H with respect to M (notation: $H \cdot M = (H \cdot M_t)_{t\geq 0}$) is the $\mathcal{H}^1$-limit of the sequence $(H^n \cdot M)_{n\geq 0}$, where H^n, $n \geq 0$, satisfy (3.69).

For the correctness of this definition we should notice that the limiting value $H \cdot M$ does not depend on the choice of approximating sequence.

Thus, the stochastic integral $H \cdot M$ is defined for the functions $H \in L^1(M)$. One can go further.

Specify the class

$$L^1_{\text{loc}}(M) = \{H : H = (H_t, \mathcal{F}_t) \text{ is a predictable process such that } \left(\int_0^{\cdot} H_t^2\, d[M]_t\right)^{1/2} \in \mathcal{A}_{\text{loc}}\}.$$

Let $H \in L^1_{\text{loc}}(M)$. Then, by definition of the class $\mathcal{A}_{\text{loc}}$, there exists an increasing sequence of Markov times $(\tau_n)_{n\geq 1}$ such that

$$\mathsf{E}\left(\int_0^{\tau_n} H_t^2\, d[M]_t\right)^{1/2} < \infty, \quad n \geq 1. \tag{3.70}$$

For every $n \geq 1$ the function $HI_{[\![0,\tau_n]\!]} \in L^1(M)$. The above constructions imply that

$$(HI_{[\![0,\tau_{n+1}]\!]} \cdot M)_{t\wedge\tau_n} = (HI_{[\![0,\tau_n]\!]} \cdot M)_t, \quad t \geq 0.$$

Consequently, there is a process (denoted by $H \cdot M$) such that

$$(H \cdot M)_{t\wedge\tau_n} = (HI_{[\![0,\tau_n]\!]} \cdot M)_t \tag{3.71}$$

for any $n \geq 1$ and $t > 0$. Therefore the process $H \cdot M$ does not depend on the choice of localizing sequence $(\tau_n)_{n\geq 1}$ and is unique (up to stochastic indistinguishability).

Definition 3.5. Let $H \in L^1_{\text{loc}}(M)$. The stochastic integral $H \cdot M$ is defined as a process satisfying the condition (3.71).

8. Let us dwell on some properties of stochastic integrals $H \cdot M$ with respect to local martingales M.

(a) If $H_1, H_2 \in L^1_{\text{loc}}(M)$ and constants $c_1, c_2 \in \mathbb{R}$, then $c_1H_1 + c_2H_2 \in L^1_{\text{loc}}(M)$ and

$$(c_1H_1 + c_2H_2) \cdot M = c_1(H_1 \cdot M) + c_2(H_2 \cdot M).$$

(b) If $H \in L^1_{\text{loc}}(M)$, then $H \cdot M \in \mathcal{M}_{\text{loc}}$ and

$$[H \cdot M]_t = \int_0^t H_s^2 \, d[M]_s. \tag{3.72}$$

(c) If τ is a Markov time, $H \in L^1_{\text{loc}}(M)$ and $(H \cdot M)^\tau$ is the "stopped integral" (*i.e.*, $(H \cdot M)^\tau_t = (H \cdot M)_{\tau \wedge t}$), then

$$(H \cdot M)^\tau = H \cdot M^\tau = HI_{[\![0,\tau]\!]} \cdot M.$$

(d) If $H \in L^1_{\text{loc}}$ and a set $D \in \mathcal{P}$, then

$$I_D \cdot (H \cdot M) = (HI_D) \cdot M.$$

(e) If $H \in L^1_{\text{loc}}$, then

$$\Delta(H \cdot M) = H \Delta M.$$

(f) If $H \in L^1(M)$, $H^n = HI(\|H\| \le n)$, then

$$H^n \cdot M \xrightarrow{\mathcal{H}^1} H \cdot M, \qquad n \to \infty.$$

(g) If $H \in L^1_{\text{loc}}(M_1) \cap L^1_{\text{loc}}(M_2)$, $M_1, M_2 \in \mathcal{M}_{\text{loc}}$ and $c_1, c_2 \in \mathbb{R}$, then $H \in L^1_{\text{loc}}(c_1 M_1 + c_2 M_2)$ and

$$H \cdot (c_1 M_1 + c_2 M_2) = c_1 (H \cdot M_1) + c_2 (H \cdot M_2).$$

The properties (a)–(c), (e), and (f) can be deduced directly from the definition of the integral $H \cdot M$ and their evident validity for the simple functions H.

To prove the property (d) consider a sequence of stopping times $(\tau_n)_{n \ge 1}$ such that $\tau_n \uparrow \infty$ and

$$\mathsf{E}[M]^{1/2}_{\tau_n} < \infty, \quad \mathsf{E}\Big(\int_0^{\tau_n} H_t^2 \, d[M]_t\Big)^{1/2} < \infty, \qquad n \ge 1.$$

For a fixed $n \ge 1$ put

$$\mathcal{D}_n = \{D \in \mathcal{P} : I_D \cdot (H \cdot M^{\tau_n}) = (HI_D) \cdot M^{\tau_n}\}.$$

By the property (c), the sets of the form $[\![0, \tau]\!]$ with the τ being Markov time belong to the class $\mathcal{D}_n$ for any $n \ge 1$. This implies that this class is monotone. It is also clear that $\mathcal{D}_n$ contains all the sets of the form $\{0\} \times B$, where $B \in \mathcal{F}_0$.

Now recalling the structure of the σ-algebra $\mathcal{P}$ and using, as was done earlier, the theorem on monotone classes, we conclude that $\mathcal{D}_n = \mathcal{P}$ for every $n \ge 1$. Applying (c) gives the proof of the required property (d).

Remark 3.7. The theorems on monotone classes (of sets and functions), based on ideas of consideration of "approaching sets" and "approaching functions" provide a powerful tool for proving assertions of type (d).

There are various versions of such theorems (see, for example, [146]). Note that when referring above to the theorem on monotone classes, we have in mind the following version.

Lemma 3.2 ("on monotone classes of sets"). *Let $\mathcal{A}$ be a certain family of subsets of the space Ω closed under finite numbers of intersections ($A, B \in \mathcal{A} \Rightarrow A \cap B \in \mathcal{A}$). Then the minimal monotone class $\mu(\mathcal{A})$ which contains all the sets from $\mathcal{A}$ coincides with the σ-algebra $\sigma(\mathcal{A})$ generated by the system $\mathcal{A}$: $\mu(\mathcal{A}) = \sigma(\mathcal{A})$.*

It is appropriate to mention here that a system $\mathfrak{M}$ of subsets of the space Ω is called a *monotone class*, if
(i) $\varnothing, \Omega \in \mathfrak{M}$;
(ii) if $A, B \in \mathfrak{M}$ and $A \subseteq B$, then $B \setminus A \in \mathfrak{M}$;
(iii) if $A, B \in \mathfrak{M}$ and $A \cap B = \varnothing$, then $A \cup B \in \mathfrak{M}$;
(iv) if $A_n \in \mathfrak{M}$, $A_n \subseteq A_{n+1}$, $n \geq 1$, then $\bigcup_{n=1}^{\infty} A_n \in \mathfrak{M}$.
Finally, let us prove the last, often used property (g).
Assume first that the function H is bounded. Then $H \in L^1_{\text{loc}}(c_1 M_1 + c_2 M_2)$, and the desired property can be established in the following way:

1) for the functions H of the form λI_D we make use of the lemma on monotone classes,

and

2) for the bounded functions H of the general form we proceed with the help of their uniform approximation by simple functions.

Now consider any $H \in L^1_{\text{loc}}(M_1) \cap L^1_{\text{loc}}(M_2)$. Let $H^n = HI(|H| \leq n)$. For such functions the property (g) is established next. Since $[c_1 M_1 + c_2 M_2] \leq 2(c_1^2 [M_1] + c_2^2 [M_2])$, we have

$$\int_0^t I(|H_s| \leq n)\, d([c_1 M_1 + c_2 M_2])_s \leq 2c_1^2 \int_0^t I(|H_s| \leq n)\, d[M_1]_s + 2c_2^2 \int_0^t I(|H_s| \leq n)\, d[M_2]_s.$$

From this we deduce that $H^n \in L^1_{\text{loc}}(c_1 M_1 + c_2 M_2)$. The required property (g) follows from its validity for bounded functions H (proved above) with the subsequent use of localization and of the property (f).

9. A Brownian motion B provides a classical example of a (continuous) local martingale. Compare the just given definition of the stochastic integral $H \cdot B$ with that introduced in Subsec. 3 (with Wiener process W = Brownian motion B). In the frame of Definition 3.5 the class $L^1_{\mathrm{loc}}(B)$ consists of the predictable functions H for which

$$\left(\int_0^{\cdot} H_t^2\,dt\right)^{1/2} \in \mathcal{A}_{\mathrm{loc}}.$$

This means that, for a certain "localizing" sequence $(\tau_n)_{n\geq 1}$,

$$\mathsf{E}\left(\int_0^{\tau_n} H_t^2\,dt\right)^{1/2} < \infty.$$

It follows from this that for any $T > 0$

$$\mathsf{E}\left(\int_0^{\tau_n \wedge T} H_t^2\,dt\right)^{1/2} < \infty,$$

and hence

$$\int_0^{\tau_n \wedge T} H_t^2\,dt < \infty \quad (\mathsf{P}\text{-a.s.}).$$

Passing to the limit as $n \to \infty$ and taking account of $\tau_n \uparrow \infty$, we find that in the case under consideration

$$H \in L^1_{\mathrm{loc}}(B) \implies \mathsf{P}\left(\int_0^t H_s^2\,ds < \infty\right) = 1,\ t > 0.$$

Actually, the reverse implication is also true. Indeed, by virtue of continuity of the process $(\int_0^t H_s^2\,ds)_{t\geq 0}$, for the times

$$S_n = \inf\left\{t : \int_0^t H_s^2\,ds \geq n\right\}$$

(where we use that $\inf(\varnothing) = \infty$), the following property is evident:

$$\mathsf{E}\int_0^{S_n} H_s^2\,ds < \infty.$$

Since $S_n \to \infty$, $n \to \infty$ (it can happen that $S_n = \infty$ already for a certain finite n), we have $\int_0^{\cdot} H_s^2\,ds \in \mathcal{A}_{\mathrm{loc}}$.

Thus, for the *predictable* functions, we have obtained the stochastic integral $H \cdot W$ for a Wiener process and $\mathcal{B}(\mathbb{R}_+) \otimes \mathcal{F}$-measurable functions H such that $\mathsf{P}(\int_0^t H_s^2\,ds < \infty) = 1$, $t > 0$, as a particular case from the general construction of stochastic integrals $H \cdot M$ for $M \in \mathcal{M}_{\mathrm{loc}}$ and $H \in L^1_{\mathrm{loc}}(M)$.

It is important to emphasize here that in reality the predictability property of H can be relaxed—it suffices to demand nothing but $\mathcal{F}_t$-measurability of the variables H_t, which follows from Remark 3.6. It is, of course, due to the specific character of the Wiener process, which allows one, in the end, to slightly weaken the measurability conditions for the process $H = (H_t)_{t \geq 0}$.

10. Case (C). Let us proceed directly to the definition of the stochastic integral $H \cdot X$ when X is a semimartingale.

Assume that X admits the decomposition

$$X = X_0 + A + M, \tag{3.73}$$

where $A \in \mathcal{V}$ and $\mathcal{M} \in \mathcal{M}_{\text{loc}}$. Certainly, it is natural to define $H \cdot X$ by the formula

$$H \cdot X = H \cdot A + H \cdot M \tag{3.74}$$

for the functions H for which both of the integrals $H \cdot A$ and $H \cdot M$ are defined.

Definition 3.6. Let $X = (X_t, \mathcal{F}_t)$ be a semimartingale with a certain decomposition (3.73), where $A_0 = M_0 = 0$, $A \in \mathcal{V}$, $M \in \mathcal{M}_{\text{loc}}$. Let

$$H \in L_{\text{var}}(A) \cap L^1_{\text{loc}}(M). \tag{3.75}$$

The *stochastic integral* $H \cdot X$ is defined by (3.74), where $H \cdot A$ is the trajectory-wise Lebesgue-Stieltjes integral and $H \cdot M$ a stochastic integral with respect to the local martingale M.

The space of X-integrable processes H will be denoted by $L(X)$.

Let us analyze the above definition.

(α) Assume a predictable process H is *locally bounded.* Then the property (3.75) will hold for *every* decomposition $X = X_0 + A + M$ with $A \in \mathcal{V}$ and $M \in \mathcal{M}_{\text{loc}}$. Indeed, the property $H \in L_{\text{var}}(A)$ is evident and the property $H \in L^1_{\text{loc}}(M)$ follows from $[M, M]^{1/2} \in \mathcal{A}_{\text{loc}}$.

(β) Let $X \in \mathcal{V} \cap \mathcal{M}_{\text{loc}}$ and $H \in L_{\text{var}}(X) \cap L^1_{\text{loc}}(X)$. Then the integral $H \cdot X$ can be understood, on one hand, as the Lebesgue–Stieltjes integral $H \cdot A$, and, on the other hand, as the stochastic integral $H \cdot M$ with respect to a local martingale M. Certainly, to confirm the correctness of the above definition, we must show that

$$H \cdot A = H \cdot M. \tag{3.76}$$

For the proof of this property, note first of all that without loss of generality we can assume that $H \in L^1(X)$ and $\mathsf{E}[X]^{1/2}_{\infty} < \infty$. (All general considerations reduce to this case with the help of a localization procedure.)

If the function H is bounded, then there is a sequence $(H^n)_{n \geq 1}$ such that $H^n \to H$ uniformly and H^n is of the form $\sum \lambda^n_k I_{D^n_k}$, where $D^n_k \in \mathcal{P}$,

$\lambda_k^n \in \mathbb{R}$. Then $H^n \cdot X = H^n \cdot A \to H \cdot A$ (uniformly in probability), but also $H^n \cdot X = H^n \cdot M \to H \cdot M$ (in the sense of $\mathcal{H}^1$-convergence and hence uniformly in probability). These two convergence results imply that $H \cdot A = H \cdot M$ for all bounded predictable functions H. The general case $H \in L_{\text{var}}(X) \cap L^1(X)$ reduces to the one considered by passing from H to bounded functions $H^n = HI(|H| \leq n)$.

(γ) Assume that $X = X_0 + A + M$ and at the same time $X = X_0 + A' + M'$ with $A, A' \in \mathcal{V}$ and $M, M' \in \mathcal{M}_{\text{loc}}$. Let

$$H \in L_{\text{var}}(A) \cap L^1_{\text{loc}}(M) \cap L_{\text{var}}(A') \cap L^1_{\text{loc}}(M').$$

Then

$$H \cdot A + H \cdot M = H \cdot A' + H \cdot M', \tag{3.77}$$

which confirms the correctness of the above-given definition of the stochastic integral $H \cdot X$, showing the independence from which decomposition ($X = X_0 + A + M$ or $X = X_0 + A' + M'$) is taken as a basis.

The proof of (3.77) follows immediately from the property (g) on page 67, the property (a') in Subsec. 1, and (3.76).

(δ) If X is a semimartingale with the decomposition $X = X_0 + A + M$, $A \in \mathcal{V}$, $M \in \mathcal{M}_{\text{loc}}$, and $H \in L_{\text{var}}(A) \cap L^1_{\text{loc}}(M)$, then the process $H \cdot X$ is a semimartingale.

The proof follows from the property (b) on page 67 and the property (a') in Subsec. 1.

11. The stochastic integral $H \cdot X$ with respect to a semimartingale X exhibits a number of features which can seem unexpected (this is related mainly to the case when the function H is *not* bounded). Consider the following example.

Example 3.1. (M. Émery.) Let τ be a random variable with exponential distribution ($\mathsf{P}(\tau > t) = e^{-t}$). Let η be a random variable such that it does not depend on τ and $\mathsf{P}(\eta = +1) = \mathsf{P}(\eta = -1) = 1/2$. Put

$$X_t = \begin{cases} 0, & t < \tau, \\ \eta/\tau, & t \geq \tau, \end{cases}$$

$\mathcal{F} = \sigma(\tau, \eta)$ and $\mathcal{F}_t = \mathcal{F}_t^X$. Let also

$$M_t = \begin{cases} 0, & t < \tau, \\ \eta, & t \geq \tau, \end{cases}$$

$H_t = 1/t$ for $t > 0$, and $H_0 = 0$.

The process $M \in \mathcal{M}_{\text{loc}}$ and $H \in L_{\text{var}}(M) \subseteq L(M)$ and it is not difficult to see that $X = H \cdot M$. It is clear that $\mathsf{E}|X_t| = \infty$, $t > 0$, and hence the process X, as an integral with respect to a local martingale M, is not a martingale. This process X is not a local martingale either, because for any stopping time T with $\mathsf{P}(T > 0) > 0$ we have $\mathsf{E}|X_T| = \infty$.

Thus, the fact that $M \in \mathcal{M}_{\text{loc}}$ and the integral $H \cdot M$ is well defined does not imply that $H \cdot M \in \mathcal{M}_{\text{loc}}$. (In [100] it is shown that in the case under consideration $H \cdot M$ is a σ-martingale.)

One can also cite instances when the process $A \in \mathcal{V}$ and the integral $H \cdot A$ is defined, but $H \cdot A \notin \mathcal{V}$.

All these facts stress that the stochastic integrals $H \cdot M$, $H \cdot A$ and $H \cdot X$ (in the case of unbounded functions H) should be dealt with quite carefully.

12. Let us dwell on some approaches to the definition of integrals on an infinite half-line $(0, \infty)$.

Recall that in classical analysis

$$\textit{the improper integral} \quad \int_0^\infty h(s)\,ds$$

of a Borel function $h = h(s)$ is defined as the following *limiting formation*:

$$\int_0^\infty h(s)\,ds = \lim_{t\to\infty} \int_{(0,t]} h(s)\,ds,$$

where $\int_{(0,t]} h(s)\,ds$ is the usual Lebesgue–Stieltjes integral over the set $(0, t]$.

But there is another definition:

$$\textit{the integral over } [0,\infty) \quad \int_{[0,\infty)} h(s)\,ds$$

is the usual Lebesgue integral over $[0, \infty)$.

If we denote

$$\begin{aligned} L &= \left\{ h : \forall t \geq 0,\ \int_{[0,t]} |h(s)|\,ds < \infty \right\}, \\ L_{\text{improp}} &= \left\{ h \in L : \exists \lim_{t\to\infty} \int_{[0,t]} h(s)\,ds < \infty \right\}, \\ L_{[0,\infty)} &= \left\{ h : \int_{[0,\infty)} |h(s)|\,ds < \infty \right\}, \end{aligned}$$

then

$$L_{[0,\infty)} \subseteq L_{\text{improp}} \subseteq L.$$

(See details in [46].)

13. So far we assumed that both the semimartingale $X = (X_t)_{t\geq 0}$ and the process $H = (H_t)_{t\geq 0}$ are one-dimensional. However there also exists a well-developed theory of the (vector) stochastic integration with respect to *multidimensional* semimartingales. Detailed presentation of this theory can be found in [100; Chap. III, § 6c] or [44]. As to our exposition, we will dwell on some particular features related to the multidimensionality of the considered processes.

Let $X = (X^1, \ldots, X^d)$ be a d-dimensional semimartingale with (semimartingale) components $X^i = (X_t^i)_{t\geq 0}$, $i = 1, \ldots, d$. Let $H = (H^1, \ldots, H^d)$ be a d-dimensional predictable process with predictable components $H^i = (H_t^i)_{t\geq 0}$. The process H is said to be *integrable with respect to the d-dimensional semimartingale* X, if there exists a representation $X = X_0 + A + M$ such that $H \in L_{\text{var}}(A) \cap L_{\text{loc}}^1(M)$. Under this assumption the *vector stochastic integral* $H \cdot X$, denoted also by $\int_0^\cdot H_s \, dX_s$ or $\int_0^\cdot (H_s, dX_s)$, is defined as in (3.74) by the formula

$$H \cdot X = H \cdot A + H \cdot M.$$

The classes $H \in L_{\text{var}}(A)$ and $L_{\text{loc}}^1(M)$ are defined in [44; §§ 3.1 and 3.2] by analogy with the afore-considered one-dimensional case taking into account however the possible "interference" between the components of the semimartingale $X = (X^1, \ldots, X^d)$. The definition of the vector stochastic integral given in [100; Chap. III, § 6c] uses somewhat different (but equivalent to the cited above) characterization of the integrability of a d-dimensional process $H = (H^1, \ldots, H^d)$.

Note that generally in the d-dimensional case $H \cdot X \neq \sum_{i=1}^d H^i \cdot X^i$.

3.4 Stochastic Differential Equations

1. In addition to the notion of *stochastic integral* (with respect to a Brownian motion, square-integrable martingales, semimartingales) considered in the preceding section, another important subject of stochastic analysis is the notion of *stochastic differential equation.*

It is well known that the modern theory of Markov processes with continuous time originates from the classical work of A. N. Kolmogorov "Analytical methods in probability theory" (1931), [114].

In this work, Kolmogorov introduces a scheme of "stochastically determined process" which describes dynamics in a phase space E of a certain system subject to random impacts. The heart of this description was the transition function $P(s, x; t, A)$ having the following meaning: it is the probability that the state of the "system" at the moment t belongs to the set A under assumption that at the moment s, $s < t$, the state x takes place.

Notice that in "Analytical methods..." Kolmogorov does not work with trajectories, nor uses the terms 'Markov process', 'diffusion process'. His main interest is to investigate the evolution of transition characteristics of "stochastically determined processes" whose primary feature is that the transition function $P(s, x; t, A)$ satisfies the equation ($s < u < t$)

$$P(s, x; t, A) = \int P(s, x; u, dy) P(u, y; t, A) \tag{3.78}$$

which now is called a *Kolmogorov–Chapman equation.*

Since we are concerned with applications of stochastic analysis to the study of financial systems, it is interesting to notice that in "Analytical methods" Kolmogorov appreciates the role of L. Bachelier—well known now as a founder of quantitative approach to the analysis of financial data: "As far back as 1900 Bachelier considered stochastic processes continuous in time", and that § 16 of "Analytical methods" is entitled "Bachelier's case".

Following [114], let us consider first the case where the phase space E is $\mathbb{R}$. Write

$$F(s, x; t, A) = P(s, x; t, (-\infty, y])$$

and

$$f(s, x; t, A) = \frac{\partial F(s, x; t, y)}{\partial y}.$$

We will assume that the transition function $P(s, x; t, A)$ is such that there exist functions $b(s, x)$ and $\sigma^2(s, x)$ for which

$$\begin{aligned}
&\lim_{\Delta \downarrow 0} \frac{1}{\Delta} \int_{-\infty}^{\infty} (y - x) f(s, x; s + \Delta, y)\, dy = b(s, x), \\
&\lim_{\Delta \downarrow 0} \frac{1}{\Delta} \int_{-\infty}^{\infty} (y - x)^2 f(s, x; s + \Delta, y)\, dy = \sigma^2(s, x), \\
&\lim_{\Delta \downarrow 0} \frac{1}{\Delta} \int_{-\infty}^{\infty} |y - x|^{2+\delta} f(s, x; s + \Delta, y)\, dy = 0 \quad \text{for some } \delta > 0.
\end{aligned} \tag{3.79}$$

The coefficients $b(s, x)$ and $\sigma^2(s, x)$ are called (*local*) *drift and diffusion coefficients*, respectively.

Under assumptions (3.79) and some additional assumptions on smoothness of $b(s,x)$ and $\sigma^2(s,x)$ Kolmogorov derived the *backward* parabolic differential equation (in (s,x)):

$$-\frac{\partial f}{\partial s} = b(s,x)\,\frac{\partial f}{\partial x} + \frac{1}{2}\,\sigma^2(s,x)\,\frac{\partial f}{\partial x}, \tag{3.80}$$

and the *forward* parabolic differential equation (in (t,y)):

$$\frac{\partial f}{\partial s} = -\frac{\partial}{\partial y}[b(s,x)f] + \frac{1}{2}\,\frac{\partial^2}{\partial y^2}[\sigma^2(t,y)f] \tag{3.81}$$

(for detail see [114], [82]).

When the phase space E is $\mathbb{R}^d$, the corresponding (backward and forward) equations (under corresponding multivariate conditions (3.79)) have the form

$$-\frac{\partial f}{\partial s} = L(s,x)f \tag{3.82}$$

and

$$\frac{\partial f}{\partial t} = L^*(t,y)f, \tag{3.83}$$

where

$$L(s,x)f = \sum_{i=1}^{d} b_i(s,x)\,\frac{\partial f}{\partial x_i} + \frac{1}{2}\sum_{i,j=1}^{d} a_{ij}(s,x)\,\frac{\partial^2 f}{\partial x_i\,\partial x_j} \tag{3.84}$$

and

$$L^*(s,x)f = -\sum_{i=1}^{d}\frac{\partial}{\partial y_i}[b_i(t,y)f] + \frac{1}{2}\sum_{i,j=1}^{d}\frac{\partial^2}{\partial y_i\,\partial y_j}\,[a_{ij}(t,y)f], \tag{3.85}$$

with

$$a_{ij} = \sum_{k=1}^{d}\sigma_{ik}\sigma_{kj}. \tag{3.86}$$

In the homogeneous case ($a_{ij} = a_{ij}(x)$, $b_i = b_i(x)$) we have $f(s,x;t,y) = f(0,x;t-s,y)$. Let $g(s;t,y) = f(0,x;t,y)$; then g solves the following parabolic equation:

$$\frac{\partial g}{\partial t} = L(x)g, \tag{3.87}$$

where

$$L(x)g = \sum_{i=1}^{d} b_i(x)\,\frac{\partial g}{\partial x_i} + \frac{1}{2}\sum_{i,j=1}^{d} a_{ij}(x)\,\frac{\partial^2 g}{\partial x_i\,\partial x_j}. \tag{3.88}$$

2. How to construct a Markov stochastic process $X = (X_t)_{t \geq 0}$ for which the conditional probability $\mathsf{P}(X_t \in A \,|\, X_s = x)$ coincides with the transition function $P(s, x; t, A)$ satisfying (3.79)? It was this question that induced K. Itô to work up the theory of stochastic differential equations with respect to the Brownian motion (and, more generally, with respect to the processes with independent increments). One of the key points of this theory was construction of a stochastic integral, as described in the previous section.

Now, for the sake of simplicity, we confine ourselves within the one-dimensional case ($E = \mathbb{R}$).

In general terms, a *stochastic differential equation* (with respect to a Brownian motion $B = (B_t)_{t \geq 0}$) for the process $X = (X_t)_{t \geq 0}$ is an equation of the form

$$dX_t = b(t, X_t)\, dt + \sigma(t, X_t)\, dB_t, \qquad X_0 = x_0. \tag{3.89}$$

This equation should be interpreted as a shortened notation for the stochastic integral relation

$$X_t = x + \int_0^t b(s, X_s)\, ds + \int_0^t \sigma(s, X_s)\, dB_s, \qquad t > 0, \tag{3.90}$$

where we assume that (P-a.s.)

$$\int_0^t |b(s, X_s)|\, ds < \infty \tag{3.91}$$

and

$$\int_0^t \sigma^2(s, X_s)\, ds < \infty \tag{3.92}$$

(conditions (3.91), (3.92) guarantee both existence of the stochastic integrals in (3.90) and the P-a.s. finiteness of the right-hand side of (3.90)).

The coefficients $b(t, x)$ and $\sigma(t, x)$ are assumed to be measurable in the pair of variables (t, x). If the identity $\mathsf{P}(X_t \in A \,|\, X_s = x) = \mathsf{P}(s, x; t, A)$ is satisfied, then we assume that $b(t, x)$ and $\sigma(t, x)$ are defined in (3.79).

In fact, the "descriptive" definition of a stochastic differential equation given above needs more precise formulation. Its necessity and substance will be clear from the definition of strong and weak solutions which is given below.

3. We begin with the definition of a strong solution.

Assume given a filtered probability space $(\Omega, \mathcal{F}, (\mathcal{F})_{t \geq 0}, \mathsf{P})$, and let a Brownian motion $B = (B_t)_{t \geq 0}$ be defined on this space. (In most cases one can think of Ω as a space of continuous functions $\omega = \omega(t)$, $t \geq 0$, $\omega(t) \in \mathbb{R}$ [or $\omega(t) \in \mathbb{R}^d$ in the d-dimensional case] endowed with σ-algebra $\mathcal{F}$ of cylindrical sets and σ-algebras $\mathcal{F}_t = \sigma(\omega : \omega(s), s \leq t)$.) The measure P can be interpreted as a Wiener measure P^W; then the process $B = (B_t)_{t\geq 0}$ can be defined canonically: $B_t(\omega) = \omega(t)$, $t \geq 0$.

In what follows, references to 'solutions' of equation (3.89) imply that two measurable functions $b = b(t,x)$ and $\sigma = \sigma(t,x)$ are given and solutions are defined and constructed starting from these functions.

Definition 3.7 (Strong solution). A process $X = (X_t)_{t\geq 0}$ such that

(a) for each $t > 0$ the random variables X_t are $\overline{\mathcal{F}}_t^B$-measurable, where $\overline{\mathcal{F}}_t^B$ is the σ-algebra generated by B_s, $s \leq t$, and by all P-null sets from the σ-algebra $\sigma(B_s, s \geq 0)$;

(b) $\mathsf{P}(X_0 = x) = 1$;

(c) for each $t > 0$ the relations (3.90), (3.91), and (3.92) hold P-a.s.,

is called a *strong solution* (on the given space $(\Omega, \mathcal{F}, (\mathcal{F}_t)_{t\geq 0}, \mathsf{P})$) to the equation (3.89).

A characteristic feature of this definition is that the Brownian motion $B = (B_t)_{t\geq 0}$ is assumed to be given *a priori* and the process $X = (X_t)_{t\geq 0}$ should be constructed upon B in an "adapted" way, in the sense that for each $t \geq 0$ the random variables $X_t(\omega)$ should be $\overline{\mathcal{F}}_t^B$-measurable.

It is clear that existence (and uniqueness) of such a process depend on the structure of the coefficients $b = b(t,x)$ and $\sigma = \sigma(t,x)$ which—together with the (Wiener) measure P^W of the Brownian motion B—determine the distribution law of the process X:

$$(b, \sigma, \mathsf{P}^W) \longrightarrow \mathsf{P}^X \equiv \mathrm{Law}(X). \tag{3.93}$$

A classical result on the existence of a strong solution is the following theorem.

Theorem 3.4 (K. Ito). *Let the coefficients $b = b(t,x)$ and $\sigma = \sigma(t,x)$ satisfy the* LIPSCHITZ CONDITION

$$|b(t,x) - b(t,y)| + |\sigma(t,x) - \sigma(t,y)| \leq C|x - y| \tag{3.94}$$

and GROWTH CONDITION

$$|b(t,x)| + |\sigma(t,x)| \leq C(1 + |x|), \tag{3.95}$$

where $x, y \in \mathbb{R}$ and $C =$ const. Then the equation (3.89) *has a strong solution $X = (X_t)_{t\geq 0}$ and the strong uniqueness takes place,* i.e., *if X and*

Y are two solutions (generated by the same Brownian motion), then these solutions are indistinguishable (i.e., $\mathsf{P}(X_t = Y_t, t \geq 0) = 1$).

The original proof of K. Ito was published in his work [94]. Now this proof is reproduced in many manuals and monographs (see, *e.g.*, [82], [125], [137], [108]).

4. Return now to the question as how to construct a Markov stochastic process $X = (X_t)_{t\geq 0}$ for which the conditional probability $\mathsf{P}(X_t \in A \mid X_s = x)$ coincides with the transition function $P(s, x; t, A)$ obeying the limiting relations (3.79) with given functions $b(t, x)$ and $\sigma(t, x)$.

The processes of such type are commonly called *diffusion processes* (in the sense of Kolmogorov) with drift and diffusion coefficients $b = b(t, x)$ and $\sigma = \sigma(t, x)$.

The following result provides an answer to the question.

Theorem 3.5 (K. Ito). (a) *Let the conditions of Theorem* 3.4 *hold, and let the process $X^{x,s} = (X_t^{x,s})_{t\geq 0}$ solve the equation*

$$X_t^{x,s} = x + \int_s^t b(u, X_u^{x,s})\, du + \int_s^t \sigma(u, X_u^{x,s})\, dB_u \tag{3.96}$$

(*for given x and s*). *Then the process $X^{x,s}$ is a Markov process whose transition function $P(s, x; t, A)$ is determined by the formula*

$$P(s, x; t, A) = \mathsf{P}(X_t^{x,s} \in A).$$

(b) *If, in addition, the functions $b = b(t, x)$ and $\sigma = \sigma(t, x)$ are continuous in t, then a solution to the equation* (3.90) *is a diffusion Markov process* (*in the sense of Kolmogorov*).

For the proof see the books cited after Theorem 3.4.

5. The assumption that the coefficients $b = b(t, x)$ and $\sigma = \sigma(t, x)$ satisfy the Lipschitz condition are in many cases too restrictive. Consider, for example, a problem of optimal control.

Let $u = u(x)$ be a function for which the Lipschitz condition is fulfilled, $|u(x) - u(y)| \leq |x - y|$).

Consider the equation

$$dX_t^u = u(X_t^u)\, dt + dB_t, \qquad X_0^u = x, \quad |x| \leq 1,$$

and let $\tau^u = \inf\{t \geq 0 : |X_t^u| = 1|\}$ is the moment when the process X^u reaches for the first time the boundary ± 1.

We look for a "control" at which

$$\sup_{u \in \mathrm{Lip}} \mathsf{E}_x \tau^u \tag{3.97}$$

is attained (the supremum is taken over the Lipschitz controls). In other words, we are interested in keeping the process between boundaries ± 1 as long as possible (in the sense of the mean value).

It is rather evident that if the process is at the point $y > 0$, then we should take $u(y) = -1$ as a control. If $y < 0$, then we should put $u(y) = 1$. The process X^* corresponding to the function

$$u^*(y) = -\operatorname{sgn} y = -\begin{cases} 1, & y > 0, \\ -1, & y \le 0, \end{cases} \tag{3.98}$$

must solve the equation

$$dX_t^* = -\operatorname{sgn} X_t^* \, dt + dB_t. \tag{3.99}$$

However, the function $u^*(y)$ does not satisfy the Lipschitz condition, and therefore there is no evident response to the question about the existence of a strong solution to equation (3.99), which would determine the "optimal" controled system generated by the Brownian motion. Nevertheless the equation (3.99) happened to have a strong solution which have lead to the more general setting (N. V. Krylov, A. N. Shiryaev) of the question on existence of strong solutions to equations with "bad" coefficients.

The following result is due to A. K. Zvonkin [176].

Theorem 3.6. *Consider the (one-dimensional) stochastic equation* (3.89), *where the functions b and σ satisfy the following conditions*:

(a) $b = b(t, x)$ *is a bounded measurable function*;

(b) $\sigma = \sigma(t, x)$ *is a continuous measurable function such that*

$$|\sigma(t,x)| \ge \varepsilon, \quad \varepsilon > 0, \quad \text{and} \quad |\sigma(t,x) - \sigma(t,y)| \le C\sqrt{|x-y|},$$

where $t \ge 0$ and $x, y \in \mathbb{R}$.

Then the equation (3.79) *has a strong solution and the strong uniqueness takes place.*

In the case of a homogeneous equation (when the coefficients b and σ do not depend on t) there exists a stronger result established by H.-J. Engelbert and W. Schmidt [68]–[70].

Theorem 3.7. *The stochastic differential equation*

$$dX_t = b(X_t)\, dt + \sigma(X_t)\, dB_t, \qquad X_0 = x,$$

where $\sigma \neq 0$ *for all* $x \in \mathbb{R}$,

$$|b(x)| + |\sigma(x)| \leq C(1 + |x|), \qquad x \in \mathbb{R},$$

$$|\sigma(x) - \sigma(y)| \leq C\sqrt{|x - y|}, \qquad x, y \in \mathbb{R},$$

and $b/\sigma^2 \in L^1_{\text{loc}}(\mathbb{R})$, i.e., $\int_{x \in K} |b(x)|/\sigma^2(x)\, dx < \infty$ *for any compact* K, *has a strong solution, and the strong uniqueness takes place.*

Remark 3.8. Initially we defined a strong solution on a *fixed* stochastic basis $(\Omega, \mathcal{F}, (\mathcal{F}_t)_{t \geq 0}, \mathsf{P}))$ with a Brownian motion $B = (B_t)_{t \geq 0}$ given on it. Let $X = (X_t)_{t \geq 0}$ be a certain strong solution (by definition of such a solution, one can represent X in the form $X = \Psi(B)$, where Ψ is an $\overline{\mathcal{F}}^B$-measurable functional (for details see [125; vol. 1, § 7.9]).

Now, if there is another filtered probability space $(\Omega', \mathcal{F}', (\mathcal{F}'_t)_{t \geq 0}, \mathsf{P}'))$ with a Brownian motion $B' = (B'_t)_{t \geq 0}$, then the process $X' = \Psi(B')$, where Ψ is the same functional as above, provides a solution on this new basis. Thus, if, on a certain stochastic basis, one knows a solution, then strong solutions on other bases with Brownian motions given on them, bring, in substance, nothing new.

We now cite several well-known examples, when there exists *no* strong solution. (In [42] one can found many other examples related to strong and weak solutions.)

Example 3.2 (Tanaka). Consider the equation

$$dX_t = \operatorname{sgn} X_t\, dB_t, \qquad X_0 = 0, \tag{3.100}$$

where $B = (B_t)_{t \geq 0}$ is a Brownian motion. Let X be a solution of this equation. Then, by the Lévy characterization theorem [146; Chap. IV, (3.6)], X is a Brownian motion. From (3.100) we find that (P-a.s.)

$$\int_0^t \operatorname{sgn} X_t\, dX_t = \int_0^t \operatorname{sgn}^2 X_t\, dB_t = B_t. \tag{3.101}$$

By Tanaka's formula [146]

$$|X_t| = \int_0^t \operatorname{sgn} X_s\, dX_s + L_t, \tag{3.102}$$

where L_t is the local time ($L_t = L_t(|X|)$) of the process X at zero.

It follows from (3.102) and (3.101) that $\mathcal{F}^B_t = \mathcal{F}^{|X|}_t$. But this conflicts with the fact that the σ-algebra $\mathcal{F}^B_t$ contains events which are designated by $\operatorname{sgn} B$ and which are absent from the σ-algebras generated by the modulus of the process X.

In Example 3.2 the function $\sigma(x) = \operatorname{sgn} x$ changes its sign. One could hope that if $\sigma(x)$ does not change sign, then a strong solution exists. However, such is not the case:

Example 3.3 (Barlow, [6]). There exists a bounded, continuous, positive function $\sigma(x)$ such that the equation

$$dX_t = \sigma(X_t)\,dB_t, \qquad X_0 = x,$$

does not have a strong solution.

6. Equally, as for stochastic differential equations (3.89) of Markov type (where the coefficients $b(t,x)$ and $\sigma(t,x)$ are functions of two variables, time $t \in \mathbb{R}_+$ and phase variable $x \in \mathbb{R}$), stochastic equations whose coefficients depend on the "past" are of great interest:

$$dX_t = b(t,X)\,dt + \sigma(t,X)\,dB_t, \tag{3.103}$$

where $b(t,x)$ and $\sigma(t,x)$ are nonanticipative functionals of $t \in \mathbb{R}_+$ and $x = (x_t,\, t \geq 0) \in C[0,\infty)$, *i.e.*, for every $t \in \mathbb{R}_+$, $b(t,x)$ and $\sigma(t,x)$ are $\mathcal{B}_t$-measurable, $\mathcal{B}_t = \sigma(x \in C[0,\infty)\colon x_s,\ s \leq t)$.

The following result [125; vol. 1, Theorem 4.6] is analogous to Theorem 3.4.

Theorem 3.8. *Let the nonanticipative functionals $b(t,x)$ and $\sigma(t,x)$, where $t \in \mathbb{R}_+$ and $x \in C[0,\infty)$, satisfy the* GLOBAL LIPSCHITZ CONDITION

$$|b(t,x)-b(t,y)|^2+|\sigma(t,x)-\sigma(t,y)|^2 \leq C_1\int_0^t |x_s-y_s|^2\,dK(s)+C_2|x_t-y_t|^2 \tag{3.104}$$

and GLOBAL GROWTH CONDITION

$$b^2(t,x)+\sigma^2 \leq C_1\int_0^t (1+x_s^2)\,dK(s)+C_2(1+x_t^2),$$

where C_1, C_2 are constants and $K = K(s)$ is a nondecreasing right-continuous function, $0 \leq K(s) < \infty$.

Then the equation (3.103) with the initial condition $X_0 = \xi$, where ξ is an $\mathcal{F}$-measurable random variable, $\mathsf{P}(|\xi| < \infty) = 1$, has a (unique) strong solution.

May the global Lipschitz condition in Theorem 3.6 be weakened? In light of Theorem 3.6 it would be natural, for example, to ask the question whether, without this condition, the equation

$$dX_t = b(t,X)\,dt + dB_t, \tag{3.105}$$

—where $b = b(t,x)$, $x \in C[0,\infty)$, is a bounded nonanticipative functional—has a strong solution. However, B. S. Tsirel'son gave the negative answer by constructing in [169] an example of a bounded functional $b = b(t,x)$, for which the equation (3.105) has no strong solution. We describe the construction of this interesting example, following [125; 2nd edn., Chap. 4, § 4].

Example 3.4 (B. S. Tsirelson). Let $(C_1, \mathcal{B}_1)$ be a measurable space of functions $x = (x_t)$, $0 \le t \le 1$, continuous on $[0,1]$ with $x_0 = 0$, where $\mathcal{B}_t = \sigma\{x\colon x_s,\ s \le t\}$.

Consider the numbers t_k, $k = 0, -1, -2, \ldots$, such that $0 < \cdots < t_{-2} < t_{-1} < t_0 = 1$ and define the function $b(t,x)$ such that $b(0,x) = 0$ and

$$b(t,x) = \left\{ \frac{x_{t_k} - x_{t_{k-1}}}{t_k - t_{k-1}} \right\}, \qquad t_k \le t < t_{k+1},$$

where $\{\alpha\}$ stands for the fractional part of α.

For such a function the equation (3.105) takes the form

$$X_{t_{k+1}} - X_{t_k} = \left\{ \frac{X_{t_k} - X_{t_{k-1}}}{t_k - t_{k-1}} \right\} (t_{k+1} - t_k) + \bigl(B_{t_{k+1}} - B_{t_k}\bigr),$$

which is equivalent to the recurrent equations

$$\eta_{k+1} = \{\eta_k\} + \varepsilon_{k+1}, \qquad k = 0, -1, -2, \ldots,$$

where

$$\eta_k = \frac{X_{t_k} - X_{t_{k-1}}}{t_k - t_{k-1}}, \qquad \varepsilon_k = \frac{B_{t_k} - B_{t_{k-1}}}{t_k - t_{k-1}}.$$

From these recurrent equations we have:

$$e^{2\pi i \eta_{k+1}} = e^{2\pi i \{\eta_{k+1}\}} e^{2\pi i \varepsilon_{k+1}} = e^{2\pi i \eta_k} e^{2\pi i \varepsilon_{k+1}}.$$

Denote $\mathsf{E} e^{2\pi i \eta_k}$ by m_k. If our equation (3.105) with the function $b(t,x)$ defined above has a strong solution, then η_k must be $\mathcal{F}^D_{t_k} = \sigma\{B_s,\ s \le t_k\}$-measurable. Hence η_k and ε_{k+1} must be independent. Consequently,

$$m_{k+1} = m_k \mathsf{E} e^{2\pi i \varepsilon_{k+1}} = m_k e^{-2\pi^2/(t_{k+1} - t_k)}$$

Therefore

$$m_{k+1} = m_{k+1-n} \exp\left\{ -2\pi^2 \left(\frac{1}{t_{k+1} - t_k} + \cdots + \frac{1}{t_{k+2-n} - t_{k+1-n}} \right) \right\},$$

thus $|m_{k+1}| \le e^{-2\pi^2 n}$ for any n, whence $m_k = 0$ for all $k = 0, -1, -2, \ldots$.

Next, from the same recurrent equations we have:

$$e^{2\pi i \eta_{k+1}} = e^{2\pi i \eta_k} e^{2\pi i \varepsilon_{k+1}} = \cdots = e^{2\pi i \eta_{k-n}} e^{2\pi i (\varepsilon_{k+1-n} + \cdots + \varepsilon_{k+1})}.$$

Under assumption of existence of a strong solution to the equation (3.105), η_{k-n} should be $\mathcal{F}^B_{t_{k-n}}$-measurable; consequently, if we denote

$$\mathcal{G}^B_{t_{k-n},t_{k+1}} = \sigma\{\omega\colon B_t - B_s,\ t_{k-n} \le s \le t_{k+1}\},$$

then the independence of the σ-algebras $\mathcal{F}^B_{t_{k-n}}$ and $\mathcal{G}^B_{t_{k-n},t_{k+1}}$ implies that

$$\mathsf{E}\Big(e^{2\pi i\eta_{k+1}} \,|\, \mathcal{G}^B_{t_{k-n},t_{k+1}}\Big) = e^{2\pi i(\varepsilon_{k+1-n}+\cdots+\varepsilon_{k+1})}\,\mathsf{E}e^{2\pi i\eta_{k-n}}.$$

Since $m_k = \mathsf{E}e^{2\pi i\eta_k} = 0$, it follows that

$$\mathsf{E}\Big(e^{2\pi i\eta_{k+1}} \,|\, \mathcal{G}^B_{t_{k-n},t_{k+1}}\Big) = 0.$$

Because $\mathcal{G}^B_{t_{k-n},t_{k+1}} \uparrow \mathcal{F}^B_{t_{k+1}}$ as $n \uparrow \infty$, we conclude ("Lévy theorem") that

$$\mathsf{E}\Big(e^{2\pi i\eta_{k+1}} \,|\, \mathcal{F}^B_{t_{k+1}}\Big) = 0.$$

If a strong solution had existed, then the variables η_{k+1} would be $\mathcal{F}^B_{t_{k+1}}$-measurable implying the identity $e^{2\pi i\eta_{k+1}} = 0$ which is clearly impossible.

The contradiction obtained shows that the equation (3.105) with the function $b(t,x)$ does not have a strong solution.

Remark 3.9. When considering strong solutions, we restricted ourselves, for the sake of simplicity, to *one-dimensional* stochastic differential equations. However Theorems 3.4, 3.5, and 3.8 remain true in the multidimensional case.

7. Now we turn to another comprehension of the concept of solution, namely the so-called weak solution. We start with justifying the need for such a modification.

Theorem 3.4 on existence of strong solutions to the equation (3.89) stipulates that the functions $b(t,x)$ and $\sigma(t,x)$, $t \in \mathbb{R}_+$, $x \in \mathbb{R}$, should satisfy the Lipschitz condition. It is rather natural that works appeared, where this condition was weakened. For example, in the monograph by A. V. Skorokhod [166] is shown that if the functions $b(t,x)$ and $\sigma(t,x)$ in (3.89) are bounded and continuous, then the equation (3.89) has a "solution".

However, it turned out that this "solution" was not a strong solution in the sense of Definition 3.7 given above.

Further, in many problems of stochastic control and nonlinear filtration the necessity arises to consider equations (3.103) with coefficients $b(t,x)$ and $\sigma(t,x)$ which depend on "past" values x_s, $s \le t$, and do not satisfy the global Lipschitz condition.

When defining a strong solution, one commonly proceed from the fact that the filtered probability space $(\Omega, \mathcal{F}, (\mathcal{F}_t)_{t\geq 0}, \mathsf{P})$ and a Brownian motion on it are given *a priori*, and one searches for a solution to (3.89) which is a nonanticipative functional $X = \Psi(B)$ of B. Sometimes the definition of a strong solution permits a filtered probability space and a Brownian motion on it to be determined and not given *a priori*.

However, one can go further and look simultaneously for both a filtered probability space and two processes $B = (B_t)_{t\geq 0}$ and $X = (X_t)_{t\geq 0}$, on that space, such that B is a Brownian motion and the stochastic integral $\int_0^t \sigma(s, X_s)\, dB_s$ is well defined for all $t \geq 0$.

There are several different (but equivalent) definitions of weak solutions. One of the first seems to be given in [125] (1974 for Russian edition, 1977 and 2001 for the English).

Definition 3.8 (Weak solution). A collection of objects

$$\mathcal{W} = (\Omega, \mathcal{F}, (\mathcal{F}_t)_{t\geq 0}, \mathsf{P}, B, X) \tag{3.106}$$

such that

(a) $(\Omega, \mathcal{F}, (\mathcal{F}_t)_{t\geq 0}, \mathsf{P})$ is a filtered probability space;
(b) $B = (B_t)_{t\geq 0}$ and $X = (X_t)_{t\geq 0}$ are stochastic processes, B_t and X_t are $\mathcal{F}_t$-measurable;
(c) $B = (B_t)_{t\geq 0}$ is a Brownian motion;
(d) $\mathsf{P}(X_0 = x) = 1$;
(e) for each $t > 0$ the relations (3.90)–(3.92) hold P-a.s.,

is called a *weak solution* to the stochastic differential equation (3.89) with given (x_0, b, σ).

Remark 3.10. It should be emphasized that, in contrast to a strong solution, a weak solution is not required to have the form $X = \Psi(B)$ (*i.e.*, to be constructed over B in a "nonanticipative" way). Moreover, it can happen that, contrariwise, B is a nonanticipative functional of X.

8. Before discussing other definitions of 'weak solutions', consider the question as how to construct the objects $\mathcal{W}$ in (3.106) for the equation (3.89), (3.79) with $\sigma \equiv 1$, *i.e*, for the equation

$$dX_t = b(t, X_t)\, dt + dB_t, \qquad X_0 = 0. \tag{3.107}$$

In what follows, it will be essential that the measure P will be constructed from considerations of "change of measure".

Let Ω be chosen as the (canonical) space $\mathbb{C} = C[0,\infty)$ of continuous functions $\omega = (\omega_t)_{t\geq 0}$, and let $\mathcal{F}_t$ and $\mathcal{F}$ be σ-algebras $\mathcal{B}_t = \sigma(\omega_s, s \leq t)$ and $\mathcal{B} = \sigma(\omega_s, s \geq 0)$. Let P^W be a Wiener measure on $(\mathbb{C}, \mathcal{B})$ which is known to be well defined. With respect to this measure, the process $W = (W_t(\omega))_{t\geq 0}$, where $W_t(\omega) = \omega_t$, is a Wiener process (Brownian motion).

Consider the process

$$B_t(\omega) = W_t(\omega) - \int_0^t b(s, W(s))\, ds \tag{3.108}$$

assuming that $\mathsf{P}^W(\int_0^t |b(s, W(s))|\, ds < \infty) = 1$, $t > 0$.

Let P_t^W be a restriction of the measure P^W onto the σ-algebra $\mathcal{F}_t$ $(= \mathcal{B}_t)$, and let

$$Z_t(\omega) = \exp\left\{\int_0^t b(s, W_s(\omega))\, dW_s(\omega) - \frac{1}{2}\int_0^t b^2(s, W_s(\omega))\, ds\right\}. \tag{3.109}$$

For the process $Z(\omega) = (Z_t(\omega))_{t\geq 0}$ to be well defined, we assume that for all $t > 0$

$$\mathsf{P}^W\left\{\int_0^t b^2(s, W_s(\omega))\, ds < \infty\right\} = 1 \tag{3.110}$$

which guarantees (see Sec. 3.3) the existence of the stochastic integral

$$\int_0^t b(s, W_s(\omega))\, dW_s(\omega).$$

In addition to (3.110), we assume that

$$\mathsf{E}_{\mathsf{P}^W} Z_t(\omega) = 1 \quad \text{for every} \quad t > 0. \tag{3.111}$$

This claim ensures that measures P_t defined by the formula

$$\mathsf{P}_t(d\omega) = Z_t(\omega)\mathsf{P}_t^W(d\omega) \tag{3.112}$$

are probability measures. These probability measures satisfy the consistency conditions

$$\mathsf{P}_t|\mathcal{F}_t = \mathsf{P}_s, \qquad s < t, \tag{3.113}$$

as immediately follows from the fact that the process $(Z_t)_{t\geq 0}$ is a martingale, $\mathsf{E}_{\mathsf{P}^W}(Z_t \mid \mathcal{F}_s) = Z_s$ (P^W-a.s.), $s < t$.

It is well known (see, *e.g.*, [168]) that if the family $\{\mathsf{P}_t, t \geq 0\}$ is consistent, then one can construct a probability measure on $(\mathbb{C}, \mathcal{B})$ such that $\mathsf{P}|\mathcal{F}_t = \mathsf{P}_t$, $t \geq 0$.

Remark 3.11. In general, one cannot construct the measure P by the formula

$$\mathsf{P}(d\omega) = Z_\infty(\omega)\,\mathsf{P}^W(d\omega), \quad \text{where } Z_\infty(\omega) = \lim_{t\to\infty} Z_t(\omega), \tag{3.114}$$

since, evidently, $\mathsf{E}_{\mathsf{P}^W} Z_\infty(\omega)$ can happen not to be equal to 1 and, moreover, it can be equal to 0. However, if for some $\varepsilon > 0$

$$\mathsf{E}_{\mathsf{P}^W} \exp\left\{\frac{1+\varepsilon}{2}\int_0^\infty b^2(s, W_w(\omega))\,ds\right\} < \infty \tag{3.115}$$

(Novikov's condition, [125]), then the measure P can be defined by (3.114), because (3.114) implies that $\mathsf{E}_{\mathsf{P}^W} Z_\infty(\omega) = 1$.

By Girsanov's theorem [125], [100], the process $B = (B_t)_{t\geq 0}$ defined in (3.108) is a Brownian motion with respect to the measure P constructed above. (Of course, with respect to this measure, the process $W = (W_t)_{t\geq 0}$, is no longer a Brownian motion.)

Let us change notation: put $X_t(\omega) = W_t(\omega)$, $t \geq 0$. Then we can rewrite (3.108) as

$$X_t = \int_0^t b(s, X_s)\,ds + B_t \tag{3.116}$$

whose differential form is exactly the equation (3.107)—for which we have just constructed a weak solution. Notice that, unlike strong solutions, for which $X = \Psi(B)$, in the present construction the process B defined by (3.108) is determined from X $(= W)$, *i.e.*, B is a nonanticipative functional of X.

Thus, the collection $\mathcal{W}$ specified in (3.110) is constructed.

It is important to notice that all our considerations were connected with the canonical space $(\Omega, \mathcal{F}) = (\mathbb{C}, \mathcal{B})$, the process X was constructed in a "coordinate" way $(X_t(\omega) = \omega_t,\ t \geq 0)$ and the process B was build by (3.108) as a functional of X. Thus, we see that the heart of the collection $\mathcal{W}$ is the measure P constructed on "change-of-measure" principle by means of the functional $Z = (Z_t)_{t\geq 0}$.

This is why this measure P—and not the whole collection $\mathcal{W}$—is often called a weak solution to the equation (3.107).

The above considerations shows that with respect to the measure P

(a) the process $B_t = X_t - \int_0^t b(s, X_s)\,ds$, $t \geq 0$, is a Brownian motion and thus a martingale ($\mathsf{E}(B_t \mid \mathcal{F}_s) = B_s$, $s \leq t$, where E stands for the expectation with respect to the measure P);

(b) the process $B_t^2 - t$, $t \geq 0$, is a martingale as well (by properties of Brownian motion).

These observations on "martingale" characterizations of the measure P led to the so-called *martingale problem*, which provides a different approach to the concept of a weak solution to the equation (3.89).

Definition 3.9. A probability measure P is said to be *solution to the martingale problem* associated to (x_0, b, σ^2) from (3.89), if for $X_t(\omega) = \omega_t$, $t \geq 0$, the following conditions are fulfilled:

(i) $\mathsf{P}(X_0 = x_0) = 1$;

(ii) $\mathsf{P}(\int_0^t (|b(s, X_s)| + \sigma^2(s, X_s))\, ds < \infty) = 1$, $t \geq 0$;

(iii) the process $M_t = X_t - \int_0^t b(s, X_s)\, ds$, $t \geq 0$, is a P-local martingale (with respect to the filtration $(\mathcal{F}_t)_{t \geq 0}$, where $\mathcal{F}_t = \mathcal{B}_t$);

(iv) the process $M_t^2 - \int_0^t \sigma^2(s, X_s)\, ds$, $t \geq 0$, is also a P-local martingale.

The following theorem establishes the connection between Definition 3.8 and Definition 3.9.

Theorem 3.9. *Let* P *be a probability measure on* $(\mathbb{C}, \mathcal{B})$. *The necessary and sufficient conditions for this measure to be a weak solution to the equation* (3.89) *associated with* (x_0, b, σ) *is that it solves the martingale problem associated with* (x_0, b, σ^2).

For a proof see, for example, [108; Theorem 18.7], [100; Chap. III, Theorem 2.33] or [42; Theorem 1.27].

9. Let us discuss the results about weak solutions. As was mentioned above, the chronologically first was the following theorem.

Theorem 3.10 (A. V. Skorokhod). *Let the coefficients* $b(t, x)$ *and* $\sigma(t, x)$ *in* (3.89) *be bounded and continuous. Then there exists a weak solution.*

In general, it can happen that there is no weak uniqueness (*i.e.*, that there is no uniqueness of the measure P solving the corresponding martingale problem). It is illustrated by the following example.

Example 3.5. Let

$$dX_t = I(X_t = 0)\, dB_t, \qquad X_0 = 0. \tag{3.117}$$

It is evident that this equation has at least two strong solutions, $X_t \equiv 0$ and $X_t \equiv B_t$. The corresponding measures P^0 and P^B provide two weak solutions of the equation (3.117).

The following example is of more interest for us.

Example 3.6 (I. V. Girsanov). Consider the differential equation

$$dX_t = |X_t|^\alpha \, dB_t, \qquad X_0 = 0, \tag{3.118}$$

with $0 < \alpha < 1/2$. It is clear that this equation has a (trivial) strong solution $X_t \equiv 0$ and the corresponding measure P^0 is a weak solution.

Now produce a nontrivial weak solution. When constructing a weak solution to the equation (3.107), we used the method based on

the change of *measure*.

To build a weak solution to the equation (3.117), we will use the method based on

the change of *time*.

(Recall the title of the book, "Change of time and change of measure".)

Let, as in Subsection 8, $(\Omega, \mathcal{F}) = (\mathbb{C}, \mathcal{B})$, and let $W = (W_t(\omega))_{t\geq 0}$ be a Wiener process (with respect to the Wiener measure P^W on $(\mathbb{C}, \mathcal{B})$).

Let $A = (A_t)_{t\geq 0}$ be a process defined by the formula

$$A_t(\omega) = \int_0^t |W_s(\omega)|^{-2\alpha} \, ds, \qquad 0 < \alpha < \frac{1}{2}, \tag{3.119}$$

and, according to (1.1),

$$\widehat{T}(\theta) = \inf\{t \geq 0 : A_t > \theta\}.$$

Since $0 < \alpha < 1/2$, the process A has P^W-a.s. continuous nondecreasing trajectories, $\mathsf{P}^W(A_t < \infty) = 1$ for each $t \geq 0$ and $A_t \to \infty$ P^W-a.s. as $t \to \infty$.

By Lemma 1.1, the family $\widehat{T} = (\widehat{T}(\theta))_{\theta\geq 0}$ is a random change of time. Starting from the process W and $\widehat{T}$, construct a new process

$$\widehat{W}_\theta(\omega) = W_{\widehat{T}(\theta)}(\omega) \tag{3.120}$$

and put

$$\widehat{\mathcal{F}}_\theta = \mathcal{F}_{\widehat{T}(\theta)}.$$

The process $\widehat{W} = (\widehat{W}_\theta(\omega))_{\theta\geq 0}$ is a $(\widehat{\mathcal{F}}_\theta)_{\theta\geq 0}$-local martingale with respect to the measure P^W. (This follows from the optional sampling theorem; for more detail see [146; Proposition 1.5].) Thus

$$\langle \widehat{W}\rangle_\theta = \widehat{T}(\theta). \tag{3.121}$$

Notice that (3.118) implies

$$\widehat{T}(\theta) = \int_0^{\hat{T}(\theta)} |W_s|^{2\alpha}\, dA_s. \tag{3.122}$$

Using (3.122), the formulae of change of variables in the Lebesgue–Stieltjes integral (see Subsec. 2 in Sec. 1.2) and (1.6), (1.10), we find that

$$\widehat{T}(\theta) = \int_0^\theta |\widehat{W}_s|^{2\alpha}\, ds. \tag{3.123}$$

Introduce the process

$$B_\theta = \int_0^\theta |\widehat{W}_s|^{-\alpha}\, d\widehat{W}_s \tag{3.124}$$

(as a stochastic integral with respect to a local martingale).

This process is an $((\widehat{\mathcal{F}}_\theta)_{\theta\geq 0}, \mathsf{P}^W)$-local martingale with the quadratic characteristic

$$\langle B\rangle_\theta = \int_0^\theta |\widehat{W}_s|^{-2\alpha}\, d\langle \widehat{X}\rangle_s. \tag{3.125}$$

Taking into account (3.120) and (3.122) we find that

$$\langle B\rangle_\theta = \int_0^\theta |\widehat{W}_s|^{-2\alpha}\, d\widehat{T}(\theta) = \int_0^\theta |\widehat{W}_s|^{-2\alpha}|\widehat{W}_s|^{2\alpha}\, ds = \theta.$$

Thus, the local martingale $B = (B_\theta)_{\theta\geq 0}$ has the quadratic characteristic $\langle B\rangle = (\langle B\rangle_\theta)_{\theta\geq 0}$ with $\langle B\rangle_\theta = \theta$; therefore, by the Lévy characterization theorem [146; Chap. IV, (3.6)], it is a P^W-Brownian motion.

As earlier, let us change notation: put

$$X_\theta = \widehat{W}_\theta.$$

Then

$$B_\theta = \int_0^\theta |X_s|^{-\alpha}\, dX_s, \tag{3.126}$$

and, consequently,

$$dX_\theta = |X_\theta|^\alpha\, dB_\theta, \qquad \theta \geq 0. \tag{3.127}$$

Thus, the pair of processes $(X_\theta, B_\theta)_{\theta\geq 0}$, where $X_\theta = W_{\widehat{T}(\theta)}$ and B_θ is defined in (3.126), provides a weak solution to the equation (3.127).

10. In the following theorem the condition on the drift coefficient $b(t,x)$ is weakened, but, due to hardened conditions on $\sigma(t,x)$, we get both weak existence and weak uniqueness.

Theorem 3.11 (D. W. Stroock, S. R. S. Varadhan). *Let the coefficient $b(t,x)$ in (3.89) be measurable and bounded, and let the coefficient $\sigma(t,x)$ be continuous, bounded and such that $\sigma(t,x) \neq 0$ for all (t,x). Then there exists a weak solution and it is unique.*

This theorem is proved in [168].

11. Questions as how to weaken conditions on drift and diffusion coefficients in the multidimensional case are paid great attention in the probabilistic literature. One often discusses stochastic differential equations, where the driving process is not only a Brownian motion but also a Poisson random measure. See, for example, [125], [124], [108], [82], [95], [166], [42], [100].

As concerns the more general martingale problem associated with the triplet of predictable characteristics of a semimartingale, see [100; Chap. III, § 2]. In particular, the martingale problem for the diffusion with jumps is discussed in [100; Chap. III, § 2c].

Chapter 4

Stochastic Exponential and Stochastic Logarithm. Cumulant Processes

4.1 Stochastic Exponential and Stochastic Logarithm

1. For the analytic studies of processes with independent increments and, in particular, of Lévy processes the well-known Kolmogorov–Lévy–Khinchin formula plays a crucial role. This formula gives a representation for the characteristic function via the cumulant function, defined by characteristic triplet of the corresponding process.

In the present chapter we give a "stochastic" version of the Kolmogorov–Lévy–Khinchin formula for a more general class of processes, namely, for semimartingales.

All the tools of stochastic calculus we need were essentially discussed in the previous chapter. For simplicity of presentation we shall consider only one-dimensional semimartingales. The corresponding extension to the multidimensional case is rather straightforward.

2. Let $X = (X_t, \mathcal{F}_t)_{t \geq 0}$ be a semimartingale given on a filtered probability space $(\Omega, \mathcal{F}, (\mathcal{F}_t)_{t \geq 0}, \mathsf{P})$ (see Sec. 3.1).

Consider the stochastic differential equation

$$dZ = Z_- \, dX, \qquad Z_0 = 1, \tag{4.1}$$

i.e.,

$$dZ_t = Z_{t-} \, dX_t, \qquad Z_0 = 1, \tag{4.2}$$

where the "solution" $Z = (Z_t, \mathcal{F}_t)_{t \geq 0}$ is assumed to be a semimartingale. We understand these equations as a symbolic form of integral equation (P-a.s.)

$$Z_t = 1 + \int_0^t Z_{s-} \, dX_s, \qquad t > 0. \tag{4.3}$$

The stochastic integral in (4.3) is well defined (Sec. 3.4) and this equation has a unique (strong) solution which is denoted usually by $\mathcal{E}(X) = (\mathcal{E}(X)_t, \mathcal{F}_t)_{t\geq 0}$ and is called the *stochastic exponential.* This name is justified by the analogy to real calculus and by the following formula:

$$\mathcal{E}(X)_t = \exp\left\{X_t - X_0 - \tfrac{1}{2}\langle X^c\rangle_t\right\} \prod_{1\leq s\leq t} (1+\Delta X_s)e^{-\Delta X_s}. \tag{4.4}$$

If $\Delta X_t > -1$ for all $t > 0$, then (4.4) can be written in the "purely exponential" form

$$\mathcal{E}(X)_t = \exp\left\{X_t - X_0 - \tfrac{1}{2}\langle X^c\rangle_t + \log((1+x) - x) * \mu_t^X\right\}. \tag{4.5}$$

The mapping $X \rightsquigarrow \mathcal{E}(X)$ can be inverted. In analogy to real calculus ($x \rightsquigarrow \exp(x)$) we call its converse the *stochastic logarithm* of X. More exactly, suppose that $Z = (Z_t, \mathcal{F}_t)_{t\geq 0}$ is a semimartingale such that Z and Z_- are $\mathbb{R}\setminus\{0\}$-valued. Consider the question about existence of a semimartingale X with $X_0 = 0$ that satisfies the equation

$$Z = Z_0\mathcal{E}(X). \tag{4.6}$$

It is easy to show (see details in [111] and [100]) that there exists a unique (up to indistinguishability) semimartingale X with $X_0 = 0$ such that $Z = Z_0\mathcal{E}(X)$. The solution is given by

$$X = \frac{1}{Z_-}\cdot Z \tag{4.7}$$

or, in detail, $X_t = \int_0^t \frac{dZ_s}{Z_{s-}}$, $t > 0$, $X_0 = 0$. As already mentioned we call the process X in (4.7) the stochastic logarithm of Z and write $X = \mathcal{L}(Z)$ $(= \frac{1}{Z_-}\cdot Z)$.

(Note that the differential form $dX_t = \frac{1}{Z_{t-}}\, dZ_t$ "follows", as one can think, from (4.2); but, of course, it should be proved because the differential equation (4.2) is only the "informal" version of the integral equation (4.3), which has a correct interpretation since the stochastic integral in the right-hand side of (4.3) is well defined.)

3. In real calculus, if $z = e^x$ then $x = \log z$. The following set of formulae summarizes the analogous properties for $Z = \mathcal{E}(X)$ and $X = \mathcal{L}(Z)$, where X and Z are real-valued semimartingales such that Z and Z_- are $\mathbb{R}\setminus\{0\}$-valued:

(a) $\mathcal{E}(X) = \exp\left\{X - X_0 - \dfrac{1}{2}\langle X^c\rangle\right\} \prod_{s\leq \cdot}(1+\Delta X_s)e^{-\Delta X_s}$

and if $\Delta X > -1$, then

$$\mathcal{E}(X) = \exp\left\{X - X_0 - \frac{1}{2}\langle X^c\rangle + \log((1+x) - x) * \mu^X\right\};$$

(b) $\mathcal{L}(Z) = \log\left|\dfrac{Z}{Z_0}\right| + \dfrac{1}{2Z_-^2}\cdot\langle Z^c\rangle - \sum_{s\le\cdot}\left(\log\left|\dfrac{Z_s}{Z_{s-}}\right| + 1 - \dfrac{Z_s}{Z_{s-}}\right)$

or

$\mathcal{L}(Z) = \log\left|\dfrac{Z}{Z_0}\right| + \dfrac{1}{2Z_-^2}\cdot\langle Z^c\rangle - \left(\log\left|1+\dfrac{x}{Z_-}\right| - \dfrac{x}{Z_-}\right) * \mu^Z;$

(c) if ΔX takes values in $\mathbb{R}\setminus\{-1\}$, then $\mathcal{L}(\mathcal{E}(X)) = X - X_0$;

(d) $\mathcal{E}(\mathcal{L}(Z)) = Z/Z_0$.

The property (a) follows by direct application of Itô's formula. For the proof of (b) it is useful to note that, again by Itô's formula,

$$\log|Z| = \log|Z_0| + \frac{1}{Z_-}\cdot Z - \frac{1}{2Z_-^2}\cdot\langle Z^c\rangle + \sum_{s\le\cdot}\left(\log|Z_s| - \log|Z_{s-}| - \frac{1}{Z_{s-}}\,\Delta Z_s\right),$$

where $\frac{1}{Z_-}\cdot Z = \mathcal{L}(Z)$ (by definition).

4. The above considerations on stochastic exponential and logarithm are useful for revealing the relationship between different ways of describing the prices $S = (S_t)_{t\ge 0}$ of financial actives.

There are two basic (multiplicative) ways to represent the positive prices $S = (S_t)_{t\ge 0}$.

Firstly — via the *compound interest* formula $S_t = S_0 e^{H_t}$,

Secondly — via the *simple interest* formula $S_t = S_0\,\mathcal{E}(\widetilde{H})_t$,

for $t > 0$ and where $H = (H_t, \mathcal{F}_t)_{t\ge 0}$ and $\widetilde{H} = (\widetilde{H}_t, \mathcal{F}_t)_{t\ge 0}$ are semimartingales. (In the sequel we put for simplicity $S_0 = 1$, $H_0 = 0$, $\widetilde{H}_0 = 0$.)

Because

$$e^{H_t} = \mathcal{E}(\widetilde{H})_t, \tag{4.8}$$

it is clear that

$$H_t = \log\mathcal{E}(\widetilde{H})_t \tag{4.9}$$

and

$$\widetilde{H}_t = \mathcal{L}(e^{H_t}). \tag{4.10}$$

The properties (a) and (b), given above, lead immediately to the following result: the processes H and $\widetilde{H}$ satisfy the relationships

$$\widetilde{H} = H + \tfrac{1}{2}\langle H^c\rangle + \sum_{0<s\le\cdot}(e^{\Delta H_s} - 1 - \Delta H_s),$$

$$H = \widetilde{H} - \tfrac{1}{2}\langle \widetilde{H}^c\rangle + \sum_{0<s\le\cdot}\left(\log(1+\Delta\widetilde{H}_s) - \Delta\widetilde{H}_s\right)$$

or, equivalently,

$$\widetilde{H} = H + \tfrac{1}{2}\langle H^c \rangle + (e^x - 1 - x) * \mu, \tag{4.11}$$

$$H = \widetilde{H} - \tfrac{1}{2}\langle \widetilde{H}^c \rangle + \big(\log(1+x) - x\big) * \widetilde{\mu}, \tag{4.12}$$

where $\mu = \mu^H$ and $\widetilde{\mu} = \mu^{\widetilde{H}}$ are the jump measures of the processes H and $\widetilde{H}$, respectively.

Note also the following formula:

$$\Delta\widetilde{H} = e^{\Delta H} - 1$$

which is useful especially for the case of discrete time: if for $n \geq 1$

$$S_n = e^{H_n}, \qquad H_n = h_1 + \cdots + h_n$$

and at the same time

$$S_n = \mathcal{E}(\widetilde{H})_n = \prod_{1 \leq k \leq n} (1 + \widetilde{h}_k)$$

then $\widetilde{h}_n = e^{h_n} - 1$ and $h_n = \log(1 + \widetilde{h}_n)$.

5. Besides the presented formulae of connection between H and $\widetilde{H}$ let us give also the corresponding relationships between the triplets $\mathbb{T} = (B, C, \nu)_h$ and $\widetilde{\mathbb{T}} = (\widetilde{B}, \widetilde{C}, \widetilde{\nu})_h$ of the processes H and $\widetilde{H}$.

Suppose that the truncation function $h = h(x)$ (see Sec. 3.2) is the same for processes H and $\widetilde{H}$. Then

$$\begin{aligned} \widetilde{B} &= B + \frac{C}{2} + \big(h(e^x - 1) - h(x)\big) * \nu, \\ \widetilde{C} &= C, \\ I_A(x) * \widetilde{\nu} &= I_A(e^x - 1) * \nu \end{aligned} \tag{4.13}$$

and

$$\begin{aligned} B &= \widetilde{B} - \frac{\widetilde{C}}{2} + \big(h(\log(1+x)) - h(x)\big) * \widetilde{\nu}, \\ C &= \widetilde{C}, \\ I_A(x) * \nu &= I_A(\log(1+x)) * \widetilde{\nu}. \end{aligned} \tag{4.14}$$

The idea of the proof of formula (4.13) (and similarly for (4.14)) is the following. Because $\Delta\widetilde{H} = e^{\Delta H} - 1$ we have $I_A(x) * \widetilde{\mu} = I_A(e^x - 1) * \mu$, where μ and $\widetilde{\mu}$ are the measures of jumps of processes H and $\widetilde{H}$. Therefore $I_A(x) * \widetilde{\nu} = I_A(e^x - 1) * \nu$ (the third formula in (4.13)). Let us consider the canonical representation of H:

$$H = B + H^c + h * (\mu - \nu) + (x - h(x)) * \mu$$

(see (3.30)). Then from (4.11)

$$\begin{aligned}\widetilde{H} &= B + \tfrac{1}{2}\langle H^c\rangle + H^c + h * (\mu - \nu) + (x - h(x)) * \mu + (e^x - 1 - x) * \mu \\ &= B + \tfrac{1}{2}\langle H^c\rangle + H^c + h * (\mu - \nu) \\ &\qquad + (e^x - 1 - h(x)) * (\mu - \nu) + (e^x - 1 - h(x)) * \nu \\ &= B + \tfrac{1}{2}\langle H^c\rangle + H^c + (e^x - 1) * (\mu - \nu) + (e^x - 1 - h(x)) * \nu. \qquad (4.15)\end{aligned}$$

On the other hand, using the canonical representation of $\widetilde{H}$ and the above formulae of connection between $\widetilde{\mu}$ and μ and between $\widetilde{\nu}$ and ν, we find

$$\begin{aligned}\widetilde{H} &= \widetilde{B} + \widetilde{H}^c + h * (\widetilde{\mu} - \widetilde{\nu}) + (x - h) * \widetilde{\mu} \\ &= \widetilde{B} + \widetilde{H}^c + h(e^x - 1) * (\mu - \nu) + (e^x - 1 - h(e^x - 1)) * \mu \\ &= \widetilde{B} + \widetilde{H}^c + h(e^x - 1) * (\mu - \nu) + (e^x - 1 - h(e^x - 1)) * (\mu - \nu) \\ &\qquad\qquad + (e^x - 1 - h(e^x - 1)) * \nu \\ &= \widetilde{B} + \widetilde{H}^c + (e^x - 1) * (\mu - \nu) + (e^x - 1 - h(e^x - 1)) * \nu. \qquad (4.16)\end{aligned}$$

From (4.15) and (4.16) we see that (with $C = \langle H^c\rangle$)

$$\widetilde{B} - \Big[\tfrac{1}{2}C + B + (e^x - 1 - h(x)) * \nu + (e^x - 1 - h(e^x - 1)) * \nu\Big] = H^c - \widetilde{H}^c. \quad (4.17)$$

Here $H^c - \widetilde{H}^c$ is a continuous (and therefore predictable) local martingale and the left-hand side is a process of bounded variation. Thus this local martingale is equal to zero (see [100; Chap. II, 3.16]), *i.e.*, $\widetilde{H}^c = H^c$ (or $\widetilde{C} = C$, that is the second formula in (4.13)). The left-hand side in (4.17) is also equal to zero. So, the first formula in (4.13) does hold.

6. From (4.13) and (4.14) we get the following useful corollaries.

Corollary 4.1. *Suppose that H and $\widetilde{H}$ are Lévy processes* (see details in Chap. 5) *with local characteristics $\mathbb{T}_{\text{loc}} = (b, c, F)_h$ and $\widetilde{\mathbb{T}}_{\text{loc}} = (\widetilde{b}, \widetilde{c}, \widetilde{F})_h$, i.e.,*

$$B_t = b\,t, \qquad C_t = c\,t, \qquad \nu(dt, dx) = dt\,F(dx)$$

and

$$\widetilde{B}_t = \widetilde{b}\,t, \qquad \widetilde{C}_t = \widetilde{c}\,t, \qquad \widetilde{\nu}(dt, dx) = dt\,\widetilde{F}(dx).$$

Then

$$\begin{aligned}&\widetilde{b} = b + \frac{c}{2} + \int \big(h(e^x - 1) - h(x)\big)\,F(dx), \\ &\widetilde{c} = c, \qquad\qquad\qquad\qquad (4.18) \\ &\widetilde{F}(G) = \int I_G(e^x - 1)\,F(dx), \qquad G \in \mathcal{B}(\mathbb{R}),\end{aligned}$$

and

$$\begin{aligned}
b &= \widetilde{b} - \frac{\widetilde{c}}{2} + \int \big[h(\log(1+x)) - h(x)\big]\, \widetilde{F}(dx), \\
c &= \widetilde{c}, \qquad\qquad\qquad\qquad\qquad\qquad\qquad (4.19)\\
F(G) &= \int I_G(\log(1+x))\, \widetilde{F}(dx), \qquad G \in \mathcal{B}(\mathbb{R}).
\end{aligned}$$

Corollary 4.2. *Suppose that H and $\widetilde{H}$ are local Lévy processes,* i.e., *semimartingales with the characteristics $\mathbb{T} = (B, C, \nu)_h$ and $\widetilde{\mathbb{T}} = (\widetilde{B}, \widetilde{C}, \widetilde{\nu})_h$, where*

$$B_t = \int_0^t b_s\, ds, \quad C_t = \int_0^t c_s\, ds, \quad \nu\big((0,t] \times G\big) = \int_0^t F_s(G)\, ds \tag{4.20}$$

and

$$\widetilde{B}_t = \int_0^t \widetilde{b}_s\, ds, \quad \widetilde{C}_t = \int_0^t \widetilde{c}_s\, ds, \quad \widetilde{\nu}\big((0,t] \times G\big) = \int_0^t \widetilde{F}_s(G)\, ds \tag{4.21}$$

with some natural integrability conditions on deterministic functions b_s, c_s, $F_s(G)$ and $\widetilde{b}_s$, $\widetilde{c}_s$, $\widetilde{F}_s(G)$. Then the following analogues of (4.18) *and* (4.19) *hold:*

$$\begin{aligned}
\widetilde{b}_s &= b_s + \frac{c_s}{2} + \int \big(h(e^x - 1) - h(x)\big)\, F_s(dx), \\
\widetilde{c}_s &= c_s, \qquad\qquad\qquad\qquad\qquad\qquad\qquad (4.22)\\
\widetilde{F}_s(G) &= \int I_G(e^x - 1)\, F_s(dx), \qquad G \in \mathcal{B}(\mathbb{R}),
\end{aligned}$$

and

$$\begin{aligned}
b_s &= \widetilde{b}_s - \frac{\widetilde{c}_s}{2} + \int \big[h(\log(1+x)) - h(x)\big]\, \widetilde{F}_s(dx), \\
c_s &= \widetilde{c}_s, \qquad\qquad\qquad\qquad\qquad\qquad\qquad (4.23)\\
F_s(G) &= \int I_G(\log(1+x))\, \widetilde{F}_s(dx), \qquad G \in \mathcal{B}(\mathbb{R}).
\end{aligned}$$

4.2 Fourier Cumulant Processes

1. Let $X = (X_t, \mathcal{F}_t)_{t \geq 0}$ be a real-valued semimartingale with the triplet

$$\mathbb{T} = (B, C, \nu)$$

of predictable characteristics (with respect to a truncation function $h = h(x)$) and such that $X_0 = 0$.

The stochastic process of *bounded variation* $\widetilde{K}(\theta) = (\widetilde{K}(\theta)_t, \mathcal{F}_t)_{t\ge 0}$ with

$$\widetilde{K}(\theta)_t = i\theta B_t - \frac{1}{2}\theta^2 C_t + \int_0^t \int_{\mathbb{R}} (e^{i\theta x} - 1 - i\theta h(x))\, \nu(\omega; ds, dx), \qquad (4.24)$$

where $\theta \in \mathbb{R}$, is called a

Fourier cumulant process

(corresponding to the value θ).

The process $K(\theta) = (K(\theta)_t, \mathcal{F}_t)_{t\ge 0}$ with

$$K(\theta)_t = \log \mathcal{E}(\widetilde{K}(\theta))_t \qquad (4.25)$$

is called a

modified Fourier cumulant process.

It is clear that

$$e^{K(\theta)} = \mathcal{E}(\widetilde{K}(\theta)) \qquad (4.26)$$

and by (4.11) and (4.12)

$$\widetilde{K}(\theta) = K(\theta) + (e^x - 1 - x) * \mu^{K(\theta)}, \qquad (4.27)$$

$$K(\theta) = \widetilde{K}(\theta) + (\log(1+x) - x) * \mu^{\widetilde{K}(\theta)}, \qquad (4.28)$$

where $\mu^{K(\theta)}$ and $\mu^{\widetilde{K}(\theta)}$ are the jump measures of the processes $K(\theta)$ and $\widetilde{K}(\theta)$, respectively. (In what follows, we shall often refer to these processes simply as *cumulant* processes; sometimes instead of $K(\theta)$ and $\widetilde{K}(\theta)$ we write $K(\theta; X)$ and $\widetilde{K}(\theta; X)$ if it is useful to emphasize that these cumulants correspond to the process X.)

2. Having cumulant processes $K(\theta)$ and $\widetilde{K}(\theta)$ let us introduce the stochastic process $M(\theta) = (M(\theta)_t, \mathcal{F}_t)_{t\ge 0}$ with

$$M(\theta)_t = \frac{e^{i\theta X_t}}{\mathcal{E}(\widetilde{K}(\theta))} \quad \left(= \frac{e^{i\theta X_t}}{e^{K(\theta)_t}}\right) \qquad (4.29)$$

assuming that $\Delta\widetilde{K}(\theta) \ne -1$ (or equivalently that $\mathcal{E}(\widetilde{K}(\theta))$ never vanishes). Note that condition $\Delta\widetilde{K}(\theta)_t \ne -1$ does hold, for example, if $\nu(\omega; \{t\}\times dx) = 0$, because in this case, as it is easy to see,

$$\Delta\widetilde{K}(\theta)_t = \mathsf{E}(e^{i\theta\Delta X_t} - 1 \,|\, \mathcal{F}_{t-}) = \int_{\mathbb{R}} (e^{i\theta x} - 1)\, \nu(\omega; \{t\} \times dx) = 0.$$

One general result of the stochastic calculus (see [100; Chap. II, Corollary 2.48]) states that (under condition $\Delta\widetilde{K}(\theta) \ne -1$) the process $M(\theta)$ is a local martingale. Even more, (under the assumption that $\Delta\widetilde{K}(\theta)_t \ne -1$

for all $\theta \in \mathbb{R}$ and $t > 0$) there exists equivalence between the following two conditions:

(i) X is a semimartingale with triplet $\mathbb{T} = (B, C, \nu)$;

(ii) the processes $M(\theta)$ are local martingales for all $\theta \in \mathbb{R}$.

In the particular case where the process X has independent increments the corresponding triplet $\mathbb{T} = (B, C, \nu)$ is *deterministic* (see [100]), the local martingale $M(\theta)$ is a *martingale* and therefore $\mathsf{E}M_t(\theta) = 1$ for each $t \geq 0$. Taking into account the formula (4.29) we find the characteristic function of the process X:

$$\mathsf{E}e^{i\theta X_t} = \mathcal{E}(\widetilde{K}(\theta))_t$$

or, in detail,

$$\mathsf{E}e^{i\theta X_t} = e^{\widetilde{K}(\theta)_t} \prod_{0<s\leq t} (1 + \Delta\widetilde{K}(\theta)_s)e^{-\Delta\widetilde{K}(\theta)_s}. \tag{4.30}$$

If a process with independent increments X is *continuous in probability* then the (deterministic) functions $B = (B_t)_{t\geq 0}$, $C = (C_t)_{t\geq 0}$, and $(\nu((0,t] \times G))_{t\geq 0}$ (where $G \in \mathcal{B}(\mathbb{R})$) are continuous and therefore $\Delta\widetilde{K}(\theta)_t = 0$ for all $\theta \in \mathbb{R}$ and $t > 0$. In this case $\widetilde{K}(\theta) = K(\theta)$ and the expression (4.30) takes the form

$$\mathsf{E}e^{i\theta X_t} = e^{K(\theta)_t} \quad (= e^{\widetilde{K}(\theta)_t})$$

which is the well-known

Kolmogorov–Lévy–Khinchin formula

for the characteristic function $\mathsf{E}e^{i\theta X_t}$ of the continuous in probability process with independent increments $X = (X_t, \mathcal{F}_t)_{t\geq 0}$.

3. In the case of Lévy processes $X = (X_t, \mathcal{F}_t)_{t\geq 0}$ (*i.e.*, processes with stationary independent increments) we have

$$B_t = bt, \qquad C_t = ct, \qquad \nu(dt, dx) = dt\, F(dx) \tag{4.31}$$

and

$$K(\theta)_t = t\, \varkappa(\theta),$$

where the *local Fourier cumulant function*

$$\varkappa(\theta) = i\theta b - \frac{\theta^2}{2} c + \int \left(e^{i\theta x} - 1 - i\theta h(x)\right) F(dx). \tag{4.32}$$

(In (4.31), (4.32) the measure $F = F(dx)$, $x \in \mathbb{R}$, is such that $F(\{0\}) = 0$ and $\int_{\mathbb{R}} (x^2 \wedge 1)\, F(dx) < \infty$.)

As a result we find the following formula:

$$\mathsf{E}e^{i\theta X_t} = e^{t\varkappa(\theta)}, \qquad t > 0, \tag{4.33}$$

or, equivalently,

$$\tfrac{1}{t} \log \mathsf{E}e^{i\theta X_t} = \varkappa(\theta), \qquad t > 0,$$

for all $\theta \in \mathbb{R}$.

4.3 Laplace Cumulant Processes

1. Along with the *complex-valued* Fourier cumulant processes $\widetilde{K}(\theta)$ and $K(\theta)$, connected by the relationship $\mathcal{E}(\widetilde{K}(\theta)) = \exp\{K(\theta)\}$ and for which the process $M(\theta) = (M(\theta)_t, \mathcal{F}_t)_{t\geq 0}$ with

$$M(\theta)_t = \frac{e^{i\theta X_t}}{\mathcal{E}(\widetilde{K}(\theta))_t} \quad \left(= \frac{e^{i\theta X_t}}{e^{K(\theta)_t}}\right)$$

is a local martingale (if $\Delta\widetilde{K}(\theta) \neq -1$), in stochastic calculus (in particular, for construction of probability measures) the *real-valued* analogues of $\widetilde{K}(\theta)$ and $K(\theta)$ prove to be useful. We denote these real-valued processes by $\widetilde{\Lambda}(\theta)$ and $\Lambda(\theta)$ and call them *Laplace cumulant process* and *modified Laplace cumulant process* (of semimartingale X with triplet $\mathbb{T} = (B, C, \nu)$).

By our definition, for each $\theta \in \mathbb{R}$

$$\widetilde{\Lambda}(\theta)_t = \theta B_t + \frac{\theta^2}{2} C_t + \int_0^t \int \big(e^{\theta x} - 1 - \theta h(x)\big)\, \nu(\omega; ds, dx). \tag{4.34}$$

(To guarantee the existence of the integrals in (4.34) we assume that $\int_0^t \int e^{\theta x} I(|x| > 1)\, \nu(\omega; ds, dx) < \infty$ for all $t > 0$, $\theta \in \mathbb{R}$.)

Having the (Laplace) cumulant process $\widetilde{\Lambda}(\theta) = (\widetilde{\Lambda}(\theta)_t, \mathcal{F}_t)_{t\geq 0}$ we define a *modified* (Laplace) cumulant process $\Lambda(\theta)$ by the formula

$$\Lambda(\theta) = \log \mathcal{E}(\widetilde{\Lambda}(\theta))$$

(cf. (4.25)).

By analogy with the process $M(\theta)$ (see (4.29)) we define now for each $\theta \in \mathbb{R}$ the process $N(\theta) = (N(\theta)_t, \mathcal{F}_t)_{t\geq 0}$ with

$$N(\theta)_t = \frac{e^{\theta X_t}}{\mathcal{E}(\widetilde{\Lambda}(\theta))_t} \quad \left(= \frac{e^{\theta X_t}}{e^{\Lambda(\theta)_t}}\right). \tag{4.35}$$

It is very natural to think that this process $N(\theta)$, like the process $M(\theta)$, has some *martingale-type* properties.

Indeed, the following result does hold: *The process $N(\theta)$ is a local martingale (for each $\theta \in \mathbb{R}$).*

The proof of this statement (given in [111]) is going along the following lines.

We have a canonical representation (see (3.30)) for process X:

$$X_t = X_0 + B_t + X_t^c + h * (\mu - \nu)_t + (x - h) * \mu_t, \qquad t \geq 0, \tag{4.36}$$

and also we know that

$$d\mathcal{E}(\widetilde{\Lambda}(\theta))_t = \mathcal{E}(\widetilde{\Lambda}(\theta))_{t-}\, d\widetilde{\Lambda}(\theta)_t. \tag{4.37}$$

Applying the Itô formula to the process $N(\theta)_t = e^{\theta X_t}/\mathcal{E}(\widetilde{\Lambda}(\theta))_t$, $t > 0$, we get that

$$N(\theta)_t = \mathcal{E}\Big(\theta X^c + \frac{e^{\theta x} - 1}{1 + \widehat{W}(\theta)} * (\mu - \nu)\Big)_t, \tag{4.38}$$

where

$$\widehat{W}(\theta)_t = \int_{\mathbb{R}} (e^{\theta x} - 1)\,\nu(\omega; \{t\} \times dx).$$

The process $\theta X^c + (e^{\theta x} - 1)/(1 + \widehat{W}(\theta)) * (\mu - \nu)$ is evidently a local martingale and therefore so is the process $N(\theta)$ (for each $\theta \in \mathbb{R}$).

2. It is important that these processes $N(\theta)$ are nonnegative and determined by the process X and the triplet $\mathbb{T} = (B, C, \nu)$ of the process X. If the process $N(\theta)$ is not only a local martingale but a *martingale*, we may use it for construction of new probability measures $\widetilde{\mathsf{P}}(\theta)_t$, $t \geq 0$, setting

$$d\widetilde{\mathsf{P}}(\theta)_t = N(\theta)_t\, d\mathsf{P}(\theta)_t,$$

where $\mathsf{P}(\theta)_t = \mathsf{P}(\theta)|\mathcal{F}_t$. This method of construction of measures $\widetilde{\mathsf{P}}(\theta)_t$, $t \geq 0$, is called an *Esscher change of measures.* In Chaps. 6 and 7 we shall give more details about this method, with applications to mathematical finance.

3. If $X = (X_t, \mathcal{F}_t)_{t\geq 0}$ is a Lévy process with the triplet of local characteristics $\mathbb{T}_{\text{loc}} = (b, c, F)$, then—similarly to the local Fourier cumulant function $\varkappa(\theta)$ (see (4.32))—one may introduce (under assumption $\int e^{\theta x} I(|x| > 1)\, F(dx) < \infty$) the *local Laplace cumulant function* $\lambda(\theta)$ by the formula

$$\lambda(\theta) = \theta b + \frac{\theta^2}{2} c + \int \big(e^{\theta x} - 1 - \theta h(x)\big)\, F(dx). \tag{4.39}$$

Thus, for Lévy processes $\Lambda(\theta)_t = t\lambda(\theta)$.

4. In the probability theory, it is well known that a lucky choice of a martingale can provide simple solution to various problems regarding the properties of first exit times for random walks.

As an example, consider a symmetric random walk $X = (X_n)_{n\geq 0}$ with $X_0 = 0$ and $X_n = \xi_1 + \dots + \xi_n$ for $n \geq 1$, where ξ_n, $n \geq 1$, are independent and identically distributed with $\mathsf{P}(\xi_n = 1) = \mathsf{P}(\xi_n = -1) = 1/2$. Let $\tau = \inf\{n \geq 1\colon X_n = 1\}$. This time of the first entry to 1 is a Markov time with respect to the filtration $(\mathcal{F}_n)_{n\geq 0}$, where $\mathcal{F}_0 = \{\varnothing, \Omega\}$ and $\mathcal{F}_n = \sigma(\xi_1, \dots, \xi_n)$ for $n \geq 1$.

Consider the sequence $(N(\theta)_n)_{n\geq 0}$, where

$$N(\theta)_n = \frac{e^{\theta X_n}}{\prod_{k=1}^{n} \mathsf{E}(e^{\theta \Delta X_k} \mid \mathcal{F}_{k-1})}, \qquad \Delta X_k = \xi_k$$

(cf. (4.35)) and, therefore,

$$N(\theta)_n = \frac{e^{\theta X_n}}{(\cosh\theta)^n} = e^{\theta X_n}(\operatorname{sech}\theta)^n.$$

It is not difficult to deduce that the sequence $(N(\theta)_n)_{n\geq 0}$ is a martingale. Therefore, by the well-known optional sampling theorem (see [160]), $\mathsf{E}N(\theta)_{n\wedge\tau} = 1$. If we take $\theta > 0$, then, by the bounded convergence theorem, $1 = \mathsf{E}N(\theta)_\tau = \mathsf{E}e^{\theta X_\tau}(\operatorname{sech}\theta)^\tau$. Letting $\theta \downarrow 0$, we find that $\mathsf{E}I(\tau < \infty) = 1$, *i. e.*, $\mathsf{P}(\tau < \infty) = 1$. Consequently, $e^{\theta X_\tau} = e^\theta$ (P-a.s.) and $1 = \mathsf{E}(\operatorname{sech}\theta)^\tau$. If we denote $\operatorname{sech}\theta$ by ρ, then

$$\mathsf{E}\rho^\tau = \sum_{n=1}^{\infty} \rho^n \mathsf{P}(\tau = n) = e^{-\theta} = \frac{1-\sqrt{1-\rho^2}}{\rho}.$$

From this we conclude that

$$\mathsf{P}(\tau = 2k-1) = (-1)^{k+1}\, C_{1/2}^k,$$

where

$$C_{1/2}^k = \frac{(\frac{1}{2})_k}{k!} = \frac{\frac{1}{2}(\frac{1}{2}-1)\cdots(\frac{1}{2}-k+1)}{k!}.$$

4.4 Cumulant Processes of Stochastic Integral Transformation $X^\varphi = \varphi \cdot X$

1. Let $X = (X_t, \mathcal{F}_t)_{t\geq 0}$ be a semimartingale, $X_0 = 0$, with the triplet $\mathbb{T} = (B, C, \nu)$, and let $\varphi = (\varphi_t, \mathcal{F}_t)_{t\geq 0}$ be a predictable process from the class $L(X)$ (see Definition 3.6 in Sec. 3.4). So, the stochastic integral $X^\varphi = \varphi \cdot X$ is well defined, and we recall that $X_t^\varphi = (\varphi \cdot X)_t$ is denoted also by $\int_0^t \varphi_s\, dX_s$ or $\int_{(0,t]} \varphi_s\, dX_s$.

The process X^φ with $\varphi \in L(X)$ is a semimartingale. Denote by

$$\mathbb{T}^\varphi = (B^\varphi, C^\varphi, \nu^\varphi)$$

its triplet of predictable characteristics (with respect to a certain truncation function h).

Instead of the expression θX_t, $t \geq 0$, used in the previous section for construction of the local martingale $N(\theta)_t$, $t \geq 0$, very often it is useful to consider more general expressions $X_t^\varphi = \int_0^t \varphi_s\, dX_s$, $t \geq 0$. (It is clear that if $\varphi_s \equiv \theta$, $s \geq 0$, then $X_t^\varphi = \theta X_t$.)

Having process φ and triplet $\mathbb{T} = (B, C, \nu)$ we want to find now the triplet $\mathbb{T}^\varphi = (B^\varphi, C^\varphi, \nu^\varphi)$ which will be useful, *e.g.*, for construction of Esscher's change of measures based on the process $N^\varphi = (N_t^\varphi, \mathcal{F}_t)_{t\geq 0}$ with

$$N_t^\varphi = \exp\left\{\int_0^t \varphi_s\, dX_s\right\}\left[\mathcal{E}(\widetilde{\Lambda}^\varphi)\right]_t^{-1}, \tag{4.40}$$

where the process $\widetilde{\Lambda}^\varphi$ (which is an analogue to $\widetilde{\Lambda}(\theta)$) is defined in (4.50) below.

2. We claim that if $X^\varphi = \varphi \cdot X$ then the corresponding triplet $(B^\varphi, C^\varphi, \nu^\varphi)$ will be given by the following formulae:

$$\begin{aligned} B^\varphi &= \varphi \cdot B + \big(h(\varphi x) - \varphi h(x)\big) * \nu, \\ C^\varphi &= \varphi^2 \cdot C, \\ I_G(x) * \nu^\varphi &= I_G(\varphi x) * \nu, \qquad G \in \mathcal{B}(\mathbb{R}). \end{aligned} \tag{4.41}$$

The proof follows the next lines.

Consider the canonical representation of X:

$$X = X_0 + B + M + (x - h(x)) * \mu^X, \tag{4.42}$$

where $M \in \mathcal{M}_{\mathrm{loc}}$, B is a predictable process of bounded variation, and μ^X is the measure of jumps of X (cf. (4.36)).

From (4.42) we find

$$X^\varphi = \varphi \cdot B + \varphi \cdot M + (\varphi x - \varphi h(x)) * \mu^X. \tag{4.43}$$

At the same time, with evident notation,

$$\begin{aligned} X^\varphi &= B^\varphi + M^\varphi + (x - h(x)) * \mu^{X^\varphi} = B^\varphi + M^\varphi + (\varphi x - h(\varphi x)) * \mu^X \\ &= B^\varphi + M^\varphi + (\varphi h(x) - h(\varphi x)) * \mu^X + (\varphi x - \varphi h(x)) * \mu^X \\ &= \big[B^\varphi + (\varphi h(x) - h(\varphi x)) * \nu^X\big] \\ &\quad + \big[M^\varphi + (\varphi h(x) - h(\varphi x)) * (\mu^X - \nu^X)\big] \\ &\quad + (\varphi x - \varphi h(x)) * \mu^X. \end{aligned} \tag{4.44}$$

Comparing (4.43) and (4.44) we find that

$$\begin{aligned} &\varphi \cdot M - [M^\varphi + (\varphi h(x) - h(\varphi x)) * (\mu^X - \nu^X)] \\ &\qquad = [B^\varphi + (\varphi h(x) - h(\varphi x)) * \nu^X - \varphi \cdot B]. \end{aligned} \tag{4.45}$$

Because the right-hand side in (4.45) is a predictable process and the left-hand side is a local martingale, we conclude (see [100; Chap. II, 3.16]) that the left- and right-hand sides are equal to zero. So $B^\varphi = \varphi \cdot B + (h(\varphi x) - \varphi h(x)) * \nu^X$, which gives the first formula in (4.41) with $\nu = \nu^X$, and $M^\varphi = \varphi \cdot M + (h(\varphi x) - \varphi h(x)) * (\mu^X - \nu^X)$. From the latter formula we obtain that $C^\varphi = \varphi^2 \cdot C$, *i.e.*, $\langle(\varphi \cdot M)^c\rangle = \varphi^2 \cdot \langle M^c\rangle$. Thus the two last formulae in (4.41) are valid (recall that $\mu^{X^\varphi}(dx) = \varphi(x)\mu^X(dx)$).

3. From (4.41) it is easy to find, for $\theta \in \mathbb{R}$, the cumulant functions $\widetilde{K}^\varphi(\theta)$, $K^\varphi(\theta)$ and $\widetilde{\Lambda}^\varphi(\theta)$, $\Lambda^\varphi(\theta)$ related to the process $X^\varphi = \varphi \cdot X$.

By definition and (4.41)

$$\begin{aligned}
\widetilde{K}^\varphi(\theta) &= i\theta B^\varphi - \frac{\theta^2}{2} C^\varphi + \big(e^{i\theta x} - 1 - i\theta h(x)\big) * \nu^\varphi \\
&= i\theta\big[\varphi \cdot B + \big(h(\varphi x) - \varphi h(x)\big) * \nu\big] \\
&\quad - \frac{\theta^2}{2}\varphi^2 \cdot C + \big(e^{i\theta\varphi x} - 1 - i\theta h(\varphi x)\big) * \nu \\
&= i\theta\varphi \cdot B - \frac{\theta^2}{2}\varphi^2 \cdot C + \big(e^{i\theta\varphi x} - 1 - i\theta\varphi h(x)\big) * \nu. \qquad (4.46)
\end{aligned}$$

Let us introduce a predictable process

$$A = \operatorname{Var} B + \operatorname{Var} C + (|x|^2 \wedge 1) * \nu$$

and represent B, C, and ν in "disintegrated" form

$$B_t = \int_0^t b_s\, dA_s, \qquad C_t = \int_0^t c_s\, dA_s, \qquad \nu(\omega; ds, dx) = F_s(dx;\omega))\, dA_s(\omega)$$

with predictable processes $b = (b_s(\omega))_{s\geq 0}$ and $c = (c_s(\omega))_{s\geq 0}$ and a transition kernel $F = (F_s(dx;\omega))_{s\geq 0}$ from $(\Omega \times \mathbb{R}_+, \mathcal{P})$ into $(\mathbb{R}, \mathcal{B}(\mathbb{R}))$. (About the possibility of such representation see [100; Chap. II, Proposition 2.9]).

These new characteristics b, c, F—which are derivative of components of the triplet $\mathbb{T} = (B, C, \nu)$—can naturally be called differential (or local) characteristics of the semimartingale X and denoted by $\partial\mathbb{T} = (b, c, F)$.

Because of different possible choices of (predictable) "disintegrators" A these differential characteristics are not uniquely defined. However it is important to emphasize that in many applications the components of the triplet $\mathbb{T}$ are absolutely continuous with respect to the Lebesgue measure and in this case one can assume $A_t = t$. (Examples are Lévy and diffusion processes; other illustrations can be found in [109].)

With introduced notation we can rewrite (4.46) in the following form:

$$\widetilde{K}^\varphi(\theta) = \Big[i\theta\varphi b - \frac{\theta^2}{2}\varphi^2 c + \big(e^{i\theta\varphi x} - 1 - i\theta\varphi h(x)\big) * F_s(dx;\omega)\Big] \cdot A. \qquad (4.47)$$

Therefore if we denote

$$\widetilde{\varkappa}(\theta\varphi)_t = i\theta b_t\varphi_t - \frac{\theta^2}{2} c_t\varphi_t^2 + \big(e^{i\theta\varphi_t x} - 1 - i\theta\varphi_t h(x)\big) * F_s(dx;\omega), \qquad (4.48)$$

then we get the following formula:

$$\widetilde{K}^{\varphi}(\theta)_t = \int_0^t \widetilde{\varkappa}(\theta\varphi)_s \, dA_s, \tag{4.49}$$

or, in short form,

$$\widetilde{K}^{\varphi}(\theta) = \widetilde{\varkappa}(\theta\varphi) \cdot A.$$

The analogous formula can be obtained for $\widetilde{\Lambda}^{\varphi}(\theta)_t$:

$$\widetilde{\Lambda}^{\varphi}(\theta)_t = \int_0^t \widetilde{\lambda}(\theta\varphi)_s \, dA_s, \tag{4.50}$$

where

$$\widetilde{\lambda}(\theta\varphi)_t = \theta b_t \varphi_t + \frac{\theta^2}{2} c_t \varphi_t^2 + \left(e^{\theta\varphi_t x} - 1 - \theta\varphi_t h(x)\right) * \nu_{\omega,t}(dx). \tag{4.51}$$

The formulae for the cumulant functions $K^{\varphi}(\theta)$ and $\Lambda^{\varphi}(\theta)$ follow from (4.49), (4.50) and (4.28), (4.35).

4. In Subsec. 2 of Sec. 4.2 we stated the necessary and sufficient conditions for a semimartingale X to have a collection (B, C, ν) as its triplet of predictable characteristics. Using the Fourier cumulant process $(\widetilde{K}_t(\theta))_{t\geq 0}$ (see (4.24)) allows us to formulate the following assertion. *The two conditions below are equivalent*:

(i) *X is a semimartingale with triplet $\mathbb{T} = (B, C, \nu)$;*
(ii) *the processes $\widetilde{M}(\theta) = (\widetilde{M}(\theta)_t)_{t\geq 0}$, $\theta \in \mathbb{R}$, with*

$$\widetilde{M}(\theta)_t = e^{i\theta X_t} - \int_0^t e^{i\theta X_{s-}} \, d\widetilde{K}(\theta)_s$$

are local martingales.

Chapter 5

Processes with Independent Increments. Lévy Processes

5.1 Processes with Independent Increments and Semimartingales

1. Suppose that $X = (X_t, \mathcal{F}_t)_{t\geq 0}$ is a càdlàg process given on a filtered probability space $(\Omega, \mathcal{F}, (\mathcal{F}_t)_{t\geq 0}, \mathsf{P})$.

We say that such a process X with $X_0 = 0$ is a *process with independent increments* ($X \in$ PII) if the random variable $X_t - X_s$ is independent of the σ-algebra $\mathcal{F}_s$ for each $0 \leq s < t < \infty$. Fundamental examples of processes from the class PII are the Wiener process $W = (W_t, \mathcal{F}_t)_{t\geq 0}$ (with $\mathsf{E}W_t = 0$, $\mathsf{E}W_t^2 = \sigma^2(t)$) and the Poisson process $N = (N_t, \mathcal{F}_t)_{t\geq 0}$ (with intensity $a(t) = \mathsf{E}N_t$).

Other interesting and important examples of the processes of the class PII can be obtained from discrete-time schemes in the following way.

Suppose that $X = (X_n, \mathcal{F}_n)_{n\geq 0}$ is a stochastic sequence, given on a discrete stochastic basis $(\Omega, \mathcal{F}, (\mathcal{F}_n)_{n\geq 0}, \mathsf{P})$, such that $X_0 = 0$, $X_n = \xi_1 + \cdots + \xi_n$, where $(\xi_n)_{n\geq 1}$ is a sequence of $\mathcal{F}_n$-measurable random variables. Define a new stochastic basis in continuous time $(\Omega, \mathcal{F}, (\overline{\mathcal{F}}_t)_{t\geq 0}, \mathsf{P})$ with $\overline{\mathcal{F}}_t = \mathcal{F}_{[t]}$ and a new stochastic process $\overline{X} = (\overline{X}_t, \overline{\mathcal{F}}_t)_{t\geq 0}$ with $\overline{X}_t = X_{[t]}$. It is easy to see that the process $\overline{X}$ is a semimartingale and at the same time a process of class PII.

2. What is the relationship of the class PII with the class of semimartingales (Sem)?

It turns out (see [100; Chap. II, Theorem 5.1]) that, essentially, processes of class PII are semimartingales. More exactly, if $X \in$ PII then X can be represented in the form $X = D + S$, where $D = (D_t)_{t\geq 0}$ is a deterministic process (maybe of infinite variation) and S is a semimartingale.

Of course, from the standpoint of probabilistic properties of X, the deterministic component D is an uninteresting object and the main interest is connected with the semimartingale component S. Following this point of view we shall assume that all processes with independent increments we consider are at the same time semimartingales and, in consequence, one may use the presented earlier results of the theory of semimartingales.

3. In view of this remark let us assume that a process $X \in \mathrm{PII} \cap \mathrm{Sem}$. Suppose that $h = h(x)$ is a truncation function and $\mathbb{T} = (B, C, \nu)$ is the triplet of predictable characteristics (with respect to h).

One very important general result of the semimartingale theory states (see [100; Chap. II, Theorem 4.15]) that if $X \in \mathrm{PII} \cap \mathrm{Sem}$ then its triplet $\mathbb{T}$ is *deterministic* (more exactly, there exists a deterministic version of predictable characteristics). This implies that the set $J = \{t : \nu(\{t\} \times \mathbb{R}) > 0\}$ consists only of fixed times of discontinuity of the compensator ν.

If this set J is empty ($J = \varnothing$) then from Sec. 4.2 (see formulae (4.26)–(4.30)) we conclude that the Fourier cumulant function of X is of the form

$$\widetilde{K}(\theta)_t = K(\theta)_t = i\theta B_t - \frac{1}{2}\theta^2 C_t + \int_0^t \int_{\mathbb{R}} (e^{i\theta x} - 1 - i\theta h(x))\, \nu(ds, dx), \quad (5.1)$$

the characteristic function being

$$\mathsf{E} e^{i\theta X_t} = e^{\widetilde{K}(\theta)_t}, \qquad t \geq 0, \quad (5.2)$$

and these two formulae constitute the well-known

Kolmogorov–Lévy–Khinchin formula,

called also the *Lévy–Khinchin formula.* (The assumption $J = \varnothing$ is equivalent to the property that the process X is continuous in probability.)

If we do not assume that $J = \varnothing$, then the corresponding formula for the characteristic functions $\mathsf{E} e^{i\theta X_t}$ has a slightly more complicated form. Indeed, if $\nu(\{t\} \times \mathbb{R}) > 0$, then

$$\Delta \widetilde{K}(\theta)_t = \mathsf{E}(e^{i\theta \Delta X_t} - 1 \,|\, \mathcal{F}_{t-}) = \int (e^{i\theta x} - 1)\, \nu(\{t\} \times dx) \neq 0, \quad (5.3)$$

so that by (4.30)

$$\mathsf{E} e^{i\theta X_t} = \exp\left\{ i\theta B_t - \frac{\theta^2}{2} C_t + \int_0^t \int_{\mathbb{R}} (e^{i\theta x} - 1 - i\theta h(x)) I_{\overline{J}}(s)\, \nu(ds, dx) \right\}$$
$$\times \prod_{0 < s \leq t} \left[e^{-i\theta \Delta B_s} \left(1 + \int (e^{i\theta x} - 1)\, \nu(\{s\} \times dx) \right) \right]. \quad (5.4)$$

4. Formulae (5.2) and (5.4) play the key role in the analytic study of the *distributional* properties of the processes from the class $\mathrm{PII} \cap \mathrm{Sem}$. From the point of view of stochastic calculus an important role belongs to the canonical representation of X.

Recall that if $X \in \text{Sem}$ then X can be represented in the following form:

$$X_t = X_0 + B_t + X_t^c + \int_0^t \int h(x)\, d(\mu - \nu) + \int_0^t \int (x - h(x))\, d\mu, \quad (5.5)$$

where μ is the jump measure of X, *i.e.*,

$$\mu(\omega; (0,t] \times A) = \sum_{0<s\leq t} I_A(\Delta X_s) \quad \text{with } A \in \mathcal{B}(\mathbb{R}\backslash\{0\}), \quad (5.6)$$

and ν is the compensator of μ.

But if additionally $X \in \text{PII} \cap \text{Sem}$ then we can say more about the structure of the canonical representation (5.5).

Indeed, in this case X^c is a continuous martingale with deterministic process $\langle X^c \rangle$. So, by the famous Lévy characterization theorem (see [146; Chap. IV, (3.6)]) the process X^c is a *Gaussian martingale.*

If $X \in \text{PII} \cap \text{Sem}$ then the jump measure μ has a very special form because its compensator ν is deterministic. More exactly, in this case the jump measure μ is an *extended Poisson random measure*, *i.e.*, an integer-valued random measure such that for any family $(A_k)_{k\leq n}$, $n \geq 1$, of pairwise-disjoint measurable sets A_k from $\mathbb{R}_+ \times \mathbb{R}$ with $\nu(A_k) < \infty$

$$\begin{aligned} \mathsf{E} \exp\Big\{i \sum_{k\leq n} \theta_k \mu(A_k)\Big\} &= \exp\Big\{\sum_{k\leq n} (e^{i\theta_k} - 1)\nu^c(A_k)\Big\} \\ &\quad \times \prod_{s>0} \Big(1 + \sum_{k\leq n} (e^{i\theta_k} - 1)\nu((\{s\} \times \mathbb{R}) \cap A_k)\Big), \end{aligned} \quad (5.7)$$

where $\nu^c = \nu - \nu^d$ and $\nu^d(dt, dx) = \nu(dt, dx) I_J(t, x)$ with $J = \{t : \nu(\{t\} \times \mathbb{R}) > 0\}$. (For details see [100; Chap. II, § 1c].)

If $\nu = \nu^c$ (continuous compensator) then the measure μ is called a *Poisson random measure.* In this case formula (5.7) takes a much simpler form:

$$\mathsf{E} \exp\Big\{i \sum_{k\leq n} \theta_k \mu(A_k)\Big\} = \exp\Big\{\sum_{k\leq n} (e^{i\theta_k} - 1)\nu(A_k)\Big\}. \quad (5.8)$$

From this formula we see that for a Poisson random measure μ the random variables $\mu(A_1), \dots, \mu(A_n)$ with pairwise-disjoint measurable sets $A_1, \dots, A_n$ in $\mathbb{R}_+ \times \mathbb{R}$ are independent random variables with Poisson distribution:

$$\mathsf{E} e^{i\theta_k \mu(A_k)} = \exp\{(e^{i\theta_k} - 1)\nu(A_k)\}, \qquad i \leq k \leq n. \quad (5.9)$$

5.2 Processes with Stationary Independent Increments (Lévy Processes)

1. Suppose that the process $X = (X_t, \mathcal{F}_t)_{t\geq 0}$ from class PII $\cap$ Sem is such that for all $0 \leq s < t$ the distribution of the random variable $X_t - X_s$ depends only on $t - s$ and therefore

$$\mathrm{Law}(X_{t+h} - X_{s+h}) = \mathrm{Law}(X_t - X_s) \quad \text{for all} \quad h > 0. \tag{5.10}$$

Such processes X are called *processes with stationary independent increments* ($X \in$ PIIS) or *Lévy processes.* From (5.10) it follows that for all $t > 0$ the distribution $\mathrm{Law}(\Delta X_t)$ of $\Delta X_t = \lim_{h\downarrow 0}(X_t - X_{t-h})$ does not depend on t. From here, because the set of fixed times of discontinuity of any càdlàg function is at most countable, we conclude that any Lévy process has no fixed time of discontinuity. In other words, $\mathsf{P}(\Delta X_t \neq 0) = 0$ for any $t > 0$. From this property it follows that Lévy processes are continuous in probability.

The property of stationarity of increments for Lévy processes leads to the following characteristic property: a process X is a PIIS- (or Lévy) process if it is a semimartingale whose characteristics (B, C, ν) (with respect to the truncation function $h = h(x)$) admit a version of the form:

$$B_t(\omega) = tb, \qquad C_t(\omega) = tc, \qquad \nu(\omega; dt, dx) = dt\, F(dx),$$

where $b \in \mathbb{R}$, $c \geq 0$, F is a positive measure on $\mathbb{R}$ which satisfies $F(\{0\}) = 0$ and integrates $x^2 \wedge 1$, *i.e.*,

$$\int_{\mathbb{R}} (x^2 \wedge 1)\, F(dx) < \infty.$$

For such processes

$$\mathsf{E}e^{i\theta X_t} = e^{t\varkappa(\theta)}, \tag{5.11}$$

for all $t \geq 0$ and $\theta \in \mathbb{R}$, where the local Fourier cumulant function

$$\varkappa(\theta) = i\theta b - \frac{\theta^2}{2} c + \int \big(e^{i\theta x} - 1 - i\theta h(x)\big)\, F(dx) \tag{5.12}$$

(see (4.32)).

Remark 5.1. We call the set (b, c, F) the *triplet of local characteristics* and denote $\mathbb{T}_{\mathrm{loc}} = (b, c, F)$. Recall that sometimes instead of (b, c, F) we write $(b, c, F)_h$ (or $\mathbb{T}_{h,\mathrm{loc}} = (b, c, F)_h$) to emphasize that the truncation function is h. Notation $(b, c, F)_0$ means that $h \equiv 0$.

2. There exists a close connection between *PIIS-processes* and *infinitely divisible random variables.*

According to a very well-known definition (see, *e.g.*, [73], [95], [100], [153], [160], [166]), we say that a random variable ξ and its probability distribution are *infinitely divisible* if for each $n \geq 1$ there exist independent identically distributed random variables $\xi_{n1}, \dots, \xi_{nn}$ such that

$$\operatorname{Law}(\xi) = \operatorname{Law}(\xi_{n1} + \cdots + \xi_{nn}).$$

Before the general theory of infinite divisibility (see, *e. g.*, [84]) was constructed, it was known that the standard Gaussian and Poisson variables, ξ_1 and ξ_2,—whose characteristic functions are of the form

$$\varphi_{\xi_1}(\theta) = \exp\Big\{i\theta b - c\frac{\theta^2}{2}\Big\}, \qquad c \geq 0,$$
$$\varphi_{\xi_2}(\theta) = \exp\{\lambda(e^{i\theta} - 1)\}, \qquad \lambda > 0,$$

respectively,—are infinitely divisible. In the 1930s, B. de Finetti unified the two cases by proving that any random variable ξ with the characteristic function

$$\varphi_\xi(\theta) = \exp\left\{i\theta b - \frac{\theta^2}{2}c + \lambda\int\left(e^{i\theta x} - 1\right)F(dx)\right\},$$

where F is a distribution function, is infinitely divisible. Finally, A. N. Kolmogorov (for the case of random variables ξ with $\mathsf{E}\xi^2 < \infty$) and then P. Lévy and A. Ya. Khinchin (for the general case) established that the characteristic function of an infinitely divisible random variable has the following form:

$$\varphi_\xi(\theta) = \exp\left\{i\theta b - \frac{\theta^2}{2}c + \int\left(e^{i\theta x} - 1 - i\theta h(x)\right)F(dx)\right\}, \tag{5.13}$$

where $b \in \mathbb{R}$, $c \geq 0$, F is a positive measure on $\mathbb{R}$ with the properties $F(\{0\}) = 0$ and $\int_{\mathbb{R}}(x^2 \wedge 1)\,F(dx) < \infty$. (This measure F is called the *Lévy measure.*) Direct comparison of (5.13) and (5.11)–(5.12) shows that if $X \in$ PIIS then $\xi = X_1$ is an infinitely divisible random variable, and *vice versa*: if ξ is an infinitely divisible random variable then there exists a process $X \in$ PIIS such that $\operatorname{Law}(X_1) = \operatorname{Law}(\xi)$.

The class of infinitely divisible distributions (and corresponding infinitely divisible random variables) is very broad. It includes, for example, the following distributions:

normal
Cauchy
Lévy–Smirnov
} see the next subsection about these distributions

Poisson
geometric
negative binomial
t-distribution (or Student distribution)
F-distribution (or Fisher distribution)
log-normal
logistic
Pareto
two-sided exponential (Laplace)
$\mathbb{G}$eneralized Inverse Gaussian ($\mathbb{G}$IG), including, in particular,
  IG (Inverse Gaussian)
  PH or H$^+$ (Positive Hyperbolic)
  Gamma
  (see Table 9.1 on page 185)
$\mathbb{G}$eneralized Hyperbolic ($\mathbb{G}$H), including in particular,
  $\mathbb{N} \circ$ IG ($\mathbb{N}$ormal Inverse Gaussian)
  $\mathbb{H}$ or $\mathbb{N} \circ$ H$^+$ ($\mathbb{H}$yperbolic or $\mathbb{N}$ormal Positive Hyperbolic)
  $\mathbb{N} \circ$ Gamma ($\mathbb{N}$ormal Gamma or $\mathbb{V}$ariance Gamma)
  (see Table 9.2 on page 193)

Note that many well-known distributions are not infinitely divisible. For example: binomial distribution, uniform distribution, each nondegenerate distribution with finite support, distributions with densities $f(x) = \text{const} \times \exp\{-|x|^\alpha\}$, $\alpha > 2$.

3. In the class PIIS there is an important subclass of so-called stable processes. Correspondingly, in the class of infinite divisible distributions a special role belongs to the class of stable laws and stable random variables defined in the following way.

The class of *stable* random variables and distributions, by definition, corresponds to the case when the random variables ξ_{nk}, $1 \le k \le n$, in the above definition of infinitely divisible random variables ξ can be constructed by means of a *special* procedure from a sequence $\eta_1, \eta_2, \ldots$ of independent and identically distributed random variables:

$$\xi_{nk} = \frac{\eta_k}{d_n} + \frac{a_n}{n}, \qquad 1 \le k \le n, \quad n \ge 1, \tag{5.14}$$

where $d_n > 0$ and $a_n \in \mathbb{R}$ are suitable constants. (See, *e.g.*, [73], [95], [100], [153], [160], [166].)

From (5.14) we see that in the *stable case*

$$\operatorname{Law}(\xi) = \operatorname{Law}\Big(\frac{\eta_1 + \cdots + \eta_n}{d_n} + a_n\Big) \quad \text{for each } n \geq 1, \tag{5.15}$$

where $\eta_1, \eta_2, \dots$ is a sequence of i.i.d. random variables.

A remarkable result of Lévy and Khinchin states that for a *stable* random variable ξ the corresponding characteristic function $\varphi_\xi(\theta)$ (given by (5.13)) can be written in the following (*Lévy–Khinchin*) form:

$$\varphi_\xi(\theta) = \begin{cases} \exp\big\{i\mu\theta - \sigma^\alpha|\theta|^\alpha\big(1 - i\beta(\operatorname{sgn}\theta)\tan\frac{\pi\alpha}{2}\big)\big\}, & \alpha \neq 1, \\ \exp\big\{i\mu\theta - \sigma|\theta|\big(1 + i\beta\frac{2}{\pi}(\operatorname{sgn}\theta)\log|\theta|\big)\big\}, & \alpha = 1, \end{cases} \tag{5.16}$$

where $0 < \alpha \leq 2$, $|\beta| \leq 1$, $\sigma > 0$, and $\mu \in \mathbb{R}$.

Here the parameters (σ, β, μ) have the following meaning:

σ is the scale parameter of the distribution density;
β is the skewness or asymmetry parameter;
μ is the location parameter.

The most important parameter α "controls" the decrease of the "tails" of the distributions: if $0 < \alpha < 2$ then

$$\lim_{x\to\infty} x^\alpha \mathsf{P}(\xi > x) = c_\alpha \frac{1+\beta}{2}\sigma^\alpha, \qquad \lim_{x\to\infty} x^\alpha \mathsf{P}(\xi < -x) = c_\alpha \frac{1-\beta}{2}\sigma^\alpha,$$

and

$$c_\alpha = \begin{cases} \dfrac{1-\alpha}{\Gamma(2-\alpha)\cos(\pi\alpha/2)}, & \alpha \neq 1, \\ 2/\pi, & \alpha = 1. \end{cases}$$

If $\alpha = 2$ then

$$\varphi_\xi(\theta) = e^{i\mu\theta - \sigma^2\theta^2} = e^{i\mu\theta - \frac{\theta^2}{2}(2\sigma^2)},$$

which shows that $\varphi_\xi(\theta)$ is the characteristic function of the normal distribution $\mathcal{N}(\mu, 2\sigma^2)$, *i.e.*, of a normally distributed random variable ξ with

$$\mathsf{E}\xi = \mu, \qquad \mathsf{D}\xi = 2\sigma^2.$$

The symmetric case ($\beta = \mu = 0$) is of a particular interest. In this case the characteristic function of ξ is

$$\varphi_\xi(\theta) = e^{-\sigma^\alpha|\theta|^\alpha}.$$

Denote by $S_\alpha(\sigma, \beta, \mu)$ the distribution of the stable random variable ξ characterized by parameters $(\alpha, \sigma, \beta, \mu)$. It is useful to note that special explicit formulae for the densities of stable distributions are not known except for some values of the parameters:

$S_2(\sigma, 0, \mu) = \mathcal{N}(\mu, 2\sigma^2)$ — the *normal* distribution with density

$$\frac{1}{2\sigma\sqrt{\pi}}\, e^{-\frac{(x-\mu)^2}{4\sigma^2}};$$

$S_1(\sigma, 0, \mu)$ — the *Cauchy* distribution with density

$$\frac{\sigma}{\pi((x-\mu)^2 + \sigma^2)};$$

$S_{1/2}(\sigma, 1, \mu)$ — the one-sided stable distribution (also called the *Lévy–Smirnov* distribution) with density

$$\sqrt{\frac{\sigma}{2\pi}}\, \frac{1}{(x-\mu)^{3/2}} \exp\left\{-\frac{\sigma}{2(x-\mu)}\right\}, \qquad x \in (\mu, \infty).$$

In particular, if $\mathrm{Law}(\xi) = S_1(\sigma, 0, 0)$ then

$$\mathsf{P}(\xi \le x) = \frac{1}{2} + \frac{1}{\pi} \arctan \frac{x}{\sigma}, \qquad x > 0,$$

and if $\mathrm{Law}(\xi) = S_{1/2}(\sigma, 1, 0)$ then

$$\mathsf{P}(\xi \le x) = 2\left(1 - \Phi\left(\sqrt{\frac{\sigma}{x}}\right)\right), \qquad x > 0.$$

4. There is another approach to the definition of a stable random variable ξ which is equivalent to that given above.

A random variable ξ is said to be *stable* if for each $n \ge 2$ there exist a positive number C_n and a number D_n such that

$$\mathrm{Law}(\xi_1 + \cdots + \xi_n) = \mathrm{Law}(C_n \xi + D_n), \tag{5.17}$$

where $\xi_1, \ldots, \xi_n$ are independent copies of ξ.

It is remarkable that, necessarily, $C_n = n^{1/\alpha}$ for some $0 < \alpha \le 2$ where α is the same as in the Lévy–Khinchin formula (5.16). (For the proof see, *e.g.*, [153].)

There are several definitions of the stable processes. One is the following. A process X from the class PIIS is a *stable process* if the characteristic functions $\mathsf{E}e^{i\theta X_t}$ have the following form (cf. (5.16)):

$$\mathsf{E}e^{i\theta X_t} = e^{t\varkappa(\theta)}, \tag{5.18}$$

where

$$\varkappa(\theta) = \begin{cases} i\mu\theta - \sigma^\alpha |\theta|^\alpha \left(1 - i\beta(\operatorname{sgn}\theta)\tan\frac{\pi\alpha}{2}\right), & \alpha \neq 1, \\ i\mu\theta - \sigma|\theta|\left(1 + i\beta\frac{2}{\pi}(\operatorname{sgn}\theta)\log|\theta|\right), & \alpha = 1. \end{cases} \tag{5.19}$$

Another definition is the natural continuous-time analogue of (5.17): a process X from the class PIIS is a *stable process* if there are numbers $0 < \alpha \leq 2$ and D such that for each $a > 0$ (D may depend on a)

$$\mathrm{Law}(X_{at}; t \geq 0) = \mathrm{Law}(a^{1/\alpha} X_t + Dt; t \geq 0). \tag{5.20}$$

(A proof of the equivalency of the two given definitions can be found, *e.g.*, in [153].) In the case $D = 0$ a stable process X is often called a *strictly α-stable Lévy process.*

Remark 5.2. The condition

$$\mathrm{Law}(X_{at}; t \geq 0) = \mathrm{Law}(a^{1/\alpha} X_t; t \geq 0) \tag{5.21}$$

is a particular case of the general property of *self-similarity* of any process X: for each $a > 0$ there exists a number $c = c(a)$ such that

$$\mathrm{Law}(X_{at}; t \geq 0) = \mathrm{Law}(cX_t; t \geq 0). \tag{5.22}$$

[The meaning of the condition (5.22) is clear: changes in the time scale ($t \to at$) produce in the sense of distributions the same results as changes in the phase scale ($x \to cx$).]

In the special case

$$c(a) = a^{\mathbb{H}}, \qquad a > 0,$$

we call the process X *self-similar with Hurst exponent* $\mathbb{H}$. (See details, *e.g.*, in [161].)

It is clear that any α-stable process X is a self-similar process with the Hurst parameter $\mathbb{H} = 1/\alpha$.

5.3 Some Properties of Sample Paths of Processes with Independent Increments

1. At present the processes with independent increments and, in particular, the Lévy processes find a lot of applications in modeling of financial data. The first continuous-time models (Bachelier, Samuelson, Black, Merton, Scholes, ...) were based on using Brownian motion which is, of course, a process with independent increments and continuous trajectories. The whole class PII and, in particular, the class PIIS give more possibilities for modeling financial data.

From this point of view it is reasonable to get some ideas (see [95], [153], [166]) about the structure of the trajectories of processes from the classes PII and PIIS. This should give a way to understand which model is compatible with the statistical data obtained from analysis of econometric information.

2. Let $X = (X_t, \mathcal{F}_t)_{t\geq 0}$ be a PII $\cap$ Sem-process which is continuous in probability, *i.e.*, $\nu(\{t\} \times \mathbb{R}) = 0$ for all $t > 0$. In this case $\nu(A) = \mathsf{E}\mu(A)$, where $A \in \mathcal{B}(\mathbb{R}_+) \otimes \mathcal{B}(\mathbb{R})$, the measure μ is a Poisson random measure (see Sec. 5.2), and a canonical representation of X has the form (with $X_0 = 0$)

$$X_t = B_t + X_t^c + \int_0^t \int h(x)\, d(\mu - \nu) + \int_0^t \int (x - h(x))\, d\mu, \tag{5.23}$$

where

$$\int_0^\infty (x^2 \wedge 1)\, \nu((0,t]\times dx) < \infty \quad \text{for all } t > 0. \tag{5.24}$$

For the processes X with $\nu(\{t\} \times \mathbb{R}) = 0$, $t > 0$, the functions $B = (B_t)_{t\geq 0}$, $C = (C_t)_{t\geq 0}$, and $\nu = (\nu((0,t]; \cdot\,))_{t\geq 0}$ are deterministic and continuous. In the canonical representation (5.23) the function $B = (B_t)_{t\geq 0}$ and the process $X^c = (X_t^c)_{t\geq 0}$ are continuous and the two last integrals give "discontinuous contributions" to the structure of the process X.

The function $B = (B_t)_{t\geq 0}$ has a meaning as a part of the "drift" of the process $X = (X_t)_{t\geq 0}$ and $X^c = (X_t^c)_{t\geq 0}$ is a continuous Gaussian process with independent increments and $\mathsf{E}X_t^c = 0$, $\mathsf{E}(X_t^c)^2 = C_t$.

Taking into account the simple structure of the components B and X^c we shall concentrate our attention on the "discontinuous" parts of X described by the two integrals in (5.23).

The structure of the sample paths of a process X with

$$X_t = \int_0^t \int h(x)\, d(\mu - \nu) + \int_0^t \int (x - h(x))\, d\mu \tag{5.25}$$

is completely defined by the properties of the compensator ν which is a σ-finite measure on $(\mathbb{R}_+ \times \mathbb{R}, \mathcal{B}(\mathbb{R}_+) \otimes \mathcal{B}(\mathbb{R}))$ with $\nu(\{0\} \times \mathbb{R}) = 0$ and with the property (5.24). Here it is useful to distinguish between the following two cases:

finite activity case, where $\nu((0,t] \times \mathbb{R}) < \infty$ for all $t > 0$,

infinite activity case, where $\nu((0,t] \times \mathbb{R}) = \infty$ for all $t > 0$.

(In the case of PIIS we have only these two possibilities, whereas for general PII there are other possibilities.)

It is clear that for all $t > 0$

$$\nu((0,t]\times\mathbb{R}) < \infty \implies \int_{\mathbb{R}} (|x| \wedge 1)\, \nu((0,t]\times dx) < \infty. \tag{5.26}$$

The condition

$$\int_{\mathbb{R}} (|x| \wedge 1)\, \nu((0,t]\times dx) < \infty \tag{5.27}$$

is necessary and sufficient for

$$\sum_{0<s\le t} |\Delta X_s| < \infty \quad (\mathsf{P}\text{-a.s.}) \tag{5.28}$$

which is (by (5.30) below) exactly the property of boundedness of variation for sample functions of X on the interval $[0,t]$. Thus

$$\int_{\mathbb{R}} (|x| \wedge 1)\, \nu((0,t]\times dx) < \infty \quad \implies \quad \begin{array}{l} X \text{ has bounded variation} \\ (\mathsf{P}\text{-a.s.}) \text{ on } [0,t]. \end{array} \tag{5.29}$$

Under condition (5.27) we can take $h(x) \equiv 0$ in the canonical representation of X. Then we find that for all $s \le t$

$$(5.27) \implies X_s = \int_0^s \int x\, d\mu \tag{5.30}$$

and

$$\mathsf{E}e^{i\theta X_s} = \exp\left\{ \int (e^{i\theta x} - 1)\, \nu((0,s] \times dx) \right\}. \tag{5.31}$$

Now we consider assumption $\nu((0,t]\times\mathbb{R}) < \infty$ which is *stronger* than the condition (5.27). In this finite activity case, provided that $\nu((0,t]\times\mathbb{R}) < \infty$ for all $t > 0$, the structure of trajectories of the process X is very simple: all sample functions are *step* functions, *i.e.*, there exist a sequence of stopping times $(\tau_k)_{k\ge 0}$, $0 = \tau_0 \le \tau_1 < \tau_2 < \cdots$, with $\tau_k \uparrow \infty$ (P-a.s.) and a sequence of random variables $(\xi_k)_{k\ge 1}$ such that ξ_k is $\mathcal{F}_k$-measurable and

$$\begin{aligned} &X_t = X_{\tau_{k-1}}, \quad \tau_{k-1} \le t < \tau_k, \qquad k \ge 1, \\ &\Delta X_{\tau_k} \; (\equiv X_{\tau_k} - X_{\tau_{k-1}}) = \xi_k. \end{aligned}$$

Therefore the process X can be completely identified with the *marked* or *multivariate* point process $(\tau_k, \xi_k)_{k\ge 1}$. It is clear that *every* point process $(\tau_k, \xi_k)_{k\ge 1}$ of the type in question (not necessarily connected with PII $\cap$ Sem-processes) is defined completely by a system of regular conditional distributions

$$G_k(\omega; dt, dx) = \mathsf{P}(\tau_k \in dt, \xi_k \in dx \,|\, \mathcal{G}_{k-1}),$$

where $\mathcal{G}_{k-1}$ is the σ-algebra $\sigma\binom{\tau_1,\dots,\tau_{k-1}}{\xi_1,\dots,\xi_{k-1}}$, *i.e.*, the σ-algebra generated by $\tau_1,\dots,\tau_{k-1}$ and $\xi_1,\dots,\xi_{k-1}$.

If μ is the jump measure of the corresponding process X (under the assumption that Ω-space is the canonical D-space of càdlàg functions) then for the compensator $\nu = \nu(\omega; dt, dx)$ we get the following formula:

$$\nu(\omega; dt, dx) = \sum_{k\ge 0} \frac{I(\tau_k < t \le \tau_{k+1})}{1 - G_{k+1}((0,t]\times\mathbb{R})} G_{k+1}(\omega; dt, dx). \tag{5.32}$$

Note that in our case (*i.e.*, when the process X has *independent* increments) the conditional distributions $G_k(\omega; dt, dx)$ are completely defined by the conditional distributions for $\tau_k - \tau_{k-1}$ and $\xi_k - \xi_{k-1}$; for the latter distributions, by "inverting" (5.32) we obtain the following formulae:

$$\mathsf{P}\Big(\tau_k - \tau_{k-1} > t \,\Big|\, \begin{matrix}\tau_1,\dots,\tau_{k-1}\\ \xi_1,\dots,\xi_{k-1}\end{matrix}\Big) = \exp\big\{-\big(\nu_{\tau_{k-1}+t}(\mathbb{R}) - \nu_{\tau_{k-1}}(\mathbb{R})\big)\big\}, \tag{5.33}$$

$$\mathsf{P}\Big(\xi_k - \xi_{k-1} \in A \,\Big|\, \begin{matrix}\tau_1,\dots,\tau_{k-1},\tau_k\\ \xi_1,\dots,\xi_{k-1}\end{matrix}\Big) = \frac{\nu_{\tau_k}(A) - \nu_{\tau_{k-1}}(A)}{\nu_{\tau_k}(\mathbb{R}) - \nu_{\tau_{k-1}}(\mathbb{R})}, \tag{5.34}$$

where $\nu_t(A) = \nu((0,t]\times A)$.

For a standard Poisson process $X = N$, $N = (N_t)_{t\ge 0}$, with parameter λ the random variables ξ_k are degenerate, $\xi_k \equiv 1$ for all $k \ge 1$, and we have $\mathsf{P}(\tau_k - \tau_{k-1} > t) = e^{-\lambda t}$ and $\nu_t(dx) = \lambda t I_{\{1\}}(dx)$.

If $(\xi_k)_{k\ge 1}$ is a sequence of i.i.d. random variables with

$$\mathsf{P}(\xi_k \in A) = \frac{\nu(A)}{\nu(\mathbb{R})},$$

where ν is a finite measure on $(\mathbb{R}, \mathcal{B}(\mathbb{R}))$, and $N = (N_t)_{t\ge 0}$ is a Poisson process with parameter $\lambda = \nu(\mathbb{R})$ independent of the sequence $(\xi_k)_{k\ge 0}$ then the process

$$X_t = \sum_{k=0}^{N_t} \xi_k$$

with $\xi_0 = 0$ is called a *compound* Poisson process. For this process a straightforward calculation shows that

$$\begin{aligned}\mathsf{E}e^{i\theta X_t} &= \sum_{k=0}^{\infty} \mathsf{E}(e^{i\theta X_t} \mid N_t = k)\mathsf{P}(N_t = k)\\ &= \sum_{k=0}^{\infty} (\mathsf{E}e^{i\theta\xi_1})^k \frac{e^{-\lambda t}(\lambda t)^k}{k!} = \exp\Big\{t\int_{\mathbb{R}} (e^{i\theta x} - 1)\,\nu(dx)\Big\}.\end{aligned} \tag{5.35}$$

So, $\nu((0,t]\times dx) = t\,\nu(dx)$.

3. The class PII∩Sem has an important subclass of *subordinators* which consists of processes with *nondecreasing* trajectories: $X_s \leq X_t$ (P-a.s.) for all $s \leq t$. The general form of such (continuous in probability) processes is the following:

$$X_t = B_t + \int_0^t \int_{(0,\infty)} x\, d\mu \qquad (X_0 = 0). \tag{5.36}$$

In this case B_t is nondecreasing, $\nu((0,t] \times (-\infty, 0)) = 0$ and

$$\mathsf{E} e^{i\theta X_t} = \exp\left\{ i\theta B_t + \int_0^\infty (e^{i\theta x} - 1)\, \nu((0,t] \times dx) \right\}, \tag{5.37}$$

where

$$\int_0^\infty (x \wedge 1)\, \nu((0,t] \times dx) < \infty. \tag{5.38}$$

(As we mentioned above, the latter condition is equivalent to the property of bounded variation of the sample functions.)

Notice that the condition (5.38) does not mean that we are in the "finite activity" case, *i.e.*, $\nu((0,t] \times \mathbb{R}_+) < \infty$ for each fixed $t > 0$. The condition (5.38) implies only that $\nu((0,t] \times A) < \infty$ for sets $A = (a, \infty)$, where $a > 0$. In other words, for subordinators we have both cases: "finite activity" and "infinite activity" in the sense that either $\nu((0,t] \times \mathbb{R}_+) < \infty$ ("finite activity case") or $\nu((0,t] \times \mathbb{R}_+) = \infty$ ("infinite activity case"). Therefore it is quite possible that a subordinator, having trajectories of bounded variations, has at the same time "a lot of small jumps", *i.e.*, $\nu((0,t] \times (0,a]) = \infty$ for $a > 0$. In the next section we shall consider other examples of processes with such properties ("infinite activity case"). For more details see [95], [166].

5.4 Some Properties of Sample Paths of Processes with Stationary Independent Increments (Lévy Processes)

The properties of the trajectories of those Lévy processes X (see Sec. 5.2) for which (for given truncation function $h = h(x)$) $B = 0$, $C = 0$ (purely discontinuous case) depend on the structure of the Lévy measure $F = F(dx)$. This σ-finite measure satisfies $F(\{0\}) = 0$ and $\int_{\mathbb{R}} (x^2 \wedge 1)\, F(dx) < \infty$. The most interesting properties of the trajectories of Lévy processes are related to the "tail" behavior of the Lévy measure $F = F(dx)$.

In the case of strictly α-stable Lévy processes the structure of the measure $F = F(dx)$ is relatively simple: if $0 < \alpha < 2$ and $x \neq 0$ then

$$F(dx) = \begin{cases} \dfrac{c_+ \, dx}{x^{\alpha+1}}, & x > 0, \\ \dfrac{c_- \, dx}{|x|^{\alpha+1}}, & x < 0, \end{cases} \tag{5.39}$$

where we exclude the trivial case $c_- = c_+ = 0$. (Recall that for $\alpha = 2$ we have $F(dx) \equiv 0$.) It explains why for description of the trajectories of Lévy processes X we shall consider only strictly α-stable Lévy processes. (For the general case of Lévy processes see, for example, [153].)

Having (5.39), note, first of all, that if X is a subordinator then $\alpha \in (0, 1)$ and $c_- = 0$ (*i.e.*, $F(dx) = 0$ for the negative x's).

The concrete structure of the Lévy measure $F = F(dx)$ given by (5.39) makes it possible to give the following rather detailed analysis of the structure of the trajectories of the strictly α-stable processes [95], [101], [153]).

A. *Case* $0 < \alpha < 1$ (*"pure jump processes of bounded variation"*). In this case the property (5.27) takes the form

$$\int_{\mathbb{R}} (|x| \wedge 1) \, F(dx) < \infty, \tag{5.40}$$

which is equivalent to the property $\sum_{s \le t} |\Delta X_s| < \infty$ (P-a.s.), $t > 0$.

Therefore, taking into account (5.39), we have the following result:

For $0 < \alpha < 1$ *the process* X *has paths with* (P-*a.s.*) *bounded variation.*

However, we have here an "infinite activity" case, because $\nu(\mathbb{R}) = \infty$ which means that $\nu((-a, a)) = \infty$ for any $a > 0$ (note that $\int_{|x|>a} d\nu < \infty$).

From the property $\nu(\mathbb{R}) = \infty$ one can conclude that for $t > 0$

$$\mathsf{P}(\text{number of jumps in } [0, t] = \infty) = 1.$$

(For a proof see, *e.g.*, [95]. For more on the behavior of the trajectories of α-stable processes see [101; p. 32], where also graphical illustrations are given.)

Condition (5.40) allows us to assume that in the canonical representation

$$X_t = bt + X_t^c + \int_0^t \int h(x) \, d(\mu - \nu) + \int_0^t \int (x - h(x)) \, d\mu, \tag{5.41}$$

where $b = b(h)$, as well as in the local Fourier cumulant function

$$\varkappa(\theta) = i\theta b(h) - \frac{\theta^2}{2} c + \int_{-\infty}^{\infty} (e^{i\theta x} - 1 - i\theta h(x)) \, F(dx) \tag{5.42}$$

the truncation function is $h(x) \equiv 0$ (*i.e.*, $h(x) = \lim_{a \downarrow 0} xI(|x| \leq a) = 0$). Then we obtain

$$X_t = b(0)t + X_t^c + \int_0^t \int x \, d\mu \tag{5.43}$$

and

$$\varkappa(\theta) = i\theta b(0) - \frac{\theta^2}{2} c + c_+ \int_0^\infty (e^{i\theta x} - 1) \frac{dx}{x^{\alpha+1}} + c_- \int_{-\infty}^0 (e^{i\theta x} - 1) \frac{dx}{|x|^{\alpha+1}}. \tag{5.44}$$

Let us show that here $b(0) = 0$ and $c = 0$. Indeed,

$$\begin{aligned}\int_0^\infty (e^{i\theta x} - 1) \frac{dx}{x^{\alpha+1}} &= \theta^\alpha \int_0^\infty (e^{ix} - 1) \frac{dx}{x^{\alpha+1}} \\ &= O(|\theta|^\alpha) = o(|\theta|) = o(|\theta^2|)\end{aligned} \tag{5.45}$$

as $|\theta| \to \infty$. Similarly

$$\int_{-\infty}^0 (e^{i\theta x} - 1) \frac{dx}{|x|^{\alpha+1}} = O(|\theta|^\alpha) = o(|\theta^2|) \qquad \text{as } |\theta| \to \infty. \tag{5.46}$$

For strictly α-stable processes we have the property of self-similarity

$$\varkappa(\theta) = \theta^\alpha \varkappa(1).$$

From here and (5.44)–(5.46) we conclude that $b(0) = c = 0$.

So, for the case $0 < \alpha < 1$ we have that X is a pure jump process:

$$X_t = \int_0^t \int_{\mathbb{R}} x \, d\mu \tag{5.47}$$

of bounded variation but with "a lot of jumps" [in the sense that P(number of jumps in $[0, t] = \infty) = 1$, $t > 0$].

B. *Case* $1 < \alpha < 2$ (*"purely discontinuous martingale of unbounded variation"*). In this "infinite activity" case

$$F(\mathbb{R}) = \infty \quad \text{and} \quad \int_{|x| \leq a} |x| \, F(dx) = \infty, \quad a > 0.$$

So, here the trajectories have unbounded variation.

Since in this case

$$\int_{\mathbb{R}} (|x| \wedge x^2) \, F(dx) < \infty, \tag{5.48}$$

one can select as a truncation function the function $h(x) = x$ which is a limit of the (usual) truncation functions $h_a(x) = xI(|x| \leq a)$ when $a \uparrow \infty$. Denote by $b(\infty)$ the value b corresponding to the truncation function $h(x) = x$. Then we obtain that

$$X_t = b(\infty)t + X_t^c + \int_0^t \int x \, d(\mu - \nu) \tag{5.49}$$

and

$$\varkappa(\theta) = i\theta b(\infty) - \frac{\theta^2}{2}c + c_+ \int_0^\infty (e^{i\theta x} - 1 - i\theta x)\,\frac{dx}{x^{\alpha+1}}$$
$$+ c_- \int_{-\infty}^0 (e^{i\theta x} - 1 - i\theta x)\,\frac{dx}{|x|^{\alpha+1}}. \tag{5.50}$$

As in the previous case, from the property (5.48) one can conclude that $\varkappa(\theta) = O(|\theta|^\alpha)$. By the self-similarity property, $\varkappa(\theta) = \theta^\alpha \varkappa(1)$. From this and the property (5.50) we obtain (letting first $|\theta| \to \infty$ and afterwards $\theta \to \infty$) that $b(\infty) = 0$ and $c = 0$. Therefore (5.49) takes the following form:

$$X_t = \int_0^t \int x\, d(\mu - \nu). \tag{5.51}$$

According to the general terminology (see [100]) this process is a "purely discontinuous local martingale" (in fact, this local martingale is a martingale because $\mathsf{E}|X_t| < \infty$ for all $t > 0$).

As we remarked above, in the case of $1 < \alpha < 2$ the process X has sample paths with unbounded variation. In fact, one can say more: for each $t > 0$

$$\mathsf{P}\Big(\sum_{s\le t} |\Delta X_s|^\beta < \infty\Big) = 1, \quad \text{if } \beta > \alpha,$$

$$\mathsf{P}\Big(\sum_{s\le t} |\Delta X_s|^\beta = \infty\Big) = 1, \quad \text{if } \beta \le \alpha.$$

C. *Case* $\alpha = 1$ (*"pure jump semimartingale of unbounded variation"*). If $c_+ = c_- = 1$ the local Fourier cumulant function is

$$\varkappa(\theta) = i\theta b + \int_{\mathbb{R}} (e^{i\theta x} - 1 - i\theta x I(|x| \le 1))\,\frac{dx}{x^2}$$

and the corresponding canonical representation is

$$X_t = bt + \int_0^t \int_{|x|\le 1} x\, d(\mu - \nu) + \int_0^t \int_{|x|>1} x\, d\mu.$$

(Cf. (5.47) for $0 < \alpha < 1$ and (5.51) for $1 < \alpha < 2$.)

If $b = 0$, then the process $X = (X_t)_{t\ge 0}$ is called a *Cauchy process*.

Chapter 6

Change of Measure. General Facts

6.1 Basic Definitions. Density Process

1. In the Introduction we said that the main aim of considering the problem of change of measures is the following. Suppose that P is a given probability measure and X is a stochastic process. The distribution of X is $\mathrm{Law}(X \mid \mathsf{P})$ and in principle it can be complicated. (For example, if X is a diffusion Markov process with $dX_t = a(X_t, t)\,dt + b(X_t, t)\,dB_t$, where $B = (B_t)_{t\geq 0}$ is a Brownian motion, then explicit formulae for $\mathrm{Law}(X \mid \mathsf{P})$ are known only for very special terms $a(X_t, t)$ and $b(X_t, t)$.) The "change of measures" task pursues an idea to construct a new measure $\widetilde{\mathsf{P}}$ equivalent to P (or only absolute continuous with respect to P) such that

$$\mathrm{Law}(X \mid \widetilde{\mathsf{P}}) = \mathrm{Law}(\widetilde{X} \mid \mathsf{P}), \tag{6.1}$$

where $\widetilde{X}$ is a process having relatively simple structure with respect to the initial measure P. (For example, $\widetilde{X}$ could be a Brownian motion with respect to P.) By the well-known "Girsanov theorems" (see, *e.g.*, [125], [100]), construction of such measures is indeed possible, for example, for a wide class of drift terms in $dX_t = a(X_t, t)\,dt + dB_t$.

2. Let us recall some basic definitions connected with *two given* measures P and $\widetilde{\mathsf{P}}$.

To make all ideas more transparent we begin with the case of discrete time; afterwards the continuous-time case will also be presented (see Subsec. 3 below). Suppose that $(\Omega, \mathcal{F}, \mathsf{P})$ is a probability space and $\widetilde{\mathsf{P}}$ is another measure on $(\Omega, \mathcal{F})$. We say that the measure $\widetilde{\mathsf{P}}$ is *absolutely continuous* with respect to the measure P (notation: $\widetilde{\mathsf{P}} \ll \mathsf{P}$) if $\widetilde{\mathsf{P}}(A) = 0$ for each $A \in \mathcal{F}$ such that $\mathsf{P}(A) = 0$. If $\widetilde{\mathsf{P}} \ll \mathsf{P}$ and $\mathsf{P} \ll \widetilde{\mathsf{P}}$ we say that measures P and $\widetilde{\mathsf{P}}$ are *equivalent* (notation: $\widetilde{\mathsf{P}} \sim \mathsf{P}$).

In the case of a filtered probability space $(\Omega, \mathcal{F}, (\mathcal{F}_n)_{n\geq 0}, \mathsf{P})$ the following notions are useful.

Let $\mathsf{P}_n = \mathsf{P}|\mathcal{F}_n$ be the restriction of the probability measure P to the σ-algebra $\mathcal{F}_n \subseteq \mathcal{F}$. We say that a measure $\widetilde{\mathsf{P}}$ is *locally absolutely continuous* with respect to P ($\widetilde{\mathsf{P}} \overset{\text{loc}}{\ll} \mathsf{P}$) if $\widetilde{\mathsf{P}}_n \ll \mathsf{P}_n$ for all $n \geq 1$. Two measures P and $\widetilde{\mathsf{P}}$ are said to be *locally equivalent* (notation: $\widetilde{\mathsf{P}} \overset{\text{loc}}{\sim} \mathsf{P}$) if $\widetilde{\mathsf{P}} \overset{\text{loc}}{\ll} \mathsf{P}$ and $\mathsf{P} \overset{\text{loc}}{\ll} \widetilde{\mathsf{P}}$.

If $\widetilde{\mathsf{P}} \overset{\text{loc}}{\ll} \mathsf{P}$ then for each $n \geq 1$ there exists the Radon–Nikodým derivative denoted by $\dfrac{d\widetilde{\mathsf{P}}_n}{d\mathsf{P}_n}$ or $\dfrac{d\widetilde{\mathsf{P}}_n}{d\mathsf{P}_n}(\omega)$ and defined as an $\mathcal{F}_n$-measurable function $Z_n = Z_n(\omega)$ such that

$$\widetilde{\mathsf{P}}_n(A) = \int_A Z_n(\omega)\, \mathsf{P}_n(d\omega) \quad \text{for each } A \in \mathcal{F}_n.$$

The process $Z = (Z_n)_{n\geq 1}$ is called the *density process* (of the measures $\widetilde{\mathsf{P}}_n$ with respect to the P_n, $n \geq 1$).

The following proposition concatenates the main properties of the density process $Z = (Z_n)_{n\geq 1}$.

Lemma 6.1. *Assume* $\widetilde{\mathsf{P}} \overset{\text{loc}}{\ll} \mathsf{P}$.

(a) *The process* $Z = (Z_n)_{n\geq 1}$ *is a nonnegative* $(\mathsf{P}, (\mathcal{F}_n)_{n\geq 1})$*-martingale.*

(b) *The following conditions are equivalent*:

(i) $\widetilde{\mathsf{P}} \ll \mathsf{P}$;

(ii) $Z = (Z_n)_{n\geq 1}$ *is a uniformly integrable martingale with respect to measure* P;

(iii) $\widetilde{\mathsf{P}}(\sup_n Z_n < \infty) = 1$.

(c) *If* $\tau = \inf\{n : Z_n = 0\}$ (*with* $\inf(\varnothing) = \infty$) *then* $\mathsf{P}\{\omega : Z_n(\omega) \neq 0$ *for some* $n \geq \tau(\omega)\} = 0$.

(d) *If* τ *is a stopping time* (*with* $\mathsf{P}(\tau < \infty) = 1$) *and* $\mathsf{P}_\tau = \mathsf{P}|\mathcal{F}_\tau$, $\widetilde{\mathsf{P}}_\tau = \widetilde{\mathsf{P}}|\mathcal{F}_\tau$, *then* $\widetilde{\mathsf{P}}_\tau \ll \mathsf{P}_\tau$ *and*

$$Z_{\tau(\omega)}(\omega) = \frac{d\widetilde{\mathsf{P}}_\tau}{d\mathsf{P}_\tau}(\omega) \quad (\mathsf{P}\text{-}a.s.).$$

(e) *If* $\mathsf{P}(Z_n > 0) = 1$ *for each* $n \geq 1$ *then* $\mathsf{P} \overset{\text{loc}}{\ll} \widetilde{\mathsf{P}}$ *and* $\widetilde{\mathsf{P}} \overset{\text{loc}}{\sim} \mathsf{P}$.

For the proof see, *e.g.*, [100], [161].

3. The given definitions (of $\widetilde{\mathsf{P}} \overset{\text{loc}}{\ll} \mathsf{P}$, $\widetilde{\mathsf{P}} \overset{\text{loc}}{\sim} \mathsf{P}$, $Z = (Z_n)_{n\geq 1}$) are easy to transfer to the case of continuous time. Indeed, if we have a filtered probability space $(\Omega, \mathcal{F}, (\mathcal{F}_t)_{t\geq 0}, \mathsf{P})$ and also a measure $\widetilde{\mathsf{P}}$ on $(\Omega, \mathcal{F})$ then we can construct restrictions $\mathsf{P}_t = \mathsf{P}|\mathcal{F}_t$, $\widetilde{\mathsf{P}}_t = \widetilde{\mathsf{P}}|\mathcal{F}_t$, $t \geq 0$, and say that $\widetilde{\mathsf{P}} \overset{\text{loc}}{\ll} \mathsf{P}$ if $\widetilde{\mathsf{P}}_t \ll \mathsf{P}_t$ for all $t \geq 0$.

If $\widetilde{\mathsf{P}} \overset{\text{loc}}{\ll} \mathsf{P}$ then the density process $Z = (Z_t)_{t\geq 1}$ is defined as a nonnegative martingale with $Z_t = \dfrac{d\widetilde{\mathsf{P}}_t}{d\mathsf{P}_t}$, $t \geq 0$. All properties listed in Lemma 6.1 can automatically be reformulated for the case of continuous time by taking t instead of n.

6.2 Discrete Version of Girsanov's Theorem

1. In this section we consider the problem of constructing a measure $\widetilde{\mathsf{P}}$ and a stochastic sequence $\widetilde{X}$ such that $\mathrm{Law}(X \mid \widetilde{\mathsf{P}}) = \mathrm{Law}(\widetilde{X} \mid \mathsf{P})$, where P and X are *a priori* given. From the point of view of applications to finance, of a particular interest are the cases, when the stochastic sequence has some martingale-like properties.

We begin with an assumption that $(\Omega, \mathcal{F}, (\mathcal{F}_n)_{n\geq 1}, \mathsf{P})$ is a given discrete filtered probability space. Let $\varepsilon = (\varepsilon_n)_{n\geq 1}$ be a sequence of $\mathcal{F}_n$-measurable i.i.d. random variables such that $\mathrm{Law}(\varepsilon_n \mid \mathcal{F}_{n-1}; \mathsf{P}) = \mathcal{N}(0,1)$. Assume also that $\mu = (\mu_n)_{n\geq 1}$ and $\sigma = (\sigma_n)_{n\geq 1}$ are predictable (*i.e.*, μ_n and σ_n are $\mathcal{F}_{n-1}$-measurable, $n \geq 1$), $\mathcal{F}_0 = \{\varnothing, \Omega\}$, and that the (*volatilities*) $\sigma_n > 0$ (P-a.s.).

Now form a sequence $x = (x_n)_{n\geq 1}$, where

$$x_n = \mu_n + \sigma_n \varepsilon_n. \tag{6.2}$$

With the notation $X_n = \sum_{k=1}^n x_k$ (*i.e.*, $\Delta X_n \equiv X_n - X_{n-1} = x_n$), $E_n = \sum_{k=1}^n \varepsilon_k$, $M_n = \sum_{k=1}^n \mu_k$, and $\Delta = 1$, one can rewrite (6.2) in the following form:

$$\Delta X_n = \mu_n \Delta + \sigma_n \Delta E_n \tag{6.3}$$

which can be regarded as a discrete counterpart to the stochastic differential equation

$$dX_t = \mu_t\, dt + \sigma_t\, dE_t \tag{6.4}$$

(called sometimes 'stochastic volatility model' with driving process $E = (E_t)_{t\geq 0}$, which is very often a Brownian motion or a Lévy process).

The sequence $x = (x_n)_{n\geq 1}$ is a conditionally Gaussian sequence (with respect to P and $(\mathcal{F}_n)_{n\geq 0}$):

$$\mathsf{P}(x_n \leq a \mid \mathcal{F}_{n-1}) = \frac{1}{\sqrt{2\pi\sigma_n^2}} \int_{-\infty}^{a} e^{-\frac{(y-\mu_n)^2}{2\sigma_n^2}}\, dy \tag{6.5}$$

or, in short form,

$$\operatorname{Law}(x_n \mid \mathcal{F}_{n-1}; \mathsf{P}) = \mathcal{N}(\mu_n, \sigma_n^2). \tag{6.6}$$

It is clear that $\mathsf{E}(x_n \mid \mathcal{F}_{n-1}) = \mu_n$ and $\mathsf{D}(x_n \mid \mathcal{F}_{n-1}) = \sigma_n^2$.

Now we want to show that (under some conditions on μ and σ) one can construct a measure $\widetilde{\mathsf{P}}$ such that with respect to $(\widetilde{\mathsf{P}}, (\mathcal{F}_n)_{n\ge 0})$ the sequence $x = (x_n)_{n\ge 1}$ will coincide in law with the sequence $(\sigma_n\varepsilon_n)_{n\ge 1}$:

$$\operatorname{Law}(x_n; n \ge 1 \mid \widetilde{\mathsf{P}}) = \operatorname{Law}(\sigma_n\varepsilon_n; n \ge 1 \mid \mathsf{P}) \tag{6.7}$$

(cf. (6.1)).

For construction of $\widetilde{\mathsf{P}}$ we consider the sequence $Z = (Z_n)_{n\ge 1}$ with

$$Z_n = \exp\biggl\{-\sum_{k=1}^{n} \frac{\mu_k}{\sigma_k}\varepsilon_k - \frac{1}{2}\sum_{k=1}^{n}\Bigl(\frac{\mu_k}{\sigma_k}\Bigr)^2\biggr\}, \qquad n \ge 1,$$

as a candidate to be a density process.

If $\mu = (\mu_n)_{n\ge 1}$ and $\sigma = (\sigma_n)_{n\ge 1}$ are such that

$$\mathsf{E}\exp\biggl\{\frac{1}{2}\sum_{k=1}^{n}\Bigl(\frac{\mu_k}{\sigma_k}\Bigr)^2\biggr\} < \infty \tag{6.8}$$

(the so-called *Novikov condition*) then $Z = (Z_n)_{n\ge 1}$ is a uniformly integrable martingale with limit (P-a.s.) $Z_\infty = \lim_n Z_n$ such that

$$\mathsf{P}(Z_\infty \ge 0) = 1, \qquad \mathsf{E}Z_\infty = 1. \tag{6.9}$$

So, we can form a new measure $\widetilde{\mathsf{P}}$ on $(\Omega, \mathcal{F})$ defining it by the formula

$$\widetilde{\mathsf{P}}(A) = \int_A Z_\infty \, d\mathsf{P} \quad (= \mathsf{E} I_A X_\infty). \tag{6.10}$$

For calculation of $\operatorname{Law}(x \mid \widetilde{\mathsf{P}})$ recall the following very useful lemma (see [161; Chap. V, § 3a]).

Lemma 6.2. (Generalized Bayes' formula or conversion formula for conditional expectations.) *Suppose that $\widetilde{\mathsf{P}}_n \ll \mathsf{P}_n$, and let Y be a bounded (or $\widetilde{\mathsf{P}}_n$-integrable) $\mathcal{F}_n$-measurable random variable. Then, for each $m \le n$, we have $\widetilde{\mathsf{P}}(Z_m > 0) = 1$ and*

$$\widetilde{\mathsf{E}}(Y \mid \mathcal{F}_m) = \frac{1}{Z_m}\,\mathsf{E}(Y Z_n \mid \mathcal{F}_m) \quad (\widetilde{\mathsf{P}}\text{-a.s.}), \tag{6.11}$$

where $Z_m = \dfrac{d\widetilde{\mathsf{P}}_m}{d\mathsf{P}_m}$ and $\widetilde{\mathsf{E}}$ stands for expectation with respect to the measure $\widetilde{\mathsf{P}}$.

Note that to prove (6.11) we need to check that for any $A \in \mathcal{F}_m$

$$\widetilde{\mathsf{E}}\Big[I_A \frac{1}{Z_m}\,\mathsf{E}(YZ_n \,|\, \mathcal{F}_m)\Big] = \widetilde{\mathsf{E}}[I_A Y] \tag{6.12}$$

which easily follows from the fact that for $\mathcal{F}_m$-measurable integrable random variables ξ

$$\widetilde{\mathsf{E}}\xi = \mathsf{E}(Z_m \xi). \tag{6.13}$$

Using this lemma we find that for every $\theta \in \mathbb{R}$

$$\begin{aligned}
\widetilde{\mathsf{E}}\big(e^{i\theta x_n} \,|\, \mathcal{F}_{n-1}\big) &= \mathsf{E}\Big[e^{i\theta x_n} \frac{Z_n}{Z_{n-1}} \,\Big|\, \mathcal{F}_{n-1}\Big] \\
&= \mathsf{E}\Big(e^{i\theta(\mu_n + \sigma_n \varepsilon_n)}\, e^{-\frac{\mu_n}{\sigma_n}\varepsilon_n - \frac{1}{2}\left(\frac{\mu_n}{\sigma_n}\right)^2} \,\Big|\, \mathcal{F}_{n-1}\Big) \\
&= e^{\frac{1}{2}\left(i\theta\sigma_n - \frac{\mu_n}{\sigma_n}\right)^2 + i\theta\mu_n - \frac{1}{2}\left(\frac{\mu_n}{\sigma_n}\right)^2}\, \mathsf{E}\Big[e^{\left(i\theta\sigma_n - \frac{\mu_n}{\sigma_n}\right)\varepsilon_n - \frac{1}{2}\left(i\theta\sigma_n - \frac{\mu_n}{\sigma_n}\right)^2}\Big] \\
&= e^{-\frac{\theta^2\sigma_n^2}{2}} \cdot 1 = e^{-\frac{\theta^2\sigma_n^2}{2}}.
\end{aligned}$$

From this it follows that

$$\mathrm{Law}(x_n \,|\, \mathcal{F}_{n-1}; \widetilde{\mathsf{P}}) = \mathcal{N}(0, \sigma_n^2) = \mathrm{Law}(\sigma_n \varepsilon_n \,|\, \mathcal{F}_{n-1}; \mathsf{P}),$$

which is the needed formula (6.7).

2. The previous case of conditionally Gaussian sequence $x = (x_n)_{n \geq 1}$ shows a way of finding the corresponding extension in a more general case.

Assume that the basic sequence $X = (X_n)_{n \geq 1}$ with $X_0 = 0$ has the representation

$$X_n = A_n + M_n, \qquad n \geq 0, \tag{6.14}$$

(Doob type decomposition), where A_n are $\mathcal{F}_{n-1}$-measurable for all $n \geq 1$ ($A_0 = 0$, $\mathcal{F}_0 = \{\varnothing, \Omega\}$) and $(M_n)_{n \geq 0}$ is a local martingale (with respect to $(\mathsf{P}, (\mathcal{F}_n)_{n \geq 1})$) with $M_0 = 0$.

Since for a local martingale $\mathsf{E}(|\Delta M_n| \,|\, \mathcal{F}_{n-1}) < \infty$ (P-a.s.), we find that $\mathsf{E}(|\Delta X_n| \,|\, \mathcal{F}_{n-1}) < \infty$ (P-a.s.) and therefore X_n has also the representation

$$X_n = \sum_{k=1}^{n} \mathsf{E}(\Delta X_k \,|\, \mathcal{F}_{k-1}) + \sum_{k=1}^{n} \big(\Delta X_k - \mathsf{E}(\Delta X_k \,|\, \mathcal{F}_{k-1})\big). \tag{6.15}$$

The representation (6.14) with predictable process $A = (A_n)_{n \geq 1}$ is unique and therefore

$$A_n = \sum_{k=1}^{n} \mathsf{E}(\Delta X_k \,|\, \mathcal{F}_{k-1}), \qquad M_n = \sum_{k=1}^{n} \big(\Delta X_k - \mathsf{E}(\Delta X_k \,|\, \mathcal{F}_{k-1})\big).$$

Assume $\widetilde{\mathsf{P}} \overset{\text{loc}}{\ll} \mathsf{P}$ and put

$$Z_n = \frac{d\widetilde{\mathsf{P}}_n}{d\mathsf{P}_n}, \qquad \alpha_n = \frac{Z_n}{Z_{n-1}} I(Z_{n-1} > 0), \quad \text{where } Z_0 \equiv 1. \tag{6.16}$$

Straightforward calculation shows that a new process $\widetilde{M} = (\widetilde{M}_n)_{n\geq 0}$ with

$$\widetilde{M}_0 = 0 \quad \text{and} \quad \widetilde{M}_n = M_n - \sum_{k=1}^{n} \mathsf{E}(\alpha_k \Delta M_k \,|\, \mathcal{F}_{k-1}) \tag{6.17}$$

is a $(\widetilde{\mathsf{P}}, (\mathcal{F}_n)_{n\geq 0})$-martingale. Note that ($\widetilde{\mathsf{P}}$-a.s.)

$$\begin{aligned} \widetilde{M}_n &= M_n - \sum_{k=1}^{n} \mathsf{E}(\alpha_k \Delta X_k \,|\, \mathcal{F}_{k-1}) + A_n \\ &= X_n - \sum_{k=1}^{n} \mathsf{E}(\alpha_k \Delta X_k \,|\, \mathcal{F}_{k-1}) = X_n - \sum_{k=1}^{n} \widetilde{\mathsf{E}}(\Delta X_k \,|\, \mathcal{F}_{k-1}). \end{aligned}$$

So, if $\widetilde{A}_n = \sum_{k=1}^{n} \widetilde{\mathsf{E}}(\Delta X_k \,|\, \mathcal{F}_{k-1})$ then ($\widetilde{\mathsf{P}}$-a.s.)

$$X_n = \widetilde{A}_n + \widetilde{M}_n, \tag{6.18}$$

where $\widetilde{M} = (\widetilde{M}_n)_{n\geq 0}$ is a local $(\widetilde{\mathsf{P}}, (\mathcal{F}_n)_{n\geq 0})$-martingale. (We interpret this result as a *discrete version* of Girsanov's theorem.)

For the model (6.2) and measure $\widetilde{\mathsf{P}}$ constructed in (6.10) we find that $\widetilde{A}_n = 0$ and $\mathrm{Law}(\widetilde{M}_n \,|\, \widetilde{\mathsf{P}}) = \mathrm{Law}(M_n \,|\, \mathsf{P}) = \mathrm{Law}(\sigma_n \varepsilon_n \,|\, \mathsf{P})$. Thus, from representation (6.18) we see that under measure $\widetilde{\mathsf{P}}$ the process $X = (X_n)_{n\geq 1}$ with $\Delta X_n = \mu_n + \sigma_n \varepsilon_n$ is a local martingale such that

$$\mathrm{Law}(\Delta X_n \,|\, \widetilde{\mathsf{P}}) = \mathrm{Law}(\sigma_n \varepsilon_n \,|\, \mathsf{P}). \tag{6.19}$$

6.3 Semimartingale Version of Girsanov's Theorem

1. In the previous section for the case of discrete time we started with the assumption that a basic process $X = (X_n)_{n\geq 1}$ has Doob type representation

$$X_n = A_n + M_n, \qquad n \geq 1, \tag{6.20}$$

with $M = (M_n)_{n\geq 1} \in \mathcal{M}_{\text{loc}}(\mathsf{P})$. After this we made a change of measure ($\mathsf{P} \to \widetilde{\mathsf{P}}$, where $\widetilde{\mathsf{P}} \overset{\text{loc}}{\ll} \mathsf{P}$) and obtained a new representation

$$X_n = \widetilde{A}_n + \widetilde{M}_n, \qquad n \geq 1,$$

where $\widetilde{M} = (\widetilde{M}_n)_{n\geq 1} \in \mathcal{M}_{\text{loc}}(\widetilde{\mathsf{P}})$.

In the present section we shall consider the case of continuous time and we assume that the basic process $X = (X_t)_{t\geq 0}$ is a semimartingale with a representation similar to (6.20):

$$X_t = A_t + M_t, \qquad t \geq 0, \tag{6.21}$$

where $A = (A_t)_{t\geq 0}$ is a process of bounded variation and $M = (M_t)_{t\geq 0}$ is a local martingale.

We assume that all these processes are considered on a filtered probability space $(\Omega, \mathcal{F}, (\mathcal{F}_t)_{t\geq 0}, \mathsf{P})$.

It is convenient to formulate the corresponding version of Girsanov's theorem for semimartingales in terms of transformation of the triplet $\mathbb{T} = (B, C, \nu)$ (of X with respect to the measure P) into a new triplet $\widetilde{\mathbb{T}} = (\widetilde{B}, \widetilde{C}, \widetilde{\nu})$ (with respect to a new measure $\widetilde{\mathsf{P}}$ such that $\widetilde{\mathsf{P}} \overset{\text{loc}}{\ll} \mathsf{P}$).

2. Consider a canonical representation of the process $X = (X_t)_{t\geq 0}$ with $X_0 = 0$:

$$X_t = B_t + X_t^c + \int_0^t \int h(x)\, d(\mu - \nu) + \int_0^t \int (x - h(x))\, d\mu. \tag{6.22}$$

Here $h = h(x)$ is a truncation function. The corresponding triplet is $\mathbb{T} = (B, C, \nu)$, where $C = \langle X^c \rangle$ and ν is the compensator of the measure μ of jumps of X. The predictable process $B = B(h)$ depends on h but the predictable characteristics C and ν do not.

Suppose now that a new measure $\widetilde{\mathsf{P}}$ is locally absolutely continuous with respect to the measure P (in short: $\widetilde{\mathsf{P}} \overset{\text{loc}}{\ll} \mathsf{P}$).

Under this measure $\widetilde{\mathsf{P}}$ the semimartingale X will be again a semimartingale with a new triplet $\widetilde{\mathbb{T}} = (\widetilde{B}, \widetilde{C}, \widetilde{\nu})$. The semimartingale version of Girsanov's theorem is exactly a statement concerning a way of finding $\widetilde{\mathbb{T}}$ *via* $\mathbb{T}$ and density process $Z = (Z_t)_{t\geq 0}$, where $Z_t = d\widetilde{\mathsf{P}}_t/d\mathsf{P}_t$.

Theorem 6.1. *If* $\widetilde{\mathsf{P}} \overset{\text{loc}}{\ll} \mathsf{P}$ *then*

$$\begin{aligned}
\widetilde{B} &= B + \beta \cdot C + h(x)(Y-1) * \nu,\\
\widetilde{C} &= C,\\
\widetilde{\nu} &= Y \cdot \nu,
\end{aligned} \tag{6.23}$$

where $|h(x)(Y-1) * \nu_t| < \infty$ *(*$\widetilde{\mathsf{P}}$*-a.s.) and* $\int_0^t \beta_s\, dC_s < \infty$ *(*$\widetilde{\mathsf{P}}$*-a.s.) for all* $t > 0$.

Let us explain the notation β and Y in these formulae.

By $\beta = (\beta_t)_{t\geq 0}$ we denote a predictable process with

$$\beta_t = \frac{d\langle Z^c, X^c\rangle_t}{d\langle X^c, X^c\rangle_t}\,\frac{I(Z_{t-} > 0)}{Z_{t-}}. \tag{6.24}$$

For definition of the process $Y = Y(\omega; t, x)$, $\omega \in \Omega$, $t \geq 0$, $x \in \mathbb{R}$, we need the following notation.

Let $m^{\mathsf{P}}_\mu(d\omega; dt, dx) = \mu(\omega; dt, dx)\,\mathsf{P}(d\omega)$ be a measure on the σ-algebra $\mathcal{F}\otimes\mathcal{B}(\mathbb{R}_+)\otimes\mathcal{B}(\mathbb{R})$ and $\widetilde{\mathcal{P}} = \mathcal{P}\otimes\mathcal{B}(\mathbb{R})$ the σ-algebra on $(\Omega, \mathbb{R}_+, \mathbb{R})$ generated by the predictable σ-algebra $\mathcal{P}$ (on $(\Omega, \mathbb{R}_+)$). We denote by M^{P}_μ the expectation with respect to the measure $m^{\mathsf{P}}_\mu = m^{\mathsf{P}}_\mu(d\omega; dt, dx)$ and by $M^{\mathsf{P}}_\mu(\cdot \,|\, \widetilde{\mathcal{P}})$ the conditional distribution with respect to the σ-algebra $\widetilde{\mathcal{P}}$. The process Y in (6.23) is defined in the following way:

$$Y = M^{\mathsf{P}}_\mu\biggl(\frac{Z}{Z_-}\,I(Z_- > 0)\,\Big|\,\widetilde{\mathcal{P}}\biggr). \tag{6.25}$$

For a proof of the formulae (6.23) we refer to [100; Chap. III, Theorem 3.24]. The meaning of formulae (6.23) can be demonstrated by the following examples.

Example 6.1. Girsanov's theorem for diffusion processes, in its original formulation, [83], states the following.

Consider an Itô process $X = (X_t, \mathcal{F}_t)_{t\geq 0}$ with differential

$$dX_t = \mu_t(\omega)\,dt + dB_t, \qquad X_0 = 0, \tag{6.26}$$

where $\mu = (\mu_t, \mathcal{F}_t)_{t\geq 0}$ is a process with $\mathsf{P}(\int_0^t |\mu_s|\,ds < \infty) = 1$, $t > 0$, and $B = (B_t, \mathcal{F}_t)_{t\geq 0}$ is a standard Brownian motion.

It is clear that the process X is a semimartingale with a triplet

$$\mathbb{T} = (B, C, 0), \tag{6.27}$$

where $B = (B_t)_{t\geq 0}$ with

$$B_t = \int_0^t \mu_s\,ds$$

and $C = (C_t)_{t\geq 0}$ is such that $C_t = t$.

Assume now that $\widetilde{\mu} = (\widetilde{\mu}_t, \mathcal{F}_t)_{t\geq 0}$ is another process such that

$$\mathsf{P}\biggl(\int_0^t (\widetilde{\mu}_s - \mu_s)^2\,ds < \infty\biggr) = 1, \qquad t > 0,$$

and define a new process $Z = (Z_t, \mathcal{F}_t)_{t\geq 0}$ such that

$$Z_t = \exp\biggl\{\int_0^t (\widetilde{\mu}_s - \mu_s)\,dB_s - \frac{1}{2}\int_0^t (\widetilde{\mu}_s - \mu_s)^2\,ds\biggr\}. \tag{6.28}$$

The latter process is a nonnegative local martingale (see, *e.g.*, Sec. 4.3). Let us assume that $\mathsf{E}Z_t = 1$ for all $t > 0$. In this case the process $Z = (Z_t, \mathcal{F}_t)_{t\geq 0}$ is a martingale. Then if $\mathsf{P}_t = \mathsf{P}|\mathcal{F}_t$, $t \geq 0$, one can introduce a probability measure $\widetilde{\mathsf{P}}_t$ by the formula

$$\widetilde{\mathsf{P}}_t(A) = \int_A Z_t \, d\mathsf{P}_t, \qquad A \in \mathcal{F}_t.$$

From the beginning we may assume that

$$(\Omega, \mathcal{F}, (\mathcal{F}_t)_{t\geq 0}) = (\mathbf{C}, \mathcal{B}(\mathbf{C}), (\mathcal{B}_t(\mathbf{C}))_{t\geq 0}),$$

where $\mathcal{B}(\mathbf{C})$ is the σ-algebra of cylindric sets in the space $\mathbf{C}$ of the continuous functions and $\mathcal{B}_t(\mathbf{C})$ is the sub-σ-algebra of $\mathcal{B}(\mathbf{C}) = \bigcap_{\varepsilon>0} \sigma(Y_s; s \leq t+\varepsilon)$, $Y \in \mathbf{C}$. It is well known (see, *e.g.*, [168]) that on the space $(\mathbf{C}, \mathcal{B}(\mathbf{C}))$ there exists a probability measure $\widetilde{\mathsf{P}}$ such that $\widetilde{\mathsf{P}}|\mathcal{F}_t = \widetilde{\mathsf{P}}_t$ for all $t > 0$.

The original *Girsanov theorem* claims that (with respect to the measure $\widetilde{\mathsf{P}}$):

(a) the process $\widetilde{B} = (\widetilde{B}_t)_{t\geq 0}$ with

$$\widetilde{B}_t = B_t - \int_0^t (\widetilde{\mu}_s - \mu_s)\, ds$$

is a standard Brownian motion;

(b) the process $X = (X_t)_{t\geq 0}$ has the stochastic differential

$$dX_t = \widetilde{\mu}_t \, dt + d\widetilde{B}_t. \tag{6.29}$$

Let us compare this Girsanov's result with the statements (6.23) in the theorem given above.

Because of (6.26) the jump measure $\mu \equiv 0$ and the formulae (6.23) have the following form:

$$\begin{aligned} \widetilde{B}_t &= \int_0^t \mu_s \, ds + \int_0^t \beta_s \, ds, \\ \widetilde{C}_t &= C_t, \\ \widetilde{\nu}_t &\equiv 0, \end{aligned} \tag{6.30}$$

where $\beta_t = \dfrac{d\langle Z, X\rangle_t}{d\langle X, X\rangle_t} \dfrac{1}{Z_t}$.

Since from (6.28)

$$dZ_t = Z_t(\widetilde{\mu}_t - \mu_t)\, dB_t$$

and $dX_t = \mu_t \, dt + dB_t$, we have $d\langle X, X\rangle_t = dt$, $d\langle Z, X\rangle_t = Z_t(\widetilde{\mu}_t - \mu_t)\, dt$ and therefore $\beta_t = \widetilde{\mu}_t - \mu_t$. So, the formula (6.30) gives that

$$\widetilde{B}_t = \int_0^t \widetilde{\mu}_s \, ds, \qquad \widetilde{C}_t = t.$$

Thus we obtained both statements (a) and (b) in Girsanov's theorem. (Recall that by Lévy's theorem a continuous martingale $M = (M_t, \mathcal{F}_t)_{t\geq 0}$ with $\mathsf{E}M_t = 0$ and $\langle M, M\rangle_t = t$ is a Brownian motion.)

Remark 6.1. For more about different kinds of Girsanov type theorems see, *e.g.*, [125], [100], [161].

Example 6.2 (Compound Poisson process). Let $X = (X_t, \mathcal{F}_t)_{t\leq T}$ be a semimartingale with a triplet $\mathbb{T} = (0, 0, \nu)$ which is such that

$$\nu((0,t] \times dx) = t\,\nu(dx),$$

where $\nu(\mathbb{R}) < \infty$ ("finite" case; see Subsec. 2 in Sec. 2.3). By (5.35) in Chap. 5

$$\mathsf{E}e^{i\theta X_t} = \exp\left\{t \int_{\mathbb{R}} (e^{i\theta x} - 1)\,\nu(dx)\right\} \tag{6.31}$$

and the process X is a compound Poisson process.

Let us form a new probability measure $\widetilde{\mathsf{P}}_T$ with

$$\widetilde{\mathsf{P}}_T(A) = \int_A Z_T\,d\mathsf{P}_T, \qquad \mathsf{P}_T = \mathsf{P}|\mathcal{F}_T,$$

where

$$Z_t = \exp\left\{t(\nu(\mathbb{R}) - \widetilde{\nu}(\mathbb{R})) + \sum_{s\leq t} \log \rho(\Delta X_s)\right\}, \qquad t \leq T, \tag{6.32}$$

with a finite measure $\widetilde{\nu}$ such that

$$\widetilde{\nu} \ll \nu \quad \text{and} \quad \rho(x) = \frac{d\widetilde{\nu}}{d\nu}(x) \quad (Y(0) = 0).$$

It is easy to check that $(Z_t, \mathcal{F}_t)_{t\leq T}$ is a P-martingale (see also Sec. 4.3) with $\mathsf{E}Z_T = 1$ and therefore the measure $\widetilde{\mathsf{P}}_T$ is a *probability* measure. Also, we can check that the right-hand side of (6.25) is exactly equal to ρ (*i.e.*, $Y = \rho$) and so, by the theorem given above we find (taking $h(x) \equiv 0$ which is possible in the case of $\nu(\mathbb{R}) < \infty$) that a new triplet is

$$\widetilde{\mathbb{T}} = (0, 0, \widetilde{\nu}). \tag{6.33}$$

Of course, the case of compound Poisson process is relatively simple and the formula (6.33) results from the following straightforward calculations based on the (continuous version of) formula (6.11) (here $\widetilde{\mathsf{E}} = \mathsf{E}_{\widetilde{\mathsf{P}}_T}$ is the expectation with respect to the measure $\widetilde{\mathsf{P}}_T$):

$$\begin{aligned}
\mathsf{E}_{\widetilde{\mathsf{P}}_T} e^{i\theta(X_t - X_s)} \,|\, \mathcal{F}_s) &= \mathsf{E}\Big(e^{i\theta(X_t-X_s)} \frac{Z_t}{Z_s} \,\Big|\, \mathcal{F}_s\Big) \\
&= \mathsf{E}\exp\Big\{ \sum_{s<u\le t} i\theta\Delta X_u + (t-s)(\nu(\mathbb{R}) - \widetilde{\nu}(\mathbb{R})) + \sum_{s<u\le t} \log\rho(\Delta X_u)\Big\} \\
&= e^{-(t-s)\widetilde{\nu}(\mathbb{R})} e^{(t-s)\nu(\mathbb{R})} \sum_{k=0}^{\infty} \frac{e^{-(t-s)\nu(\mathbb{R})}((t-s)\nu(\mathbb{R}))^k}{k!} \\
&\qquad \times \Big[\int_{\mathbb{R}} e^{i\theta x + \log\rho(x)} \frac{\nu(dx)}{\nu(\mathbb{R})}\Big]^k \\
&= e^{-(t-s)\widetilde{\nu}(\mathbb{R})} \sum_{k=0}^{\infty} \frac{((t-s)\nu(\mathbb{R}))^k}{k!} \Big[\int_{\mathbb{R}} e^{i\theta x + \log\rho(x)} \nu(dx)\Big]^k \\
&= \exp\Big\{ -(t-s)\widetilde{\nu}(\mathbb{R}) + (t-s)\int_{\mathbb{R}} e^{i\theta x + \log\rho(x)}\nu(dx)\Big\} \\
&= \exp\Big\{ (t-s)\int_{\mathbb{R}} (e^{i\theta x} - 1)\,\widetilde{\nu}(dx)\Big\}.
\end{aligned}$$

So, with respect to the measure $\widetilde{\mathsf{P}}_T$ the increments $X_t - X_s$ do not depend on the σ-algebra $\mathcal{F}_s$ and the process $X = (X_t, \mathcal{F}_t)_{t\le T}$ is a compound Poisson process (cf. (5.35)) with triplet $\widetilde{\mathbb{T}} = (0, 0, \widetilde{\nu})$.

In particular, if the measure $\nu = \nu(dx)$ is concentrated only in the point $x = 1$ then the compound Poisson process is a standard Poisson process with intensity $\lambda = \nu(\{1\})$. If the measure $\widetilde{\nu} = \widetilde{\nu}(dx)$ is also concentrated in the point $x = 1$ and if we put $\widetilde{\lambda} = \widetilde{\nu}(\{1\})$ then we see from (6.32) that

$$Z_t = \exp\Big\{ X_t \log\frac{\widetilde{\lambda}}{\lambda} - t(\widetilde{\lambda} - \lambda)\Big\}, \qquad t \le T.$$

The process $X = (X_t)_{t\le T}$ with respect to the measure $\widetilde{\mathsf{P}}_T$ with

$$\widetilde{\mathsf{P}}_T(A) = \int_A Z_T \, d\mathsf{P}, \qquad A \in \mathcal{F}_T,$$

is then a standard Poisson process with intensity $\widetilde{\lambda} = \widetilde{\nu}(\{1\})$.

3. One can observe that both of the considered examples are particular cases of the following scheme.

Let $X = (X_t)_{t\ge 0}$ be a semimartingale with differential (local) characteristics $\partial\mathbb{T} = (b, c, F)$. Let $\widetilde{\mathsf{P}} \overset{\text{loc}}{\ll} \mathsf{P}$ and let the derivative $Z = d\widetilde{\mathsf{P}}/d\mathsf{P}$ have the following structure:

$$Z = \mathcal{E}\big(H \cdot X^c + W * (\mu^X - \nu^X)\big)$$

(for notation see Sec. 2.1 and Sec. 3.4), where X^c is a continuous martingale part of X, μ^X is the measure of jumps and ν^X is its compensator, and where H and W meet some integrability conditions (see [100; Chap. III, Theorem 4.34]). Under these assumptions on the structure of the process Z, Theorem 6.1 can be restated in the following way (see [109]).

Theorem 6.2. *The differential characteristics $\partial\widetilde{\mathbb{T}} = (\tilde{b}, \tilde{c}, \widetilde{F})$ of the process X with respect to the measure $\widetilde{\mathsf{P}}$, where $d\widetilde{\mathsf{P}} = Z\, d\mathsf{P}$, are given by the formulae*

$$\begin{aligned} \tilde{b}_t &= b_t + H_t c_t + \int W(t,x) h(x)\, F_t(dx), \\ \tilde{c}_T &= c_t, \\ \widetilde{F}_t(G) &= \int I_G(x)\big(1 + W(t,x)\big)\, F_t(dx), \qquad G \in \mathcal{B}(\mathbb{R}). \end{aligned}$$

(The "random" argument ω in all the functions here is omitted for brevity.)

6.4 Esscher's Change of Measure

1. Suppose that $(\Omega, \mathcal{F}, (\mathcal{F}_t)_{t\geq 0}, \mathsf{P})$ is a given filtered probability space and $X = (X_t, \mathcal{F}_t)_{t\geq 0}$ is a semimartingale (with respect to measure P) with the triplet $\mathbb{T} = (B, C, \nu)$ of predictable characteristics.

In the Introduction we wrote that one of the motivations for constructing a new measure $\widetilde{\mathsf{P}}$ (usually with property $\widetilde{\mathsf{P}} \overset{\text{loc}}{\ll} \mathsf{P}$ or $\widetilde{\mathsf{P}} \overset{\text{loc}}{\sim} \mathsf{P}$) is our desire to get some "good" properties of the process X under the measure $\widetilde{\mathsf{P}}$ (for example, to have a martingale property of X with respect to $\widetilde{\mathsf{P}}$).

We mentioned above several times (*e.g.*, in Sec. 4.4) that one reasonable procedure of constructing the density processes $Z = (Z_t)_{t\geq 0}$, which can be used for construction of a new measure $\widetilde{\mathsf{P}}$, is based on the so-called Esscher transformation (invented by F. Esscher in 1932 [72]). Let us explain by a simple example the ideas of this transformation.

Assume that $(\Omega, \mathcal{F}, \mathsf{P})$ is a probability space and $\xi = \xi(\omega)$ is a random variable with (for simplicity) $\varphi(\theta) = \mathsf{E}e^{\theta\xi} < \infty$ for all $\theta \in \mathbb{R}$.

The transformation

$$\mathsf{P}(dx) \rightsquigarrow \mathsf{P}^\theta(dx) = \frac{e^{\theta x}}{\varphi(\theta)}\, \mathsf{P}(dx), \qquad \theta \in \mathbb{R}, \quad x \in \mathbb{R},$$

is well known in the actuarial practice as the *Esscher transform.* Let us show that, with respect to new measures P^θ constructed *via* the Esscher transform, the random variable can have some "good" properties. Indeed, define, for a given θ,

$$Z^\theta(x) = \frac{e^{\theta x}}{\varphi(\theta)}. \tag{6.34}$$

Because $\mathsf{E}Z^\theta(\xi) = 1$ we can construct a new measure P^θ on $\mathcal{F}$ with

$$\mathsf{P}^\theta(A) = \int_A Z^\theta(\xi(\omega))\,\mathsf{P}(d\omega), \qquad A \in \mathcal{F}.$$

We see that (under assumption that $\varphi(\theta) < \infty$ for all $\theta \in \mathbb{R}$)

$$\mathsf{E}_{\mathsf{P}^\theta}\xi = \mathsf{E}\xi Z^\theta(\xi) = \frac{\varphi'(\theta)}{\varphi(\theta)}. \tag{6.35}$$

The function $\varphi(\theta)$ is strictly convex since $\varphi''(\theta) > 0$, and if additionally (to the assumption $\varphi(\theta) < \infty$, $\theta \in \mathbb{R}$)

$$\mathsf{P}(\xi > 0) > 0 \quad \text{and} \quad \mathsf{P}(\xi < 0) > 0$$

then there exists a $\theta^* \in \mathbb{R}$ such that $\varphi'(\theta^*) = 0$.

Then from (6.35)

$$\mathsf{E}_{\mathsf{P}^{\theta^*}}\xi = 0. \tag{6.36}$$

In other words, with respect to the measure P^{θ^*}, constructed by means of an Esscher transform, the random variable ξ has "equilibrium" behavior in the sense that the expectation $\mathsf{E}_{\mathsf{P}^{\theta^*}}\xi$ is equal to zero.

If we put $X_0 \equiv 0$, $X_1 = \xi$ and $\mathcal{F}_0 = \{\varnothing, \Omega\}$, $\mathcal{F}_1 = \mathcal{F}$ then we note that (P-a.s.)

$$\mathsf{E}_{\mathsf{P}^{\theta^*}}(X_1 \,|\, \mathcal{F}_0) = X_0 \;\; (= 0),$$

which is exactly a martingale property (with respect to P^{θ^*}) of the sequence (X_0, X_1).

This simple "one-step" model is a good illustration of the construction of a "martingale" measure P^{θ^*} by Esscher's procedure which works also for "multi-step" models and in continuous-time models used in mathematical finance, and actuarial mathematics.

2. Taking into account the previous considerations let us form, for a semimartingale $X = (X_t, \mathcal{F}_t)_{t\geq 0}$ and a predictable process $\varphi = (\varphi_t)_{t\geq 0}$ from the class $L(X)$ (see Sec. 3.4), a new process $X^\varphi = (X_t^\varphi, \mathcal{F}_t)_{t\geq 0}$ and a process $Z^\varphi = (Z_t^\varphi, \mathcal{F}_t)_{t\geq 0}$ with

$$Z_t^\varphi = \frac{\exp\{X_t^\varphi\}}{\mathcal{E}(\widetilde{\mathcal{K}}^\varphi)_t} \tag{6.37}$$

(in (4.40) this process was denoted by $\mathcal{M}_t^\varphi$), where $\widetilde{\mathcal{K}}^\varphi = \widetilde{\mathcal{K}}^\varphi(1)$ with $\widetilde{\mathcal{K}}^\varphi(\theta)$ defined in (4.47)–(4.49).

The process $Z^\varphi = (Z_t^\varphi)_{t\geq 0}$ is a local martingale (with respect to P). The problem about the martingale property and uniform martingale property was investigated in [111], and we refer to this paper for all details.

So, let us *assume* that the process $(Z_t^\varphi)_{t\geq T}$ is a martingale. As a result this process has the property $\mathsf{E}Z_T^\varphi = 1$, and because $Z_T^\varphi \geq 0$ we may define an (Esscher) measure $\widetilde{\mathsf{P}}_T^\varphi = \widetilde{\mathsf{P}}_T^\varphi(A)$, $A \in \mathcal{F}_T$, with

$$\widetilde{\mathsf{P}}_T^\varphi(A) = \int_A Z_T^\varphi(\omega)\, \mathsf{P}(d\omega). \tag{6.38}$$

With respect to this measure $\widetilde{\mathsf{P}}_T^\varphi$ the semimartingale $X = (X_t, \mathcal{F}_t)_{t\geq T}$ will be again a semimartingale with the triplet $\widetilde{\mathbb{T}}^\varphi = (B_t^\varphi, C_t^\varphi, \nu_t^\varphi)_{t\leq T}$ that was described in (4.41).

Note that if the process $(Z_t^\varphi)_{t\geq 0}$ is a *uniformly* integrable martingale then there exists a limit (P-a.s.) $Z_\infty^\varphi = \lim_{t\to\infty} Z_t^\varphi$ and we can construct also a measure $\widetilde{\mathsf{P}}_\infty^\varphi$ taking in (6.38) $T = \infty$. The corresponding triplet $\widetilde{\mathbb{T}}^\varphi = (B^\varphi, C^\varphi, \nu^\varphi)$ is given by the same formulae (4.41).

In the next chapter we give a systematic exposition of the problems of construction—by means of Esscher transforms—of a new measure (with "good" properties) for the Lévy processes.

Finally we make the following remark, which explains why we pay much attention to the Esscher transform which is only a *special* way to construct a new measure. Suppose that we have two measures, P and $\widetilde{\mathsf{P}}$ ($\widetilde{\mathsf{P}} \overset{\text{loc}}{\ll} \mathsf{P}$), and $Z_t = \dfrac{d\widetilde{\mathsf{P}}_t}{d\mathsf{P}_t}$. Assuming the canonical setting (*i.e.*, Ω is the space of càdlàg functions) we may say that the process $Z = (Z_t)_{t\geq 0}$ with nonanticipating functionals $Z_t = Z_t(X)$, $t \geq 0$, where X is a canonical process on Ω, necessarily should be a martingale. So, if we knew the structure of all such positive martingales then it would be no problem to describe all possible measures $\widetilde{\mathsf{P}}$ which are absolutely continuous or locally absolutely continuous with respect to the measure P. However, there are few general results in this direction. (Some of them can be found in [100; Chap. III, §§ 4c, 5a–c].) This is one explanation why we use so often Esscher transforms (see, *e.g.*, (6.36)): we know definitively that this leads, at least, to positive local martingales.

Chapter 7

Change of Measure in Models Based on Lévy Processes

7.1 Linear and Exponential Lévy Models under Change of Measure

1. Let $X = (X_t, \mathcal{F}_t)_{t\geq 0}$ be a Lévy process with local characteristics (b, c, F) (relative to the fixed truncation function) and with the characteristic function

$$\mathsf{E}e^{i\theta X_t} = e^{t\varkappa(\theta)}, \tag{7.1}$$

where the local Fourier cumulant function is

$$\varkappa(\theta) = i\theta b - \frac{\theta^2}{2}c + \int \bigl(e^{i\theta x} - 1 - i\theta h(x)\bigr)\, F(dx). \tag{7.2}$$

(See Sec. 4.2.)

In *Mathematical Finance* and other fields of applications the following two models (in particular, for the financial prices $S = (S_t)_{t\geq 0}$), based on Lévy processes, are popular.

I. *Linear model*

$$S_t = S_0 + X_t, \qquad t \geq 0. \tag{7.3}$$

II. *Exponential model*

$$S_t = S_0 e^{X_t}, \qquad t \geq 0, \tag{7.4}$$

or, equivalently,

$$S_t = S_0 \mathcal{E}(\widetilde{X})_t, \qquad t \geq 0, \tag{7.5}$$

where $\widetilde{X} = \mathcal{L}(e^X)$ is the stochastic logarithm of e^X:

$$\widetilde{X} = X + \frac{1}{2}\langle X^c \rangle + \sum_{s\leq \cdot} \bigl(e^{\Delta X_s} - 1 - \Delta X_s\bigr). \tag{7.6}$$

Remark 7.1. In Financial Economics the representations (7.4) and (7.5) are called *compound return* and *simple return* representations of prices, respectively.

We intend to consider below the following questions:

(a) when is the process S a local martingale or martingale with respect to the initial measure P?

and, in the case S is not a (local) martingale (with respect to P)

(b) how to construct a measure $\widetilde{\mathsf{P}}$ with the property $\widetilde{\mathsf{P}} \overset{\text{loc}}{\ll} \mathsf{P}$ or $\widetilde{\mathsf{P}} \overset{\text{loc}}{\sim} \mathsf{P}$, relative to which the process S will be a local martingale or martingale?

2. To answer the question (a) for the *linear* model (7.3) we use a general result from Subsec. 3 of Sec. 3.2. This result states that a semimartingale X with the triplet $\mathbb{T} = (B, C, \nu)$ is a local martingale if and only if (up to an evanescent set)

$$B + (x - h) * \nu = 0. \tag{7.7}$$

For Lévy process it gives the following condition:

$$b + \int_{\mathbb{R}} (x - h(x))\, F(dx) = 0. \tag{7.8}$$

For the *exponential* model (7.4) the corresponding condition for "local martingality" of $S = e^X$ follows easily from the representation $S = \mathcal{E}(\widetilde{X})$, where $\widetilde{X} = \mathcal{L}(e^X)$ (Sec. 4.1). Indeed, the stochastic exponential $\mathcal{E}(\widetilde{X})$ is a local martingale if and only if so is $\widetilde{X}$. The process $\widetilde{X}$ is also a semimartingale with the triplet $\widetilde{\mathbb{T}} = (\widetilde{B}, \widetilde{C}, \widetilde{F})$. So, by (7.8) the process $S = e^X$ $(= \mathcal{E}(\widetilde{X}))$ is a local martingale if and only if

$$\tilde{b} + \int_{\mathbb{R}} (y - h(y))\, \widetilde{F}(dy) = 0. \tag{7.9}$$

By (4.18)

$$\tilde{b} = b + \frac{c}{2} + \int_{\mathbb{R}} \big(h(e^x - 1) - h(x)\big)\, F(dx), \tag{7.10}$$

$$\widetilde{F}(G) = \int I_G(e^x - 1)\, F(dx), \qquad G \in \mathcal{B}(\mathbb{R}). \tag{7.11}$$

From (7.10)–(7.11) we find that

$$\begin{aligned}\widetilde{b}+\int_{\mathbb{R}}(y-h(y))\,\widetilde{F}(dy)&=b+\frac{c}{2}+\int_{\mathbb{R}}\big(h(e^x-1)-h(x)\big)\,F(dx)\\&\quad+\int_{\mathbb{R}}\big((e^x-1)-h(e^x-1)\big)\,F(dx)\\&=b+\frac{c}{2}+\int_{\mathbb{R}}\big(e^x-1-h(x)\big)\,F(dx).\end{aligned}\tag{7.12}$$

From here, (7.9), and (7.8) we get the following theorem.

Theorem 7.1. *Let $X=(X_t,\mathcal{F}_t)_{t\ge0}$ be a Lévy process with the local characteristics (b,c,F).*

In the linear model $S_t=S_0+X_t$, $t\ge0$, the process $S=(S_t)_{t\ge0}$ is a local martingale (with respect to P*) if and only if*

$$b+\int_{\mathbb{R}}(x-h(x))\,F(dx)=0\tag{7.13}$$

(i.e., (7.8) *is fulfilled*).

In the exponential model $S_t=S_0e^{X_t}$, $t\ge0$, the process $S=(S_t)_{t\ge0}$ is a local martingale (with respect to P*) if and only if*

$$b+\frac{c}{2}+\int_{\mathbb{R}}\big(e^x-1-h(x)\big)\,F(dx)=0.\tag{7.14}$$

(Here we assume that $\int_{|x|>1}e^x\,F(dx)<\infty$.)

Example 7.1. Assume that the process $X=(X_t,\mathcal{F}_t)_{t\ge0}$ is a Poisson difference process with drift, *i.e.*,

$$X_t=\mu t+\alpha N_t^{(1)}-\beta N_t^{(2)},$$

where $\alpha>0$, $\beta>0$, and $N^{(1)}=(N_t^{(1)})_{t\ge0}$ and $N^{(2)}=(N_t^{(2)})_{t\ge0}$ are two independent Poison processes with intensity parameters $\lambda_1>0$ and λ_2, respectively.

It is clear that

$$\begin{aligned}\mathsf{E}X_t&=(\mu+\alpha\lambda_1-\beta\lambda_2)t,\\\mathsf{D}X_t&=(\alpha^2\lambda_1^2-\beta^2\lambda_2^2)t.\end{aligned}$$

Taking the function $h=h(x)$ to be identically null, we find that Laplace's cumulant function $\lambda(\theta)$ (see (4.39)) is given by

$$\lambda(\theta)=\mu\theta+\lambda_1(e^{\alpha\theta}-1)+\lambda_2(e^{-\beta\theta}-1),$$

since for the process X, evidently,

$$F(dx)=\lambda_1I_{\{\alpha\}}(dx)+\lambda_2I_{\{-\beta\}}(dx).$$

Note that the process X can be defined alternatively as a compound Poisson process with drift:

$$X_t = \mu t + \sum_{k=1}^{N_t} \xi_k,$$

where $N = (N_t)_{t\geq 0}$ is a Poisson process with parameter $\lambda = \lambda_1 + \lambda_2$ and $(\xi_k)_{k\geq 1}$ is i.i.d. sequence independent of N and such that

$$\mathsf{P}(\xi_k = \alpha) = \frac{\lambda_1}{\lambda_1 + \lambda_2}, \qquad \mathsf{P}(\xi_k = -\beta) = \frac{\lambda_2}{\lambda_1 + \lambda_2}.$$

We conclude from (7.13) that in a *linear* model $S_t = S_0 + X_t$ with the considered process $X = (X_t)_{t\geq 0}$ the process $S = (S_t)_{t\geq 0}$ is a martingale if and only if

$$\mu + \alpha\lambda_1 - \beta\lambda_2 = 0,$$

which is an easy consequence of $\mathsf{E}(X_t - X_s) = (\mu + \alpha\lambda_1 - \beta\lambda_2)(t - s)$.

From (7.14) we see that in the exponential model $S_t = S_0 e^{X_t}$, $S_0 > 0$, the process $S = (S_t)_{t\geq 0}$ is a martingale if and only if

$$\mu + \alpha\lambda_1(e^{\alpha} - 1) + \lambda_2(e^{-\beta} - 1) = 0.$$

It is interesting to note that for the difference $N_1^{(1)} - N_1^{(2)}$ of two independent random variables which have Poisson distributions with parameters λ_1 and λ_2 the explicit formula for probabilities $\mathsf{P}(N_1^{(1)} - N_1^{(2)} = k)$, $k = 0, \pm 1, \pm 2, \ldots$, can be given. (Clearly, the distribution of the sum $N_1^{(1)} + N_1^{(2)}$ is Poissonian with parameter $\lambda_1 + \lambda_2$.)

To this end consider the generating function

$$G_{N_1^{(i)}}(s) = \mathsf{E} s^{N_1^{(i)}}, \qquad i = 1, 2.$$

It is clear that because

$$\lambda^{(i)}(\theta) = \mathsf{E} e^{\theta N_1^{(i)}} = e^{\lambda_i(e^{\theta} - 1)},$$

we have

$$G_{N_1^{(i)}}(s) = e^{-\lambda_i(1-s)}.$$

Therefore

$$G_{N_1^{(1)} - N_1^{(2)}}(s) = G_{N_1^{(1)}}(s) \cdot G_{N_1^{(2)}}\left(\tfrac{1}{s}\right) = e^{-(\lambda_1+\lambda_2)} \cdot e^{\sqrt{\lambda_1\lambda_2}(t+1/t)},$$

where $t = s\sqrt{\lambda_1/\lambda_2}$.

It is a known fact from analysis that

$$e^{x(t+1/t)} = \sum_{k=-\infty}^{\infty} t^k I_k(2x),$$

where $I_k(2x)$ is the modified Bessel function of the first kind with index k (see, *i.e.*, [67; v. 1, p. 373], [173]), which admits the following representation:

$$I_k(2x) = x^k \sum_{r=0}^{\infty} \frac{x^{2r}}{r!\,\Gamma(k+r+1)}, \qquad k = 0, \pm 1, \pm 2, \ldots .$$

The above formulae imply

$$G_{N_1^{(1)} - N_1^{(2)}}(s) = e^{-(\lambda_1+\lambda_2)}\, e^{\sqrt{\lambda_1\lambda_2}(t+1/t)} = e^{-(\lambda_1+\lambda_2)} \sum_{k=-\infty}^{\infty} t^k I_k(2\sqrt{\lambda_1\lambda_2})$$

$$= e^{-(\lambda_1+\lambda_2)} \sum_{k=-\infty}^{\infty} s^k \left(\sqrt{\frac{\lambda_1}{\lambda_2}}\right)^k I_k\big(2\sqrt{\lambda_1\lambda_2}\big).$$

Whence we find that

$$\mathsf{P}\big(N_1^{(1)} - N_1^{(2)} = k\big) = e^{-(\lambda_1+\lambda_2)} \left(\frac{\lambda_1}{\lambda_2}\right)^{k/2} I_k\big(2\sqrt{\lambda_1\lambda_2}\big).$$

3. Suppose now that neither condition (7.13) nor (7.14) holds. We want to give at least sufficient conditions guaranteeing the possibility to construct new measures $\widetilde{\mathsf{P}}$ with respect to which the processes $S_t = S_0 + X_t$ and $S_t = S_0 e^{X_t}$, $t \geq 0$, are again local martingales.

In both cases we shall try to construct the corresponding measures by means of Esscher transforms.

From Sec. 4.3 we know that for all $a \in \mathbb{R}$ the processes $Z^{(a)} = (Z_t^{(a)})_{t \leq T}$ with

$$Z_t^{(a)} = \frac{e^{aX_t}}{e^{\Lambda_t(a)}} = e^{aX_t - t\lambda(a)} \tag{7.15}$$

under assumption $\int e^{ax} I(|x| > 1)\, F(dx) < \infty$, $a \in \mathbb{R}$, are positive local martingales. From now on we assume that these processes $Z^{(a)}$ are martingales. Then evidently $\mathsf{E} Z_T^{(a)} = 1$ and for each $a \in \mathbb{R}$ we may form a new probability (Esscher) measure $\mathsf{P}_T^{(a)}$ with

$$\mathsf{P}_T^{(a)}(A) = \int_A Z_T^{(a)}\, \mathsf{P}_T(d\omega), \tag{7.16}$$

where $A \in \mathcal{F}_T$, $\mathsf{P}_T = \mathsf{P}|\mathcal{F}_T$.

It is clear that under the formulated assumptions

$$\mathsf{E}_{\mathsf{P}_T^{(a)}} e^{\theta X_t} = e^{t\lambda^{(a)}(\theta)} \tag{7.17}$$

with $\lambda^{(a)}(\theta) = \lambda(a+\theta) - \lambda(a)$, and that

$$\mathsf{E}_{\mathsf{P}_T^{(a)}} \big(e^{\theta(X_t - X_s)} \,|\, \mathcal{F}_s\big) = e^{(t-s)\lambda^{(a)}(\theta)}. \tag{7.18}$$

So, the process $X = (X_t)_{t\le T}$ with respect to the new measure $\mathsf{P}_T^{(a)}$ is again a Lévy process. Denote by $(b^{(a)}, c^{(a)}, F^{(a)})$ the local characteristics of this process.

Comparing the expressions

$$\mathsf{E}_{\mathsf{P}_T^{(a)}} e^{\theta X_t} = \exp\{t[\lambda(a+\theta) - \lambda(a)]\}$$

and

$$\mathsf{E}_{\mathsf{P}_T^{(a)}} e^{\theta X_t} = \exp\bigg\{t\bigg[\theta b^{(a)} + \frac{\theta^2}{2} c^{(a)} + \int_{\mathbb{R}} \big(e^{\theta x} - 1 - \theta h(x)\big)\, F^{(a)}(dx)\bigg]\bigg\}, \tag{7.19}$$

we find that

$$\begin{aligned} b^{(a)} &= b + ac + \int_{\mathbb{R}} h(x)(e^{ax} - 1)\, F(dx), \\ c^{(a)} &= c, \\ F^{(a)}(dx) &= e^{ax}\, F(dx). \end{aligned} \tag{7.20}$$

From (7.8) we conclude that if $\tilde{a}$ is such that

$$b^{(\tilde{a})} + \int_{\mathbb{R}} (x - h(x))\, F^{(\tilde{a})}(dx) = 0 \tag{7.21}$$

then the process $X = (X_t)_{t\le T}$ with respect to the measure $\widetilde{\mathsf{P}}_T = \mathsf{P}_T^{(\tilde{a})}$ is a local martingale. Together with (7.20) this yields the equivalence of (7.21) to the following condition: there exists $\tilde{a}$ such that

$$b + \tilde{a}c + \int_{\mathbb{R}} \big(x e^{\tilde{a}x} - h(x)\big)\, F(dx) = 0. \tag{7.22}$$

Finally, consider the question about existence of a measure $\widetilde{\mathsf{P}}_T$ such that the process $S_t = S_0 e^{X_t}$ is a local martingale when condition (7.14) does not hold.

From (7.18) it follows that in exponential models $S = (S_t)_{t\le T}$ with $S_t = S_0 e^{X_t}$ the process S is a local martingale with respect to the Esscher measure $\widetilde{\mathsf{P}}_T = \mathsf{P}_T^{(\tilde{a})}$ (see (7.16)) if $\tilde{a}$ solves the equation $\lambda(\tilde{a}+1) - \lambda(\tilde{a}) = 0$, which is equivalent to condition that

$$b + \left(\tilde{a} + \frac{1}{2}\right)c + \int_{\mathbb{R}} \left(e^{\tilde{a}x}(e^x - 1) - h(x)\right) F(dx) = 0 \tag{7.23}$$

(under the assumption that $\int_{\mathbb{R}} \left|e^{\tilde{a}x}(e^x - 1) - h(x)\right| F(dx) < \infty$).

Remark 7.2. If $\tilde{a} = 0$ then condition (7.22) coincides with (7.8) and condition (7.23) turns into condition (7.14).

Thus, we have the following theorem.

Theorem 7.2. *In the linear model $S_t = S_0 + X_t$, $t \le T$, the process $S = (S_t)_{t \le T}$ is a local martingale with respect to the Esscher measure $\mathsf{P}_T^{(\tilde{a})}$ if $\tilde{a}$ is such that condition* (7.22) *does hold.*

In the exponential model $S_t = S_0 e^{X_t}$, $t \le T$, the process $S = (S_t)_{t \le T}$ is a local martingale with respect to the Esscher measure $\mathsf{P}_T^{(\tilde{a})}$ if $\tilde{a}$ satisfies (7.23).

Example 7.2. Turn again to the process $X_t = \mu t + \alpha N_t^{(1)} - \beta N_t^{(2)}$, $t \ge 0$, considered in Example 7.1.

Assume that the condition $\mu + \alpha\lambda_1 - \beta\lambda_2 = 0$ fails. Then the linear process $X = (X_t)_{t \ge 0}$ is not a martingale (with respect to the initial probability measure P). By the first part of Theorem 7.2 we see that if the parameter $\tilde{a}$ is such that

$$\mu + \lambda_1 \alpha e^{\tilde{a}\alpha} - \lambda_2 \beta e^{-\tilde{a}\beta} = 0, \tag{7.24}$$

then the process $X = (X_t)_{t \le T}$ is a martingale with respect to the measure $\mathsf{P}_T^{(\tilde{a})}$ constructed by means of Esscher's transformation; see (7.16) with $a = \tilde{a}$. It is easy to see that $\tilde{a}$ solves (7.24).)

In the case of exponential model $S_t = S_0 e^{X_t}$, $t \le T$, the process $S = (S_t)_{t \le T}$ is (by the second part of Theorem 7.2) a martingale with respect to the measure $\mathsf{P}_T^{(\tilde{a})}$ if the parameter $\tilde{a}$ is chosen to be such that

$$\mu + \lambda_1 e^{\tilde{a}\alpha}(e^{\alpha} - 1) + \lambda_2 e^{-\tilde{a}\beta}(e^{-\beta} - 1) = 0.$$

Considering this relation as an equation in $\tilde{a}$, we see that it has a unique solution. If $\mu = 0$, then $\tilde{a}$ is easy to find:

$$\tilde{a} = \frac{1}{\alpha + \beta} \log \frac{\lambda_2(1 - e^{-\beta})}{\lambda_1(e^{\alpha} - 1)}.$$

7.2 On the Criteria of Local Absolute Continuity of Two Measures of Lévy Processes

1. The main problems considered in the previous Sec. 7.1 and in Chap. 6 were related with the construction of a measure $\widetilde{\mathsf{P}} \overset{\text{loc}}{\ll} \mathsf{P}$ such that the process $X = (X_t, \mathcal{F}_t)_{t\geq 0}$, given on a filtered probability space $(\Omega, \mathcal{F}, (\mathcal{F}_t)_{t\geq 0}, \mathsf{P})$, has some "good" properties with respect to the measure $\widetilde{\mathsf{P}}$.

It is very natural also to change a little bit the problem, asking the following question.

Suppose that $X = (X_t, \mathcal{F}_t)_{t\geq 0}$ is a stochastic process given on the filtered probabilistic-statistical space $(\Omega, \mathcal{F}, (\mathcal{F}_t)_{t\geq 0}, \mathsf{P}$ and $\widetilde{\mathsf{P}})$, where P and $\widetilde{\mathsf{P}}$ are two probability measures. We shall assume also that we deal with the "canonical setting", *i.e.*, Ω is the canonical space of all càdlàg functions $\omega = \omega(t)$, $t \geq 0$, and $(X_t(\omega))_{t\geq 0}$ is the canonical process defined by $X_t(\omega) = \omega(t)$.

Under such assumption $\mathrm{Law}(X \mid \mathsf{P}) = \mathsf{P}$ and $\mathrm{Law}(X \mid \widetilde{\mathsf{P}}) = \widetilde{\mathsf{P}}$.

In [100; Chap. IV] one can find a detailed investigation (based on the notions of Hellinger–Kakutani distance, Hellinger process) of the problems about absolute continuity, equivalency, and singularity of the probability distributions P and $\widetilde{\mathsf{P}}$ of the process X. Here we shall use some of those results in the case where the process X is a Lévy process with respect to both measures P and $\widetilde{\mathsf{P}}$. For this case an important general result is the following analog of the "Kakutani alternative" for two product measures P and $\widetilde{\mathsf{P}}$ on $(\mathbb{R}^\infty, \mathcal{B}(\mathbb{R}^\infty))$ (see [100; introduction to Chap. IV] or [160; Chap. VII, § 6]): *If* P *and* $\widetilde{\mathsf{P}}$ *are distributions of a Lévy process* X *then*

$$\text{either} \quad \widetilde{\mathsf{P}} = \mathsf{P} \quad \text{or} \quad \widetilde{\mathsf{P}} \perp \mathsf{P}. \tag{7.25}$$

For example, if $\mathsf{P} = \mathsf{P}^0$ and $\widetilde{\mathsf{P}} = \mathsf{P}^\mu$ are distributions of Brownian motions $X_t = B_t$ and $X_t = \mu t + B_t$, $t \geq 0$, respectively, then $\widetilde{\mathsf{P}} \perp \mathsf{P}$ if $\mu \neq 0$. [Note that $\lim_{t\to\infty} X_t/t = \mu$ (P^μ-a.s.).]

If $\mathsf{P} = \mathsf{P}^{\lambda_0}$ and $\widetilde{\mathsf{P}} = \mathsf{P}^{\lambda_1}$, $\lambda_0 \neq \lambda_1$, are distributions of two Poisson processes with parameters λ_0 and λ_1, then $\widetilde{\mathsf{P}} \perp \mathsf{P}$.

In many cases we are interested not in the measures P and $\widetilde{\mathsf{P}}$ but in their restrictions $\mathsf{P}_t = \mathsf{P}|\mathcal{F}_t$ and $\widetilde{\mathsf{P}}_t = \widetilde{\mathsf{P}}|\mathcal{F}_t$ for $t > 0$.

In this case the corresponding result (cf. (7.25)) has the following from:

$$\begin{aligned} &\text{either} \quad \text{for all } t > 0, \ \mathsf{P}_t \text{ and } \widetilde{\mathsf{P}}_t \text{ are not singular,} \\ &\text{or} \quad \text{for all } t > 0, \ \mathsf{P}_t \perp \widetilde{\mathsf{P}}_t. \end{aligned} \tag{7.26}$$

(For the proof see [100; Chap. IV, Theorem 4.39 a), b)].)

2. Now assume that P and $\widetilde{\mathsf{P}}$ are distributions of Lévy processes with triplets (b, c, F) and $(\widetilde{b}, \widetilde{c}, \widetilde{F})$ (relative to a truncation function $h = h(x)$). The basic result here is the following theorem (see [100; Chap. IV, Theorem 4.39 c)]).

Theorem 7.3. *The property* $\widetilde{\mathsf{P}} \overset{\text{loc}}{\ll} \mathsf{P}$ *is valid if and only if the following set of conditions hold:*

$$\text{(a)}\quad \widetilde{F} \ll F \ \ \left(\textit{with } K(x) = \frac{d\widetilde{F}}{dF}(x)\right);$$

$$\text{(b)}\quad \int \big|h(x)[1 - K(x)]\big|\, F(dx) < \infty;$$

$$\text{(c)}\quad \widetilde{b} = b + \int h(x)[K(x) - 1]\, F(dx) + c\beta \ \ (\textit{for some } \beta \in \mathbb{R});$$

$$\text{(d)}\quad \widetilde{c} = c;$$

$$\text{(e)}\quad \int \big(1 - \sqrt{K(x)}\big)^2 F(dx) < \infty.$$

Remark 7.3. One can show that (e) $\Rightarrow$ (b).

Corollary 7.1. *Take* $h(x) = xI(|x| \le a)$ *for some* $a \in \mathbb{R}$. *Suppose that* $\widetilde{F} = F$ *on the set* $\{x : |x| \le a\}$ *(in this case* $K(x) = 1$ *for* $|x| \le a$*) and* $\widetilde{F} \ll F$ *on the set* $\{x : |x| > a\}$.

Assume also that $\widetilde{b} = b$ *and* $\widetilde{c} = c$. *Then condition* (c) *holds with* $\beta = 0$. *Condition* (b) *evidently does hold. Condition* (e) *also does hold (because* $F(\{|x| \ge a\}) < \infty$ *and* $\widetilde{F}(\{|x| \ge a\}) < \infty$*). Thus,* $\widetilde{\mathsf{P}} \overset{\text{loc}}{\ll} \mathsf{P}$ *and it is not difficult to show that the density process* $Z_t = d\widetilde{\mathsf{P}}_t/d\mathsf{P}_t$, $t \ge 0$, *is defined by*

$$Z_t = \exp\Bigg\{t\Bigg[\sum_{s \le t} \log K\big((\Delta X_s) I(|\Delta X_s| > a)\big) \\ + F(\{|x| > a\}) - \widetilde{F}(\{|x| > a\})\Bigg]\Bigg\}.$$

Corollary 7.2. *For distributions* P *and* $\widetilde{\mathsf{P}}$ *of Lévy processes the property* $\widetilde{\mathsf{P}} \overset{\text{loc}}{\sim} \mathsf{P}$ *is valid if and only if*

(α) $\widetilde{F} \sim F$ *and* $\int (1 - (d\widetilde{F}/dF)^{1/2})^2\, dF < \infty$;

(β) $\widetilde{c} = c$;

(γ) *either* $c > 0$ *or*

$$c = 0 \quad \textit{and} \quad \widetilde{b} = b + \int h(x)\, (\widetilde{F} - F)(dx).$$

The proof is nothing but a straightforward verification of the conditions of Theorem 7.3. Indeed, note, first of all, that from (α) we get

(δ) $\int h(x)\, d\mathrm{Var}\,(F-\widetilde{F}) < \infty$.

Then it is easy to see that

(α) $\Longrightarrow$ (a) and ($\widetilde{\text{a}}$): $F \ll \widetilde{F}$;

(α) $\Longrightarrow$ (e);

(α) $\Longrightarrow$ (δ) $\Longrightarrow$ ($\widetilde{\text{e}}$): $\displaystyle\int \big(1-[\widetilde{K}(x)]^{1/2}\big)^2\, \widetilde{F}(dx) < \infty$;

(β) $\Longrightarrow$ (d);

(γ) $\Longrightarrow$ (c) and ($\widetilde{\text{c}}$): $b = \widetilde{b} + \displaystyle\int h(x)[\widetilde{K}(x)-1]\, \widetilde{F}(dx) + \widetilde{c}\widetilde{\beta}$ for some $\widetilde{\beta} \in \mathbb{R}$;

(α) $\Longrightarrow$ (δ) $\Longrightarrow$ (b) and ($\widetilde{\text{b}}$): $\displaystyle\int \big|h(x)[\widetilde{K}(x)-1]\big|\, \widetilde{F}(dx) < \infty$.

So, all conditions (a)–(e) and also the corresponding conditions ($\widetilde{\text{a}}$)–($\widetilde{\text{e}}$) hold, and one may apply the conclusion of Theorem 7.2 (about validity at the same time of both properties $\widetilde{\mathsf{P}} \overset{\text{loc}}{\ll} \mathsf{P}$ and $\mathsf{P} \overset{\text{loc}}{\ll} \widetilde{\mathsf{P}}$, *i.e.*, $\widetilde{\mathsf{P}} \overset{\text{loc}}{\sim} \mathsf{P}$).

7.3 On the Uniqueness of Locally Equivalent Martingale-type Measures for the Exponential Lévy Models

1. Let $S = (S_t)_{t\le T}$ be an exponential Lévy model,

$$S_t = S_0 e^{X_t}, \tag{7.27}$$

where $X = (X_t)_{t\le T}$ is a Lévy process given on the canonical filtered probability space $(\Omega, \mathcal{F}, (\mathcal{F}_t)_{t\le T}, \mathsf{P})$, $\mathcal{F} = \mathcal{F}_T$. Now we shall denote by $(b, c, F)_h$ the triplet of local characteristics of the process X with respect to the truncation function $h = h(x)$. The following four classes of measures $\widetilde{\mathsf{P}}$ with the property $\widetilde{\mathsf{P}} \sim \mathsf{P}$ play an essential role in mathematical finance, in particular in the problems about the "absence of arbitrage" and the "completeness" of markets (see Chap. 10), where the exponential Lévy processes are used for the description of the evolution of the prices of financial indices:

$$\begin{aligned}
\mathcal{M}_{\mathrm{U}} &= \{\widetilde{\mathsf{P}} \sim \mathsf{P} : S = e^X \text{ is a uniformly integrable martingale} \\
&\qquad\qquad\qquad\qquad \text{with respect to } \widetilde{\mathsf{P}}\}, \\
\mathcal{M} &= \{\widetilde{\mathsf{P}} \sim \mathsf{P} : S = e^X \text{ is a } \widetilde{\mathsf{P}}\text{-martingale}\}, \\
\mathcal{M}_{\mathrm{loc}} &= \{\widetilde{\mathsf{P}} \sim \mathsf{P} : S = e^X \text{ is a } \widetilde{\mathsf{P}}\text{-local martingale}\}, \\
\mathcal{M}_{\sigma} &= \{\widetilde{\mathsf{P}} \sim \mathsf{P} : S = e^X \text{ is a } \widetilde{\mathsf{P}}\text{-}\sigma\text{-martingale}\}.
\end{aligned}$$

(For description of the class $\mathcal{M}_\sigma$ see Sec. 10.1.)

For these four classes we have the following inclusions:

$$\mathcal{M}_{\mathrm{U}} \subseteq \mathcal{M} \subseteq \mathcal{M}_{\mathrm{loc}} \subseteq \mathcal{M}_\sigma.$$

2. The following two models are the classical models based on a standard Brownian motion $B = (B_t)_{t\le T}$ and a Poisson process $N = (N_t)_{t\le T}$ with the parameter $\lambda > 0$:

$$(B) \qquad S_t = S_0 e^{X_t}, \quad X_t = \mu t + \sigma B_t,$$
$$(N) \qquad S_t = S_0 e^{X_t}, \quad X_t = \mu t + a N_t.$$

(In both cases we assume that S_0 is a positive constant and $\sigma > 0$, $a > 0$.)

It is clear that in the model (B)

$$(b, c, \nu)_0 = (\mu, \sigma^2, 0)$$

and in the model (N)

$$(b, c, \nu(dx))_0 = (\mu, 0, \lambda\delta_{\{a\}}(dx)).$$

(Note that $(aN_t)_{t\le T}$ is a compound Poisson process, see Sec. 5.3; for notation $(b, c, \nu)_0$ see Remark 5.1 on page 108.)

From (7.14) (see Theorem 7.1) we have that the process in (B) is a P-local martingale if and only if

$$\mu + \frac{\sigma^2}{2} \ \left(= b + \frac{c}{2}\right) \ = 0.$$

In other words, if

$$X_t = \sigma B_t - \frac{\sigma^2}{2} t \tag{7.28}$$

then the process $S = (S_t)_{t\le T}$ with

$$S_t = S_0 e^{\sigma B_t - \frac{\sigma^2}{2} t} \tag{7.29}$$

is a P-local martingale (in fact, it is a P-martingale).

Now consider the model (N). Again by Theorem 7.1 we get that the process $S = S_0 e^{X_t}$ is a P-local martingale if and only if (see (7.14))

$$\mu + \int_{\mathbb{R}} (e^x - 1)\lambda\delta_{\{a\}}(dx) = 0,$$

that is,

$$\mu + \lambda(e^a - 1) = 0. \tag{7.30}$$

In other words, for the model (N) the process

$$S_t = S_0 e^{-\lambda(e^a-1)t+aN_t} \tag{7.31}$$

is a P-local martingale. (Here also this process is a P-martingale.)

For the cases under consideration we can obtain even more:

Theorem 7.4. (a) *In the exponential model* (B) *with* $\mu = -\sigma^2/2$, i.e.,

$$S_t = S_0 e^{\sigma B_t - \frac{\sigma^2}{2}t}, \qquad t \le T,$$

the initial ("physical") measure P *is a unique martingale (and local martingale) measure* (notation: $|\mathcal{M}| = 1$ and $|\mathcal{M}_{\text{loc}}| = 1$).

(b) *In the exponential model* (N) *with* $\mu = -\lambda(e^a - 1)$, i.e.,

$$S_t = S_0 e^{-\lambda(e^a-1)+aN_t},$$

the initial ("physical") measure P *is unique in the class* $\mathcal{M}$ *and in the class* $\mathcal{M}_{\text{loc}}$ (i.e., $P \in \mathcal{M} \subseteq \mathcal{M}_{\text{loc}}$ *and* $|\mathcal{M}| = 1$ *and* $|\mathcal{M}_{\text{loc}}| = 1$).

Proof. We have to prove only uniqueness of the measure in the classes $\mathcal{M}$ and $\mathcal{M}_{\text{loc}}$. For case (B) a proof of uniqueness of measure P is provided in [161; Chap. VII, § 4a]. Another—simpler—proof is given in [157], [158] and is based on the following idea.

Suppose that there exists a measure $\overline{\mathsf{P}}$ such that the process S is a $\overline{\mathsf{P}}$-local martingale. Take for simplicity $\sigma = 1$. Then because $dS_t = S_t\,dB_t$ we see that B is a stochastic logarithm of S (see Sec. 4.1), *i.e.*,

$$B_t = \int_0^t \frac{dS_u}{S_u}.$$

From here it follows that the process $B = (B_t)_{t \le T}$ is a continuous $\overline{\mathsf{P}}$-local martingale. Because $\overline{\mathsf{P}} \sim \mathsf{P}$ we have $\langle B\rangle_t^{\overline{\mathsf{P}}} = \langle B\rangle_t^{\mathsf{P}} = t$. So, by the Lévy characterization theorem [146; Chap. IV, (3.6)] the process $B = (B_t)_{t \le T}$ with respect to $\overline{\mathsf{P}}$ is a Brownian motion. Taking into account that we are now in canonical setting, we conclude that $\overline{\mathsf{P}}$ is a Wiener measure and therefore $\overline{\mathsf{P}} = \mathsf{P}$.

In case (N) our proof of uniqueness is going in a similar way. Suppose that a measure $\overline{\mathsf{P}}$ is such that $\overline{\mathsf{P}} \sim \mathsf{P}$ and the process $S = (S_t)_{t \le T}$ is a $\overline{\mathsf{P}}$-local martingale. Then from

$$dS_t = (e^a - 1)S_{t-}\,d(N_t - \lambda t),$$

using again the notion of stochastic logarithm, we get

$$N_t - \lambda t = \int_0^t \frac{dS_t}{(e^a - 1)S_{t-}}.$$

So, the process $(N_t - \lambda t)_{t \le T}$ is a $\overline{\mathsf{P}}$-local martingale with $\overline{\mathsf{P}}$-compensator λt, $t \le T$. This implies (see [100; Chap. II, Theorem 4.5]) that the process $N = (N_t)_{t \le T}$ is a Poisson process with parameter λ and $\overline{\mathsf{P}}$ is the corresponding Poissonian measure. Because P is as well the Poissonian measure (with parameter λ), we get that $\overline{\mathsf{P}} = \mathsf{P}$. The proof of uniqueness in models (B) and (N) is done. □

7.4 On the Construction of Martingale Measures with Minimal Entropy in the Exponential Lévy Models

1. As in the previous section, we consider a Lévy process $X = (X_t, \mathcal{F}_t)_{t \le T}$ with local characteristic (b, c, F) and local Laplace cumulant function $\lambda(\theta)$ (see Subsec. 3 of Sec. 4.3). The method, suggested in Sec. 7.1, of construction of (Esscher) martingale measures $\mathsf{P}_T^{(a)}$ by means of the formula

$$d\mathsf{P}_T^{(a)} = Z_T^{(a)}\, d\mathsf{P}_T, \tag{7.32}$$

where $Z_T^{(a)}$ is defined in (7.15), not only is an convenient tool for constructing such measures but proves to be very effective for determining the so-called rational prices of options on arbitrage-free markets. (General questions of the arbitrage theory will be reviewed later in Chap. 10.)

In *complete* arbitrage-free financial markets each contingent claim can be replicated under the right choice of a portfolio of securities and the uniquely determined rational price of the considered derivative (*e.g.*, option). This price is calculated with respect to that unique martingale measure (equivalent to the initial one), which is determined by the conditions of no arbitrage and completeness.

However, in *incomplete* markets an exact replication is not possible. And what is more, the no-arbitrage arguments do not guarantee that the rational price is uniquely determined. In fact, there exists a whole spectrum of martingales measures that leads to a spectrum of rational prices.

It is clear that this results in different degrees of RISK from possessing the corresponding derivative. And the choice of one or other martingale measure (and then of the rational price of the derivative) depends on economical advisability of the admitted degree of risk, on various preference considerations.

One of the well-known methods of choosing an "optimal" martingale measure is grounded on ideas of the utility-based derivative pricing (see [54]).

In many interesting cases it was shown (see, for example, [27]) that the problems of utility *maximization* admit a *dual* formulation, which consists in finding the equivalent martingale measure *minimizing* certain "distances" to the initial (physical) probability measure.

Different choice of utility function leads to different "distances". For example, Schweizer [156] deals with L^2-metrics; Keller [112] works with Hellinger's distance.

In the present section the measure of closeness between the probability measures will be chosen to be either *relative entropy* (which is a particular case of so-called f-divergence; see Definition 7.1 further), or the corresponding utility function $u = u(x)$ of the form $u(x) = 1 - e^{-px}$.

Out main interest will be related not to interconnection between the relative entropy and the utility function but to how the minimum-entropy martingale measure can be found by means of measure constructions based on the Esscher transform.

In conclusion of this introductory excursus, let us cite an interesting recent approach to the pricing in incomplete markets, based on an idea to consider on the securities market only the strategies which are not "profitable operations" (see in this connection [79]–[81]).

2. Let $(\Omega, \mathcal{F}, \mathsf{P})$ be a probability space and let Q be another probability measure on $(\Omega, \mathcal{F})$. Assume that $\mathsf{Q} \ll \mathsf{P}$, and let $f = f(x)$ be a convex function, defined on $(0, \infty)$ and taking values in $\mathbb{R}$.

The following definitions are given in [123] and [170].

Definition 7.1. *f-divergence $f(\mathsf{Q}\|\mathsf{P})$ between Q and P* is

$$f(\mathsf{Q}\|\mathsf{P}) = \begin{cases} \displaystyle\int f\Big(\frac{d\mathsf{Q}}{d\mathsf{P}}\Big)\, d\mathsf{P} & \text{if the integral exists,} \\ \infty & \text{else,} \end{cases} \tag{7.33}$$

where $f(0) = \lim_{x\downarrow 0} f(x)$.

Examples of f-divergences.

Relative entropy (or entropy distance, or Kullback–Leibler distance):

$$f(x) = x \log x;$$

total variation distance:

$$f(x) = |x - 1|;$$

Hellinger's distance:

$$f(x) = -\sqrt{x}.$$

3. Of a special interest for us is the *relative entropy* (or simply *entropy*), for which we shall use the particular notation

$$H(\mathsf{Q}\|\mathsf{P}) = \begin{cases} \mathsf{E}_{\mathsf{Q}} \log \dfrac{d\mathsf{Q}}{d\mathsf{P}} & \text{if } \mathsf{Q} \ll \mathsf{P}, \\ \infty & \text{else,} \end{cases} \tag{7.34}$$

where

$$\begin{aligned} \mathsf{E}_{\mathsf{Q}} \log \frac{d\mathsf{Q}}{d\mathsf{P}} &= \int_\Omega \log\Big(\frac{d\mathsf{Q}}{d\mathsf{P}}\Big)\, d\mathsf{Q} \\ &= \int_\Omega \log\Big(\frac{d\mathsf{Q}}{d\mathsf{P}}\Big) \cdot \frac{d\mathsf{Q}}{d\mathsf{P}}\, d\mathsf{P} = \mathsf{E}\frac{d\mathsf{Q}}{d\mathsf{P}} \log \frac{d\mathsf{Q}}{d\mathsf{P}}. \end{aligned} \tag{7.35}$$

Let $S = (S_t)_{t\geq 0}$ be a stochastic process given on a filtered probability space $(\Omega, \mathcal{F}, (\mathcal{F}_t)_{t\geq 0}, \mathsf{P})$ and modeling the price process of certain financial instruments. Let

$$\mathbb{Q} = \big\{\mathsf{Q} : \mathsf{Q} \ll \mathsf{P} \text{ and } S \text{ is a } \mathsf{Q}\text{-local martingale}\big\}.$$

Definition 7.2. A measure $\widetilde{\mathsf{Q}} \in \mathbb{Q}$ is called *minimal entropy martingale measure* (MEMM) if

$$H\big(\widetilde{\mathsf{Q}}\|\mathsf{P}\big) = \inf_{\mathsf{Q}\in\mathbb{Q}} H(\mathsf{Q}\|\mathsf{P}). \tag{7.36}$$

MEMMs for the Lévy processes were studied in many works (see, for example, [77], [85], [92] and references therein).

The main result on the connection between minimal entropy martingale measures (MEMM) and martingale measures constructed by means of Esscher transformations consists in the following.

Let $X = (X_t)_{t\leq T}$ be a Lévy process with local characteristics (b, c, F) and Laplace cumulant function

$$\lambda(\theta) = b\theta + \frac{c}{2}\theta^2 + \int_{\mathbb{R}} \big(e^{\theta x} - 1 - h(x)\theta\big)\, F(dx). \tag{7.37}$$

Theorem 7.5. 1. *Suppose that for a (linear) process $\widetilde{X} = (\widetilde{X}_t)_{t\leq T}$ (such that $\mathcal{E}(\widetilde{X}) = e^X$) there exists an Esscher martingale measure $\widetilde{\mathsf{Q}}_T$. Then this measure is a minimal entropy measure for the exponential Lévy process $e^X = (e^{X_t})_{t\leq T}$.*

2. *Conversely, if a minimal entropy martingale measure $\widetilde{\mathsf{Q}}_T$ for the exponential Lévy process $e^X = (e^{X_t})_{t\leq T}$ exists, then this measure is an Esscher martingale measure for the process $\widetilde{X} = (\widetilde{X}_t)_{t\leq T}$ (such that $\mathcal{E}(\widetilde{X}) = e^X$).*

For the detailed proof of this theorem see [92].

According to Theorem 7.2, the Esscher martingale measure for the process e^X (and thus for the processes $\mathcal{E}(\widetilde{X})$ and $\widetilde{X}$) exists if the parameters (b, c, F) are such that there exists $\tilde{a} \in \mathbb{R}$ with

$$b + \left(\tilde{a} + \frac{1}{2}\right)c + \int \left[e^{\tilde{a}x}(e^x - 1) - h(x)\right] F(dx) = 0.$$

Thus Theorem 7.2 provides an effective method allowing one to determine when the exponential process e^X has minimal entropy martingale measure (which coincides—if it exists—with the Esscher martingale measure for the process $\widetilde{X}$).

Note also that if

$$\widetilde{Z}_t^{(a)} = e^{a\widetilde{X}_t - t\widetilde{\lambda}(a)},$$

where $\widetilde{\lambda}(a)$ is the cumulant of the process $\widetilde{X}$, and the measure $\widetilde{\mathsf{P}}_T^{(a)}$ is specified by

$$d\widetilde{\mathsf{P}}_T^{(a)} = \widetilde{Z}_t^{(a)}\, d\mathsf{P}_T,$$

then

$$H\big(\widetilde{\mathsf{P}}_T^{(a)} \| \mathsf{P}_T\big) = \mathsf{E}_{\widetilde{\mathsf{P}}_T^{(a)}} \log \frac{d\widetilde{\mathsf{P}}_T^{(a)}}{d\mathsf{P}_T} = \mathsf{E}_{\widetilde{\mathsf{P}}_T^{(a)}} \big(a\widetilde{X}_T - T\widetilde{\lambda}(a)\big). \tag{7.38}$$

If the parameter $\tilde{a}$ is chosen so that with respect to the measure $\widetilde{\mathsf{P}}_T^{(a)}$ the process $\widetilde{X} = (\widetilde{X}_t)_{t \le T}$ with $\widetilde{X}_0 = 0$ is a martingale, then (7.38) yields that

$$H\big(\widetilde{\mathsf{P}}_T^{(a)} \| \mathsf{P}_T\big) = -\widetilde{\lambda}(a)T,$$

where $\widetilde{\lambda}(\theta)$ is the Laplace cumulant function of the process $\widetilde{X}$:

$$\widetilde{\lambda}(\theta) = \tilde{b}\theta + \frac{\tilde{c}}{2}\theta^2 + \int_{\mathbb{R}} \left(e^{\theta x} - 1 - h(x)\theta\right) \widetilde{F}(dx),$$

where the triplet $(\tilde{b}, \tilde{c}, \widetilde{F})$ is determined from the triplet (b, c, F) by (7.18).

Chapter 8

Change of Time in Semimartingale Models and Models Based on Brownian Motion and Lévy Processes

8.1 Some General Facts about Change of Time for Semimartingale Models

1. The general principles of "change of time" were discussed in detail in Chap. 1. There we formulated the properties of such changes (from "old" time t to "new" time θ and vice versa) and cited examples of processes which can be obtained from "simple" processes by means of change of time.

In the present chapter we will develop this theme of "change of time" by considering new aspects due, *e.g.*, to different choices of filtration, or of initial "simple" processes, and we shall revise the results of Chap. 1 taking into account this new context.

2. All our considerations will deal with the processes which are semimartingales. We start by recalling some of their properties that are related to change of filtration and change of time and will be useful in the subsequent exposition.

As usual, assume that we are given a stochastic basis, *i.e.*, a filtered probability space $(\Omega, \mathcal{F}, (\mathcal{F}_t)_{t\geq 0}, \mathsf{P})$, and let $X = (X_t)_{t\geq 0}$ be a semimartingale defined on this space (see [100]).

Together with the filtration $\mathbb{F} = (\mathcal{F}_t)_{t\geq 0}$ consider filtrations

$$\mathbb{F}^X = (\mathcal{F}_t^X)_{t\geq 0} \quad \text{and} \quad \mathbb{G} = (\mathcal{G}_t)_{t\geq 0}$$

where

$$\mathcal{F}_t^X = \bigcap_{\varepsilon>0} \sigma(X_s; s \leq t+\varepsilon)$$

and $\mathcal{G}_t$ is a sub-σ-algebra of $\mathcal{F}_t$ (*i.e.*, $\mathcal{G}_t \subseteq \mathcal{F}_t$), $t \geq 0$. (The filtration $\mathbb{F}^X$ is called the *natural* filtration generated by the process X.) Since $\mathbb{G}$ is a filtration, we have $\mathcal{G}_s \subseteq \mathcal{G}_t$, $s \leq t$.

The first property of semimartingales to be pointed out here is the following.

Theorem 8.1. *A semimartingale $X = (X_t)_{t \geq 0}$ considered with respect to the filtration $\mathbb{F} = (\mathcal{F}_t)_{t \geq 0}$ is a semimartingale relative both to its natural filtration $\mathbb{F}^X = (\mathcal{F}_t^X)_{t \geq 0}$ and any filtration $\mathbb{G} = (\mathcal{G}_t)_{t \geq 0}$ such that $\mathcal{F}_t^X \subseteq \mathcal{G}_t \subseteq \mathcal{F}_t$, $t \geq 0$.*

For the proof of this property see [124; Chap. 4, § 6].

It should be noticed that the triplets of predictable characteristics of the semimartingale X relative to these filtrations are not the same. For example, let

$$X_t = \int_0^t a_s(\omega)\, ds + B_t,$$

where a_s is $\mathcal{F}_s$-measurable, $|a_s| \leq c$, $s \geq 0$, and $B = (B_t, \mathcal{F}_t)_{t \geq 0}$ is a Brownian motion. The triplet (B, C, ν) with respect to $\mathbb{F}$ is

$$\Big(\int_0^t a_s(\omega)\, ds, t, 0 \Big)_{t \geq 0},$$

whereas relative to $\mathbb{F}^X$ the triplet is

$$\Big(\int_0^t \mathsf{E}(a_s \,|\, \mathcal{F}_s^X)\, ds, t, 0 \Big)_{t \geq 0}.$$

The latter follows from the fact that the process X admits the following *innovation* representation [125; Theorem 7.12]:

$$X_t = \int_0^t \mathsf{E}\big(a_s \,|\, \mathcal{F}_s^X\big)\, ds + \overline{B}_t,$$

where $\overline{B} = (\overline{B}, \mathcal{F}_t^X)_{t \geq 0}$ is also a Brownian motion (*innovation* process).

Let $\widehat{T} = (\widehat{T}(\theta))_{\theta \geq 0}$ be a random change of time ($\{\widehat{T}(\theta) \leq t\} \in \mathcal{F}_t$, $t \geq 0$, $\theta \geq 0$, and $\widehat{T}(0) = 0$) and $\widehat{\mathbb{F}} = (\widehat{\mathcal{F}}_\theta)_{\theta \geq 0}$, where $\widehat{\mathcal{F}}_\theta = \mathcal{F}_{\widehat{T}(\theta)}$.

The second property of semimartingales which will be useful in the sequel consists in the following.

Theorem 8.2. *If $X = (X_t)_{t \geq 0}$ is a semimartingale with respect to a filtration $\mathbb{F}$, then in the "new" θ-time the process $\widehat{X} = (\widehat{X}_\theta)_{\theta \geq 0}$, where $\widehat{X}_\theta = X_{\widehat{T}(\theta)}$, is (with respect to the filtration $\widehat{\mathbb{F}} = (\widehat{\mathcal{F}}_\theta)_{\theta \geq 0}$) a semimartingale too.*

For the proof see [124; Chap. 4, § 7].)

This property implies, in particular, that if $X = B$ is a Brownian motion (which is, of course, a semimartingale), then the process $\widehat{B} = (\widehat{B}_t)_{t \geq 0}$ with $\widehat{B}_\theta = B_{\widehat{T}(\theta)}$ is a semimartingale as well.

Let the semimartingale $X = (X_t, \mathcal{F}_t)_{t \geq 0}$ have the triplet $\mathbb{T} = (B, C, \nu)$ of predictable characteristics. As we said, the process $\widehat{X}_\theta = (\widehat{X}_\theta, \widehat{\mathcal{F}}_\theta)_{\theta \geq 0}$ with $\widehat{X}_\theta = X_{\widehat{T}(\theta)}$ and $\widehat{\mathcal{F}}_\theta = \mathcal{F}_{\widehat{T}(\theta)}$, where $\widehat{T} = (\widehat{T}(\theta))_{\theta \geq 0}$ is a finite change of time, is a semimartingale. Denote the triplet of its predictable characteristics by $\widehat{\mathbb{T}} = (\widehat{B}, \widehat{C}, \widehat{\nu})$.

The third property of semimartingales which is useful in many respects is described in the following theorem.

Theorem 8.3. *The triplet $\widehat{\mathbb{T}} = (\widehat{B}, \widehat{C}, \widehat{\nu})$ can be determined from the triplet $\mathbb{T} = (B, C, \nu)$ by*

$$\widehat{B}_\theta = B_{\widehat{T}(\theta)}, \qquad \widehat{C}_\theta = C_{\widehat{T}(\theta)}, \qquad I_G * \widehat{\nu}_\theta = I_G * \nu_{\widehat{T}(\theta)}.$$

The Fourier cumulant processes $\widetilde{K}^X(u) = \big(\widetilde{K}^X(u)_t\big)_{t \geq 0}$ and $\widetilde{K}^{\widehat{X}}(u) = \big(\widetilde{K}^{\widehat{X}}(u)_\theta\big)_{\theta \geq 0}$ (see Sec. 4.2) *are connected by*

$$\widetilde{K}^{\widehat{X}}(u)_\theta = \widetilde{K}^X(u)_{\widehat{T}(\theta)}.$$

For the proof see, *e.g.*, [110].

Is is worth mentioning how the triplet of differential characteristics $\partial\mathbb{T} = (b, c, F)$ alters under an *absolute continuous* change of time.

Namely, we will assume that the random time-change $\widehat{T} = (\widehat{T}(\theta))_{\theta \geq 0}$ is such that $\widehat{T}(\theta)$ for every $\theta \geq 0$ is a finite stopping time and

$$\widehat{T}(\theta) = \int_0^\theta \hat{t}(u)\, du, \qquad \theta \geq 0,$$

where the "derivative" $\hat{t} = (\hat{t}(u))_{u \geq 0}$ is nonnegative. Then we have

Theorem 8.4. *The triplet $\partial\widehat{\mathbb{T}} = (\hat{b}, \hat{c}, \widehat{F})$ of differential characteristics of the time-changed process*

$$\widehat{X}_\theta = X_{\widehat{T}(\theta)}, \qquad \theta \geq 0,$$

which is a semimartingale with respect to the time-changed filtration $(\widehat{\mathcal{F}}_\theta)_{\theta \geq 0} \equiv (\mathcal{F}_{\widehat{T}(\theta)})_{\theta \geq 0}$, is given by

$$\begin{aligned} \hat{b}_\theta &= b_{\widehat{T}(\theta)} \hat{t}(\theta), \\ \hat{c}_\theta &= c_{\widehat{T}(\theta)} \hat{t}(\theta), \\ \widehat{F}_\theta(G) &= F_{\widehat{T}(\theta)}(G) \hat{t}(\theta), \qquad G \in \mathcal{B}(\mathbb{R}). \end{aligned}$$

Finally, the fourth property of interest (which was already referred to in Sec. 1.3) is the "Monroe theorem":

> *If* $X = (X_t)_{t\geq 0}$ *is a semimartingale (with respect to the natural filtration* $\mathbb{F}^X$*), then there exists a filtered probability space with a Brownian motion* $\widehat{B} = (\widehat{B}_\theta)_{\theta\geq 0}$ *and a change of time* $T = (T(t))_{t\geq 0}$ *defined on it such that*
>
> $$X \stackrel{\text{law}}{=} \widehat{B} \circ T, \tag{8.1}$$
>
> *where* $\stackrel{\text{law}}{=}$ *means that the processes* X *and* $\widehat{B} \circ T$ *coincide in distribution.*

(For the proof see the paper by I. Monroe [131].)

8.2 Change of Time in Brownian Motion. Different Formulations

1. The Monroe theorem formulated above states in principle the possibility to obtain the representation (8.1) with a Brownian motion $\widehat{B}$ and a change of time T. However it does not answer the question about the *concrete* structure of this change of time. Moreover, the representation (8.1), in general, is not unique, and the connection between $\widehat{B}$ and T depends on the filtered probability space on which these processes are defined.

Certainly, it would be natural to have, for a given semimartingale, the representation (8.1) with "simple" change of time. Thus it is interesting to answer the following questions [45]:

(I) Which processes X admit the representation with T a *continuous* process?

(II) When does there exist a representation with *independent* processes $\widehat{B}$ and T?

(III) When does there exist a representation with *independent* processes $\widehat{B}$ and T, the process T being *continuous*?

(IV) When does there exist a representation with the process T being a *subordinator*, *i.e.*, a nondecreasing Lévy process? What are the predictable characteristics of this process T?

Questions of such kind are interesting, of course, not only in the Brownian motion case but also for other processes, *e.g.*, Lévy processes, diffusion processes, semimartingales,. . . .

2. It is clear that in the case (I)—when the process T has continuous trajectories—the process $\widehat{B} \circ T$ has continuous trajectories as well. Let us show that this process is a local martingale.

Let

$$\sigma_n = \inf\{t \geq 0 : |\widehat{B}_{T(t)}| \geq n\}.$$

The "stopped" process

$$(\widehat{B} \circ T)^{\sigma_n} = (\widehat{B}_{T(t \wedge \sigma_n)})_{t \geq 0} \tag{8.2}$$

admits the representation

$$\widehat{B}^{\rho_n} \circ T = (\widehat{B}_{\rho_n \wedge T(t)})_{t \geq 0} \tag{8.3}$$

with $\rho_n = \inf\{\theta : |\widehat{B}_\theta| \geq n\}$.

By Doob's optional sampling (or stopping) theorem [100; Chap. I, Theorem 1.39],

$$\mathsf{E}(\widehat{B}^{\rho_n}_{T(t)} \,|\, \widehat{\mathcal{F}}_{T(s)}) = \widehat{B}^{\rho_n}_{T(s)}, \qquad s \leq t,$$

i.e.,

$$\mathsf{E}(\widehat{B}^{\rho_n}_{T(t)} \,|\, \mathcal{F}^{\widehat{B} \circ T}_s) = \widehat{B}^{\rho_n}_{T(s)}.$$

This and (8.1), (8.3) yield that the process $\widehat{B} \circ T$ is a local martingale, and it is easy to conclude that so is the process X.

The Dambis–Dubins–Schwarz theorem (Theorem 1.1 on page 17) implies that in fact any *continuous* local martingale X admits a representation $X \overset{\text{law}}{=} \widehat{B} \circ T$. Thus we can provide the complete answer to (I):

> *A semimartingale X can be represented as $X \overset{\text{law}}{=} \widehat{B} \circ T$ with a continuous process T if and only if the process X is a continuous local martingale.*

There is, to our knowledge, no so satisfactory answer to the question (II). Concrete examples of processes which can be represented as $\widehat{B} \circ T$ with independent $\widehat{B}$ and T were cited in Sec. 1.3 (Examples 1.4–1.6). See also Sec. 12.3 about L($\mathbb{G}$IG)- and L($\mathbb{G}$H)-processes.

In the setting (III), *i.e.*, if the processes $\widehat{B}$ and T are independent and the process T is continuous, the following result by D. L. Ocone is known.

The conditions below are equivalent:

(i) $X \overset{\text{law}}{=} \widehat{B} \circ T$;

(ii) X is a continuous local martingale such that

$$\left(\int_0^t H_s\, dX_s; t \geq 0\right) \stackrel{\text{law}}{=} (X_t; t \geq 0)$$

for any $\mathbb{F}^X$-predictable process $H = (H_t)_{t \geq 0}$ such that $|H| = 1$.

(The proof as well as other conditions equivalent to (i) and (ii) can be found in [136]; see also [45].) It is remarkable that in the case under consideration not only the representation $X \stackrel{\text{law}}{=} \widehat{B} \circ T$ but also the representation $X \stackrel{\text{a.s.}}{=} \widehat{B} \circ T$ holds, maybe on an extension of an initial filtered probability space.

The case (IV) is discussed in the next sections.

8.3 Change of Time Given by Subordinators. I. Some Examples

1. If $T = (T(t))_{t \geq 0}$ is a subordinator, *i.e.*, a Lévy process with nondecreasing trajectories, then its triplet of local characteristics is of the form $(\beta, 0, \rho)$ as follows from the equality

$$\mathsf{E} e^{-uT(t)} = e^{-t\chi(u)}, \tag{8.4}$$

where $t \geq 0$, $u \geq 0$, and

$$\chi(u) = u\beta + \int_0^\infty (1 - e^{-ux})\, \rho(dx) \tag{8.5}$$

with $\beta \geq 0$ and $\int_0^\infty (x \wedge 1)\, \rho(dx) < \infty$. (Cf. (5.37) on page 117.)

Remark 8.1. If T is a subordinator having properties (8.4) and (8.5), then for any *complex* w such that $\operatorname{Re} w \leq 0$ we have

$$\mathsf{E} e^{wT(t)} = e^{t\psi(w)}, \tag{8.4$'$}$$

where

$$\psi(w) = \beta w + \int_0^\infty (e^{wx} - 1)\, \rho(dx). \tag{8.5$'$}$$

Thus for real $w = -u \leq 0$

$$\psi(-u) = -\chi(u).$$

2. In Sec. 1.3 there were given concrete examples of processes X, having representation of the form $X = \widehat{B} \circ T$, where $\widehat{B}$ is a Brownian motion and T is a subordinator which does not depend on $\widehat{B}$. Let us recall some of them.

Cauchy process. Let

$$T^\circ(t) = \inf\{\theta \geq 0 : \widehat{\beta}_\theta > t\}, \qquad t \geq 0, \tag{8.6}$$

where $\widehat{\beta} = (\widehat{\beta}_\theta)_{\theta \geq 0}$ is a standard Brownian motion independent of $\widehat{B}$.

For any $\lambda \geq 0$ and $t \geq 0$ the process

$$\left(\exp\left\{\lambda\widehat{\beta}_{T^\circ(t)\wedge\theta} - \tfrac{\lambda^2}{2}(T^\circ(t) \wedge \theta)\right\}\right)_{\theta \geq 0}$$

is a bounded martingale. Since $T^\circ(t) < \infty$ a.s., the Doob optional sampling theorem implies that for $\lambda \geq 0$

$$\mathsf{E}\exp\left\{\lambda\widehat{\beta}_{T^\circ(t)} - \frac{\lambda^2}{2}T^\circ(t)\right\} = 1. \tag{8.7}$$

Hence

$$\mathsf{E}e^{-\frac{\lambda^2}{2}T^\circ(t)} = e^{-\lambda t},$$

i.e., for any $u \geq 0$

$$\mathsf{E}e^{-uT^\circ(t)} = e^{-\sqrt{2u}t}. \tag{8.8}$$

The Lévy–Khinchin representation for stable processes (see (5.16) on page 111) and (8.22) below allow one to conclude that the subordinator T° is a $\frac{1}{2}$-stable process with the triplet $(0, 0, \rho)$ and Lévy measure

$$\rho_0(dx) = (2\pi)^{-1/2}x^{-3/2}\,dx, \tag{8.9}$$

which results from the easy-to-establish equality (cf. (8.5))

$$\int_0^\infty (1 - e^{-ux})(2\pi)^{-1/2}x^{-3/2}\,dx = \sqrt{2u}.$$

As we assume that $\widehat{B}$ and T° are independent we have for any $\lambda \in \mathbb{R}$

$$\mathsf{E}e^{i\lambda\widehat{B}_{T^\circ(t)}} = \mathsf{E}e^{-\lambda^2T^\circ(t)/2} = e^{-|\lambda|t}.$$

This means (in accordance with the Lévy–Khinchin representation) that the process $X^\circ = \widehat{B} \circ T^\circ$ is a 1-stable symmetric process, *i.e.*, the standard Cauchy process.

Normal Inverse Gaussian and Hyperbolic Lévy processes $\mathrm{L}(\mathbb{N} \circ \mathrm{IG})$ **and** $\mathrm{L}(\mathbb{H})$**.** These processes (let us denote them by X^1 and X^2, respectively) were introduced in Sec. 1.3 by

$$X_t^i = \mu_i t + \beta_i T^i(t) + \widehat{B}_{T^i(t)}, \qquad i = 1, 2, \tag{8.10}$$

where $T^i = (T^i(t))_{t \geq 0}$ are subordinators constructed in a special way. The subordinator T^1 (under the appropriate choice of the parameters a and c) can be realized in the form

$$T^1(t) = \inf\{\theta \geq 0 : a\theta + \widehat{\beta}_\theta \geq ct\} \tag{8.11}$$

(cf. (8.6)), where $\widehat{\beta} = (\widehat{\beta}_\theta)_{\theta \geq 0}$ is a Brownian motion independent of $\widehat{B}$. (See [7]–[9], [51], [58]–[64], [30]; in more detail these processes will be discussed in Chap. 12.)

3. Generalized Hyperbolic Lévy processes $\mathrm{L}(\mathbb{GH})$.The above processes $\mathrm{L}(\mathbb{N} \circ \mathbb{IG})$ and $\mathrm{L}(\mathbb{H})$ are particular cases of the class of $\mathrm{L}(\mathbb{GH})$ processes. As in the case of processes $\mathrm{L}(\mathbb{N} \circ \mathbb{IG})$ and $\mathrm{L}(\mathbb{H})$, the $\mathrm{L}(\mathbb{GH})$ process X can be realized in the form

$$X_t = \mu t + \beta T(t) + \widehat{B}_{T(t)}, \tag{8.12}$$

where the subordinator $T = (T(t))_{t \geq 0}$ is generated by the nonnegative infinitely divisible random variable τ $(= T(1))$ having *Generalized Inverse Gaussian* ($\mathbb{GIG}$) distribution, *i.e.*, the distribution $\mathrm{Law}(\tau)$ whose density $p(s) = p(s; a, b, \nu)$ is given by

$$p(s; a, b, \nu) = c(a, b, \nu) s^{\nu - 1} e^{-(as + b/s)/2} \tag{8.13}$$

where $c(a, b, \nu)$ is a normalizing constant, and the parameters a, b, ν satisfy conditions (9.43) in Chap. 9. See details in Sec. 12.2.

Comparison of (8.13) with (1.36) and (1.37) on page 13 and with Table 9.1 on page 185 shows that

> if $\nu = -1/2$ then $\mathbb{GIG}$ gives Normal Inverse Gaussian distribution (1.36);
> if $\nu = 1$ then $\mathbb{GIG}$ gives Hyperbolic distribution.

Thus the $\mathrm{L}(\mathbb{N} \circ \mathbb{IG})$ and $\mathrm{L}(\mathbb{H})$ processes form subclasses in $\mathrm{L}(\mathbb{GH})$. Notice also that (8.13) as compared with (1.38) shows that the Gamma distribution is a distribution of the class $\mathbb{GIG}$, and the Variance Gamma ($\mathbb{VG}$) Lévy process introduced in Sec. 1.3 belongs to the class of $\mathrm{L}(\mathbb{GH})$ processes.

For more detail on $\mathrm{L}(\mathbb{GH})$ processes and their applications in Finance see [7]–[9], [23], [26], [30], [58]–[64] and Chap. 12.

8.4 Change of Time Given by Subordinators. II. Structure of the Triplets of Predictable Characteristics

1. Before we formulate results on the structure of the triplets of local characteristics of the processes $\widehat{B} \circ T$ and $\widehat{L} \circ T$, where $\widehat{B}$ is a Brownian motion and $\widehat{L}$ is a Lévy process, let us recall some notation due to the choice of truncation function in the representations of the characteristic functions.

If the characteristic function $\mathsf{E}e^{i\lambda X_t}$ of the Lévy process $X = (X_t)_{t\geq 0}$ is of the form

$$\mathsf{E}e^{i\lambda X_t} = \exp\left\{t\left[i\lambda b - \frac{\lambda^2}{2}c + \int_{\mathbb{R}} \left(e^{i\lambda x} - 1 - \lambda h(x)\right) F(dx)\right]\right\} \tag{8.14}$$

with the truncation function $h = h(x)$, we write

$$\mathbb{T}_{h,\mathrm{loc}} = (b, c, F)_h$$

and call this formation the triplet of local characteristics with respect to the truncation function h (see Sec. 3.2 and Remark 5.1 on page 108).

If the condition

$$\int \min(1, |x|)\, F(dx) < \infty \tag{8.15}$$

is fulfilled then in (8.14) one can take $h \equiv 0$ as a truncation function. (Under the assumption (8.15) the integral in (8.14) is well defined for $h \equiv 0$.)

In what follows, the canonical truncation function $xI(|x| \leq 1)$ will be denoted by $\widetilde{h} = \widetilde{h}(x)$.

2. The following assertion is true.

Theorem 8.5. *Let $\widehat{B} = (\widehat{B}_\theta)_{\theta\geq 0}$ be a Brownian motion, and let $T = (T(t))_{t\geq 0}$ be a subordinator independent of $\widehat{B}$, with local characteristics $(\beta, 0, \rho)_0$.*

The process $X = \widehat{B} \circ T$ is a Lévy process with the triplet of local characteristics

$$\mathbb{T}_{\widetilde{h},\mathrm{loc}} = (0, \beta, \nu)_{\widetilde{h}}, \tag{8.16}$$

where

$$\nu(A) = \int_0^\infty Q_x(A)\, \rho(dx), \qquad A \in \mathcal{B}(\mathbb{R}), \tag{8.17}$$

and $Q_x(\,\cdot\,)$ is the normal distribution with zero mean and variance equal to x.

Proof. We find from (8.4) and (8.5) that for $\lambda \in \mathbb{R}$

$$\begin{aligned} \mathsf{E}e^{i\lambda(\widehat{B}\circ T)_1} = \mathsf{E}e^{-\frac{\lambda^2}{2}T(1)} &= \exp\left\{-\frac{\lambda^2}{2}\beta + \int_0^\infty \left(e^{-\lambda^2 x/2} - 1\right) \rho(dx)\right\} \\ &= \exp\left\{-\frac{\lambda^2}{2}\beta + \int_0^\infty \mathsf{E}\left(e^{i\lambda\xi(x)} - 1\right) \rho(dx)\right\}, \end{aligned} \tag{8.18}$$

where $\xi(x)$ is the normally distributed random variable with $\mathsf{E}\xi(x) = 0$ and $\mathsf{E}\xi^2(x) = x$ for $x \geq 0$.

It is clear that

$$\mathsf{E}(e^{i\lambda\xi(x)} - 1) = \mathsf{E}(e^{i\lambda\xi(x)} - 1 - i\lambda\xi(x)I(|\xi(x)| \le 1)). \tag{8.19}$$

With the notation $Q_x(dz)$ for the normal distribution of the random variable $\xi(x)$, we find from (8.18), (8.19), and the Fubini theorem that

$$\begin{aligned}\mathsf{E}e^{i\lambda(\widehat{B}\circ T)_1} &= \exp\left\{ -\frac{\lambda^2}{2}\beta + \int_0^\infty \left[\int_{\mathbb{R}} \Big(e^{i\lambda z} - 1 \right.\right. \\ &\qquad\qquad \left.\left. - i\lambda z I(|z| \le 1) \Big) Q_x(dz) \right] \rho(dx) \right\} \\ &= \exp\left\{ -\frac{\lambda^2}{2}\beta + \int_{\mathbb{R}} \Big(e^{i\lambda z} - 1 - i\lambda z I(|z| \le 1) \Big)\, \nu(dz) \right\},\end{aligned}$$

where the measure $\nu = \nu(A)$ for $A \in \mathcal{B}(\mathbb{R})$ is defined by (8.17). □

3. Theorem 8.5 implies that if we start with a Brownian motion and a subordinator T independent of it and having local characteristics $(\beta, 0, \rho)_0$ then the resulting process $X = \widehat{B} \circ T$ will be the Lévy process with local characteristics $(0, \beta, \nu)_{\widetilde{h}}$, where ν is given by (8.17). The reverse question naturally arises (see the setting (IV) on page 154): under what assumptions on the triplet $\mathbb{T}_{h,\mathrm{loc}} = (b, c, \nu)_h$ of the process X does there exist a representation in the form $X \overset{\text{law}}{=} \widehat{B} \circ T$, and what are the corresponding characteristics of the subordinator (independent of $\widehat{B}$)? According to Theorem 8.5, the coefficient b in the triplet $(b, c, \nu)_h$ is necessarily equal to zero. It is less evident what c and ν must be and what is the triplet of subordinator T (note that this subordinator is *uniquely* defined by the distribution of the process X). (If $\widehat{B}^1 \circ T^1 \overset{\text{law}}{=} \widehat{B}^2 \circ T^2$, where $\widehat{B}^i$ and T^i are independent, $i = 1, 2$, then $T^1 \overset{\text{law}}{=} T^2$, see [45; Exercise 3.24].)

Theorem 8.6. *If a Lévy process X with the triplet $(0, c, \nu)_h$ admits the representation $X \overset{\text{law}}{=} \widehat{B} \circ T$, then*

(a) *the measure $\nu = \nu(dz)$ is symmetric, absolutely continuous relative to the Lebesgue measure with density $q(z) = \frac{\nu(dz)}{dz}$, having the property that the function $q(\sqrt{z})$, $z > 0$, is completely monotone*[1];

(b) *there exists a unique positive measure μ on $(0, \infty)$ such that*

$$q(\sqrt{z}) = \int_0^\infty e^{-zy}\, \mu(dy), \qquad z > 0; \tag{8.20}$$

[1] A function φ is said to be *completely monotone*, if it is infinitely differentiable and $\varphi' < 0$, $\varphi'' > 0$, $\varphi''' < 0$ and so on.

(c) *the triplet* $(\beta, 0, \rho)_0$ *of the subordinator* T *in the representation* $X \overset{\text{law}}{=} \widehat{B} \circ T$ *is such that*

$$\beta = c \tag{8.21}$$

and

$$\rho(dx) = \sqrt{2\pi x}\,(\mu \circ \theta^{-1})(dx), \tag{8.22}$$

where $\theta\colon x \rightsquigarrow (1/2)x$.

Proof. First of all let us notice that if ρ is a positive measure on $(0, \infty)$ and the measure $\nu = \nu(A)$, $A \in \mathcal{B}(\mathbb{R})$, is defined by

$$\nu(A) = \int_0^\infty Q_x(A)\,\rho(dx) \tag{8.23}$$

(cf. (8.17)) and satisfies (as a Lévy measure) the condition

$$\int_{\mathbb{R}} \min(1, z^2)\,\nu(dz) < \infty \tag{8.24}$$

then by the properties of the normal distribution $Q_x(\,\cdot\,)$ the following condition for ρ holds:

$$\int_0^\infty \min(1, x)\,\rho(dx) < \infty. \tag{8.25}$$

Thus the possibility to represent $X \overset{\text{law}}{=} \widehat{B} \circ T$ guarantees that for ν the properties (8.23) and (8.24) are fulfilled.

The representation (8.23) shows that the measure ν has density

$$q(z) = \frac{\nu(dz)}{dz}$$

such that

$$q(z) = \int_0^\infty \frac{1}{\sqrt{2\pi x}}\, e^{-z^2/(2x)}\,\rho(dx), \tag{8.26}$$

where $\rho = \rho(dx)$ is a positive measure on $(0, \infty)$.

This yields that the measure ν is a symmetric measure on $(\mathbb{R}, \mathcal{B}(\mathbb{R}))$ and for $z \geq 0$

$$q(\sqrt{z}) = \int_0^\infty e^{-zy}\,\mu(dy), \tag{8.27}$$

where μ is a positive measure on $[0, \infty)$.

The following result of S. N. Bernstein is well known (see, *e.g.*, [73; Vol. II, Chap. XIII, § 4]: If for a nonnegative function $\varphi = \varphi(z)$, $z > 0$, the representation

$$\varphi(z) = \int_0^\infty e^{-zy}\,\mu(dy)$$

holds with a positive measure μ on $[0, \infty)$, then the function $\varphi = \varphi(z)$ is completely monotone.

The last assertions (8.21) and (8.22) in (c) follow from the constructions described above and the scheme

$$\nu \xrightarrow{(8.23)} q \xrightarrow{(8.27)} \mu \xrightarrow{(8.26)} \rho. \qquad \square$$

Chapter 9

Conditionally Gaussian Distributions and Stochastic Volatility Models for the Discrete-time Case

9.1 Deviation from the Gaussian Property of the Returns of the Prices

1. As one can see from the preceding chapters, our main interest is related to different models of stochastic processes in *continuous time.* From the point of view of analysis of financial data, such choice is well justified, because the modern technique of trading (especially, electronic) results in data that comes "almost continuously" in time. On the other hand, the real statistical analysis operates with discrete data which come, *e.g.*, at times $n = 0, 1, 2, \dots$.

Taking the probabilistic point of view on financial markets, we assume here that the prices (of stocks, for example) are described by a random sequence $S = (S_n)_{n\geq 0}$ of positive random variables S_n, $n \geq 0$.

To make a probabilistic analysis of such sequences, we assume that all considerations are underlaid by a filtered probability space

$$(\Omega, \mathcal{F}, (\mathcal{F}_n)_{n\geq 0}, \mathsf{P}),$$

where $(\mathcal{F}_n)_{n\geq 0}$ is a filtration, $\mathcal{F}_0 \subseteq \mathcal{F}_1 \subseteq \cdots$ and the σ-algebra $\mathcal{F}_n$ is interpreted as a collection of events observable until the time n inclusive. In this sense $\mathcal{F}_n$ is an "information" available at time n.

Assume that

$$S_n = S_0 e^{H_n} \tag{9.1}$$

with $H_n = h_1 + h_2 + \cdots + h_n$ for $n \geq 1$ and $H_0 = 0$, where evidently

$$h_n = \log \frac{S_n}{S_{n-1}}, \qquad n \geq 1. \tag{9.2}$$

(We assume $h_0 = 0$.)

In Sec. 4.1 we mentioned that for the sequence $(S_n)_{n\geq 0}$ the following representation proves also useful:

$$S_n = S_0 \mathcal{E}(\widetilde{H})_n, \tag{9.3}$$

where $\widetilde{H}_n = \tilde{h}_1 + \tilde{h}_2 + \cdots + \tilde{h}_n$ with $\tilde{h}_k = e^{h_k} - 1$ and $\mathcal{E}(\widetilde{H})_n$ is the stochastic exponential:

$$\mathcal{E}(\widetilde{H})_n = \prod_{0<k\leq n} (1 + \Delta\widetilde{H}_k), \tag{9.4}$$

where $\Delta\widetilde{H}_k = \widetilde{H}_k - \widetilde{H}_{k-1}$, $\widetilde{H}_0 = 0$. Since $\Delta\widetilde{H}_k = \tilde{h}_k$, it follows that

$$\mathcal{E}(\widetilde{H})_n = \prod_{0<k\leq n} (1 + \tilde{h}_k). \tag{9.5}$$

It is the representation (9.1) which is convenient to operate with when making *statistical* analysis of the sequence $(S_n)_{n\geq 0}$, because this representation reduces the problem to investigation of the statistical properties of the sequence $(h_n)_{n\geq 0}$. As for *probabilistic* analysis, there the representation (9.3) proves to be much more convenient, since the stochastic exponential have, *e.g.*, the remarkable property that if the sequence $\widetilde{H} = (\widetilde{H}_n)_{n\geq 0}$ is a local martingale ($\widetilde{H} \in \mathcal{M}_{\text{loc}}$), then so is the sequence $(\mathcal{E}(\widetilde{H}))_{n\geq 0}$ ($\mathcal{E}(\widetilde{H}) \in \mathcal{M}_{\text{loc}}$); but if $H \in \mathcal{M}_{\text{loc}}$, then e^H is not necessarily a local martingale.

2. From the point of view of classical statistics, it would be very attractive that the sequence $h = (h_n)_{n\geq 0}$ of returns be a Gaussian (normally distributed) random sequence.

Such a sequence is uniquely determined by the means

$$\mu_n = \mathsf{E}h_n, \qquad n \geq 1,$$

and covariances

$$\operatorname{cov}(h_n, h_m) = \mathsf{E}h_n h_m - \mathsf{E}h_n\, \mathsf{E}h_m.$$

Let

$$\sigma_n^2 = \mathsf{D}h_n \; (= \operatorname{cov}(h_n, h_n)).$$

If we assume that $h = (h_n)_{n\geq 1}$ is a Gaussian sequence of independent random variables, then there exists a sequence $\varepsilon = (\varepsilon_n)_{n\geq 1}$ of independent standard normally distributed, $\mathcal{N}(0,1)$, random variables such that the following representations hold:

$$h_n = \mu_n + \sigma_n \varepsilon_n. \tag{9.6}$$

3. However, the statistical analysis of financial data reveals that the attractive hypothesis of *normality, independency, and identical distribution* of random variables $h_1, h_2, \ldots$ could not be accepted except with a great prudence. Moreover, these assumptions should be revised, if we are looking for simplest possible models adequately fitting the relevant features of the financial data—which is important for the effective risk management, for correct pricing of contracts (*e.g.*, of options), for creation of hedging strategies (see Chap. 11), and so on.

Consider the sample mean calculated upon the observed data $h_1, \ldots, h_n$:

$$\bar{h}_n = \frac{1}{n} \sum_{i=1}^{n} h_i$$

and the sample standard deviation

$$\hat{\sigma}_n = \sqrt{\frac{1}{n-1} \sum_{i=1}^{n} (h_i - \bar{h}_i)^2}.$$

Form the confidence intervals

$$\big[\bar{h}_n - k\hat{\sigma}_n, \bar{h}_n + k\hat{\sigma}_n\big],$$

say, for $k = 1, 2, 3$.

It turns out that the fraction of the variables $h_1, \ldots, h_n$ whose values lie beyond these confidence intervals is substantially greater than that expected under hypothesis of normality, independency, and identical distribution of $h_1, \ldots, h_n$. This fact can be interpreted as saying that the empirical densities $\hat{p}_{h_n}(x)$ have "heavy tails", *i.e.*, as $|x| \to \infty$, they decrease more slowly than the density

$$\frac{1}{\sqrt{2\pi\sigma^2}} \exp\left\{-\frac{(x-\mu)^2}{2\sigma^2}\right\}$$

of the normal (Gaussian), $\mathcal{N}(\mu, \sigma^2)$, distribution.

Moreover, for a standard Gaussian random variable $\varepsilon \sim \mathcal{N}(0,1)$ its kurtosis defined by

$$\kappa = \frac{\mathsf{E}\varepsilon^4}{(\mathsf{E}\varepsilon^2)^2} - 3$$

is equal to zero. However, in many cases the empirical kurtosis

$$\hat{\kappa}_n = \frac{\widehat{m}_n^{(4)}}{(\widehat{m}_n^{(2)})^2} - 3,$$

where $\widehat{m}_n^{(4)}$ and $\widehat{m}_n^{(2)}$ are fourth and second empirical moments calculated upon $\varepsilon_1, \dots, \varepsilon_n$ with $\varepsilon_k = (h_k - \mu_k)/\sigma_k$, is significantly greater than zero. This again tells on digression of distribution of the variables $h_1, \dots, h_n$ from normality. In fact, numerous researches show that the empirical densities $\hat{p}_{h_n}(x)$ have a strongly pronounced peak in the neighborhood of central values as well as having tails that are substantially heavier than those of a fitted normal distribution.

These arguments testify for the $(h_n)_{n\geq 1}$ having non-Gaussian distribution.

Moreover, the analysis of empirical financial time series show that generally there is a very significant time-wise dependence, often bordering on long range dependence (although the autocorrelations may be essentially 0. (See also discussion in Sec. 12.1.)

In the next section we will consider a modification of the model (9.6) whose characteristic feature is that, further to ε_n, the quantities μ_n and σ_n are also *random* variables.

To formulate "right" models (in which, on the one hand, the appearance of the random ε_n, μ_n, and σ_n would be natural and which, on the other hand, are easy to extend for the continuous time case), it is expedient to begin with description of the martingale approach to studying random sequences.

9.2 Martingale Approach to the Study of the Returns of the Prices

1. Let $S_n = S_0 e^{H_n}$, where $H_n = h_1 + \cdots + h_n$, $n \geq 1$. The random sequences $H = (H_n)_{n\geq 1}$ and $h = (h_n)_{n\geq 1}$ can be studied by different methods elaborated in the probability theory. For example, by the methods of the theory of "sums of independent random variables", by Markovian methods or by means of techniques of stationary sequences.

In mathematical finance the martingale methods proved to be most useful; among them one should mention first the Doob decomposition of stochastic sequences (briefly reviewed in Subsec. 8 of Sec. 3.1, in connection with the Doob–Meyer decomposition).

Assume given a filtered probability space $(\Omega, \mathcal{F}, (\mathcal{F}_n)_{n\geq 0}, \mathsf{P})$, where $\mathcal{F}_0 = \{\varnothing, \Omega\}$. Let the sequence $H = (H_n)_{n\geq 0}$ with $H_0 = 0$ be such that H_n is $\mathcal{F}_n$-measurable and $\mathsf{E}|H_n| < \infty$, $n \geq 1$. With the notation

$h_n = H_n - H_{n-1}$ we have

$$H_n = \sum_{k \le n} \mathsf{E}(h_k \mid \mathcal{F}_{k-1}) + \sum_{k \le n} \big[h_k - \mathsf{E}(h_k \mid \mathcal{F}_{k-1})\big], \tag{9.7}$$

where the conditional expectation $\mathsf{E}(h_k \mid \mathcal{F}_{k-1})$ is well defined, because $\mathsf{E}|h_k| < \infty$, $k \ge 1$.

Using the notation

$$A_n = \sum_{k \le n} \mathsf{E}(h_k \mid \mathcal{F}_{k-1}) \tag{9.8}$$

and

$$M_n = \sum_{k \le n} \big[h_k - \mathsf{E}(h_k \mid \mathcal{F}_{k-1})\big], \tag{9.9}$$

one can rewrite (9.7) in the form

$$H_n = A_n + M_n, \tag{9.10}$$

where

(a) the sequence $A = (A_n)_{n \ge 0}$ with $A_0 = 0$ is *predictable*, *i.e.*, the random variables A_n are $\mathcal{F}_{n-1}$-measurable, $n \ge 1$;

(b) the sequence $M = (M_n)_{n \ge 0}$ is a *martingale*, *i.e.*, the random variables M_n are $\mathcal{F}_n$-measurable, $\mathsf{E}|M_n| < \infty$, $n \ge 0$, and the following martingale property holds:

$$\mathsf{E}(M_n \mid \mathcal{F}_{n-1}) = M_{n-1} \quad \text{for all} \quad n \ge 1. \tag{9.11}$$

The representation (9.10) is called a *Doob decomposition* of the sequence $H = (H_n)_{n \ge 1}$ into predictable (A) and martingale (M) components:

$$H = A + M. \tag{9.12}$$

It is remarkable that if there exists another representation $H = A' + M'$ with a predictable sequence $A' = (A'_n)_{n \ge 0}$, $A'_0 = 0$, then $A' = A$ and $M' = M$. In other words, the Doob decomposition with a predictable process A is unique. (Cf. Subsec. 8 in Sec. 3.1.)

2. Above we assumed that $\mathsf{E}|H_n| < \infty$ for all $n \ge 1$ (or, equivalently, that $\mathsf{E}|h_n| < \infty$ for all $n \ge 1$). If this condition is not satisfied, then, to obtain a Doob decomposition, one should proceed as follows.

Choose an $a > 0$ and write h_k in the form

$$h_k = h_k I(|h_k| \le a) + h_k I(|h_k| > a). \tag{9.13}$$

Applying the Doob decomposition (9.10) to the random variables

$$H_n^{(\leq a)} = \sum_{k \leq n} h_k I(|h_k| \leq a), \qquad n \geq 1,$$

and using (9.13), we get *generalized Doob decomposition*

$$H_n = A_n^{(\leq a)} + M_n^{(\leq a)} + \sum_{k \leq n} h_k I(|h_k| > a), \tag{9.14}$$

where

$$A_n^{(\leq a)} = \sum_{k \leq n} \mathsf{E}\big[h_k I(|h_k| \leq a)\big] \tag{9.15}$$

and

$$M_n^{(\leq a)} = \sum_{k \leq n} \Big(h_k I(|h_k| \leq a) - \mathsf{E}\big[h_k I(|h_k| \leq a) \,|\, \mathcal{F}_{k-1}\big]\Big). \tag{9.16}$$

It is sometimes helpful to rewrite (9.14) in a different form, using the notion of integer-valued measure of jumps of the sequence $H = (H_n)_{n \geq 1}$.

Namely, put

$$\mu_n(A; \omega) = I_A(\Delta H_n(\omega)), \qquad A \in \mathcal{B}(\mathbb{R} \setminus \{0\}).$$

In other words, let

$$\mu_n(A; \omega) = \begin{cases} 1 & \text{if } \Delta H_n(\omega) \in A, \\ 0 & \text{if } \Delta H_n(\omega) \notin A. \end{cases}$$

The sequence $\mu = (\mu_n(\cdot))_{n \geq 1}$ is called a *measure of jumps* of the sequence $H = (H_n)_{n \geq 1}$.

Further, let $\nu = (\nu_n(\cdot))_{n \geq 1}$ be the sequence of *regular* $\mathcal{F}_{n-1}$-conditional distributions $\nu_n(\cdots)$ of ΔH_n, *i.e.*, the sequence of functions $\nu_n(A; \omega)$, $A \in \mathcal{B}(\mathbb{R} \setminus \{0\})$, $\omega \in \Omega$, such that for each $n \geq 1$:

(a) $\nu_n(\cdot; \omega)$ is a probability measure on $(\mathbb{R} \setminus \{0\}, \mathcal{B}(\mathbb{R} \setminus \{0\}))$ for every $\omega \in \Omega$;

(b) for fixed $A \in \mathcal{B}(\mathbb{R} \setminus \{0\})$, $\nu_n(A; \omega)$ is a realization of the conditional probability $\mathsf{P}(\Delta H_n \in A \,|\, \mathcal{F}_{n-1})(\omega)$, *i.e.*,

$$\nu_n(A; \omega) = \mathsf{P}(\Delta H_n \in A \,|\, \mathcal{F}_{n-1})(\omega) \quad (\mathsf{P}\text{-a.s.}).$$

(See, *e.g.*, [160; Chap. II, § 7].) The sequence $\nu = (\nu_n(\cdot))_{n \geq 1}$ is referred to as the compensator of the measure of jumps $\mu = (\mu_n(\cdot))_{n \geq 1}$. This is motivated by the fact that ν compensates μ to a martingale: for each

$A \in \mathcal{B}(\mathbb{R} \setminus \{0\})$ the sequence of random variables $(\mu_n(A;\omega) - \nu_n(A;\omega))_{n\geq 0}$ is a martingale-difference. Equivalently, if

$$\mu_{(0,n]}(A;\omega) = \sum_{k=1}^{n} \mu_k(A;\omega) \quad \text{and} \quad \nu_{(0,n]}(A;\omega) = \sum_{k=1}^{n} \nu_k(A;\omega),$$

then the sequence

$$(\mu_n(A;\omega) - \nu_n(A;\omega))_{n\geq 1}$$

forms a martingale (for each $A \in \mathcal{B}(\mathbb{R} \setminus \{0\})$).

In terms of the introduced notions, the Doob decomposition can be rewritten in the form:

$$\begin{aligned} H_n = {} & \sum_{k\leq n} \int_{\mathbb{R}\setminus\{0\}} x\, \nu_k(dx;\omega) \\ & + \sum_{k\leq n} \int_{\mathbb{R}\setminus\{0\}} x\, \big(\mu_k(dx;\omega) - \nu_k(dx;\omega)\big). \end{aligned} \tag{9.17}$$

Denoting, for brevity, the first and second summands in (9.17) by

$$(x * \nu)_n \quad \text{and} \quad (x * (\mu - \nu))_n,$$

respectively, we find that

$$H_n = (x * \nu)_n + (x * (\mu - \nu))_n, \tag{9.18}$$

or, in a concise form,

$$H = x * \nu + x * (\mu - \nu). \tag{9.19}$$

If we do not assume that $\mathsf{E}|h_n| < \infty$, $n \geq 1$, then instead of (9.19) we have, by (9.14), the generalized Doob decomposition in the following form:

$$H = \varphi * \nu + \varphi * (\mu - \nu) + (x - \varphi) * \mu,$$

or, in more details,

$$H_n = (\varphi * \nu)_n + (\varphi * (\mu - \nu))_n + ((x - \varphi) * \mu)_n,$$

where $\varphi(x) = xI(|x| \leq 1)$ is a truncation function (in Chap. 3 denoted by $h(x)$) and

$$(\varphi * \nu)_n, \ (\varphi * (\mu - \nu))_n, \ \text{and} \ ((x - \varphi) * \mu)_n$$

are defined by analogy with the case $\varphi(x) = x$ considered above (*i.e.*, one should replace x by $\varphi(x)$ in the previous formulae).

3. In the subsequent two sections we shall see that identification of the sequences with their Doob's decomposition is very effective in constructing models for $H = (H_n)_{n \geq 1}$ which would fit adequately the statistics of financial data.

We conclude with an example which shows that, however simple the derivation of Doob's decomposition may be, it allows one to deduce conclusions which are far from trivial.

Example 9.1. Let $X_0 = 0$ and

$$X_n = \xi_1 + \cdots + \xi_n, \qquad n \geq 1,$$

where $(\xi_n)_{n \geq 1}$ is a sequence of independent identically distributed random variables such that

$$\mathsf{P}(\xi_n = 1) = \mathsf{P}(\xi_n = -1) = \frac{1}{2}.$$

Let $H_n = |X_n|$, $n \geq 1$. Then

$$h_n = \Delta H_n = \Delta |X_n| = |X_n| - |X_{n-1}| = |X_{n-1} + \xi_n| - |X_{n-1}|.$$

It is not difficult to find (for more detail see [161; Chap. II, § 1b]) that the Doob decomposition here has the form

$$|X_n| = \sum_{1 \leq k \leq n} (\operatorname{sgn} X_{k-1}) \Delta X_k + L_n(0), \tag{9.20}$$

where $L_n(0) = \#\{1 \leq k \leq n \colon X_{k-1} = 0\}$ is the number of k, $1 \leq k \leq n$, such that $X_{k-1} = 0$, and $(\sum_{1 \leq k \leq n} \operatorname{sgn} X_{k-1} \Delta X_k)_{n \geq 1}$ is a martingale. The formula (9.20) is a discrete-time analog of the well-known Tanaka formula for the Brownian motion $B = (B_t)_{t \geq 0}$:

$$|B_t| = \int_0^t \operatorname{sgn} B_s \, dB_s + L_t(0),$$

where $L_t(0)$ is the local time of Brownian motion at zero:

$$L_t(0) = \lim_{\varepsilon \downarrow 0} \frac{1}{2\varepsilon} \int_0^t I(|B_s| \leq \varepsilon) \, ds.$$

It follows from (9.20) that

$$\mathsf{E} L_n(0) = \mathsf{E}|X_n|.$$

By the central limit theorem, $X_n / \sqrt{n} \sim \mathcal{N}(0,1)$ as $n \to \infty$. One can deduce from here that asymptotically, as $n \to \infty$,

$$\mathsf{E}|X_n| \sim \sqrt{\frac{2}{\pi} n}.$$

Thus,

$$\mathsf{E}L_n(0) \sim \sqrt{\frac{2}{\pi}n},$$

in other words, for a symmetrical Brownian motion the averaged quantity of zeroes ($X_k = 0$) increases as $\sqrt{n}$ with $n \to \infty$ (rather than, as might be expected, a linear increase with n).

9.3 Conditionally Gaussian Models. I. Linear (AR, MA, ARMA) and Nonlinear (ARCH, GARCH) Models for Returns

1. The aforementioned (see Sec. 9.1) deviation of distributions of returns $h_n = \log(S_n/S_{n-1})$ from normality—which results in the effect of "heavy tails" and "leptokurtosis"—stimulates one to search for adequate distributions outside the class of Gaussian distributions (which are characterized by two parameters, mean and dispersion). The required distributions should be not too complicated and have the smallest possible number of parameters which, in addition, should be easy to estimate statistically. In other words, it is desirable to have parsimonious *parametric* models of distributions.

One can go further and consider *functional parametric* models such as, *e.g.*, infinitely divisible distributions, determined by the triplet $(b, c, F(dx))$, where $F(dx)$ is a Lévy measure (see Sec. 5.1).

A classical example of a parametric family of distributions is the *Karl Pearson system.* In this system, the distributions are described by their densities $f = f(x)$ which depend on *four parameters* (a, b_0, b_1, b_2) and satisfy the differential equation

$$f'(x) = \frac{(x-a)f(x)}{b_0 + b_1 x + b_2 x^2}. \tag{9.21}$$

(For the origine of this equation as a limit of hypergeometric distributions see [113; 5.28 and 6.2].)

By Pearson's classification, the Beta distribution belongs to Type I, Gamma distribution belongs to Type III, Student distribution belongs to Type VII, *etc.* (see [113; Chap. 6] and [138]).

2. The development of probability theory and, in particular, the creation of the theory of infinitely divisible distributions (aimed to describe the limiting distributions for sums of independent random variables) as well as significant progress in the theory of stochastic processes (Brownian motion, Lévy's processes, stochastic calculus) favored the contrivance of new models for families of probability distributions.

Below we dwell upon two classes of *conditionally Gaussian models*, which proved helpful in construction of distributions fitting well the empirical data.

3. Let $S = (S_n)_{n\geq 0}$ be a sequence of prices $S_n = S_0 e^{H_n}$, where $H_n = h_1 + \cdots + h_n$, $n \geq 1$, and $S_0 = 1$, $H_0 = 0$.

If the random variables h_n, $n \geq 1$, are independent and Gaussian, *i.e.*, have normal distribution $\mathcal{N}(\mu_n, \sigma_n^2)$, then they can be represented in the form

$$h_n = \mu_n + \sigma_n \varepsilon_n, \tag{9.22}$$

where $(\varepsilon_n)_{n\geq 1}$ is a sequence of standard normal, $\mathcal{N}(0,1)$, independent random variables.

However, as was mentioned in Sec. 9.1, it would be inconsiderate to count on h_n to have a normal (Gaussian) distribution; as a result, $\mathrm{Law}(h_1, \ldots, h_n)$ may not be the multivariate Gaussian distribution.

Suppose that the considered price process $S = (S_n)_{n\geq 0}$ is given on a filtered probability space $(\Omega, \mathcal{F}, (\mathcal{F}_n)_{n\geq 0}, \mathsf{P})$, where the "information" flow $(\mathcal{F}_n)_{n\geq 0}$ is such that S_n is $\mathcal{F}_n$-measurable (see [161; Chap. I, § 2a]). Then, having in mind the Doob decomposition considered above (see Sec. 9.2) and the tendency to not deviate too much from the Gaussianity assumption, it is natural to assume that, by analogy with (9.22), $h_n = h_n(\omega)$ admits the following representation:

$$h_n(\omega) = \mu_n(\omega) + \sigma_n(\omega)\varepsilon_n(\omega), \tag{9.23}$$

where $\mu_n = \mu_n(\omega)$ and $\sigma_n = \sigma_n(\omega)$ are $\mathcal{F}_{n-1}$-measurable random variables. The random variables $\varepsilon_n = \varepsilon_n(\omega)$ are assumed $\mathcal{F}_n$-measurable with the conditional distribution $\mathrm{Law}(\varepsilon_n \mid \mathcal{F}_{n-1}) = \mathcal{N}(0,1)$. This implies that, just as in (9.11), the sequence $(\varepsilon_n)_{n\geq 1}$ is a sequence of standard normal, $\mathcal{N}(0,1)$, independent random variables.

It is important to emphasize that ε_n is assumed $\mathcal{F}_n$-measurable for each $n \geq 1$.

It follows directly from (9.23) that

$$\mathrm{Law}(h_n \mid \mathcal{F}_{n-1})(\omega) = \mathcal{N}(\mu_n(\omega), \sigma_n^2(\omega)), \tag{9.24}$$

which justifies the appellation of the sequence $(h_n)_{n\geq 1}$ as *conditionally Gaussian*.

Let us emphasize that it follows from (9.24) that Law(h_n) is a *mixture* of Gaussian distributions obtained from $\mathcal{N}(\mu_n(\omega), \sigma_n^2(\omega))$ by averaging upon the joint distribution of $(\mu_n(\omega), \sigma_n^2(\omega))$.

From mathematical statistics we know that the distribution densities which are *mixtures* of Gaussian densities can both have "heavy tails" and imitate the "peak" form in the central domain. All this, together with deflection of behavior of financial assets from Gaussianity discussed in Sec. 9.1, allows one to hope that the *conditionally Gaussian distributions* would approximate well the empirical distributions of h_n.

4. The following well-known *linear models* are particular cases of the conditionally Gaussian model (9.23).

AR(p): AutoRegressive model of order p. In this linear model

$$\begin{aligned}
\mathcal{F}_n &= \sigma(\varepsilon_1, \dots, \varepsilon_n),\\
\mu_n &= a_0 + a_1 h_{n-1} + \cdots + a_p h_{n-p},\\
\sigma_n &\equiv \sigma = \text{const}, \qquad h_k = \text{const} \quad \text{for} \quad k \le 0.
\end{aligned}$$

Thus,

$$h_n = \mu_n + \sigma_n \varepsilon_n = a_0 + a_1 h_{n-1} + \cdots + a_p h_{n-p} + \sigma \varepsilon_n.$$

As usual, $(\varepsilon_n)_{n \ge 1}$ is a sequence of independent and identically distributed, $\mathcal{N}(0,1)$, random variables.

MA(q): Moving Average model of order q. In this model the "initial values" $(\varepsilon_{1-q}, \dots, \varepsilon_{-1}, \varepsilon_0)$ are assumed given, and

$$\begin{aligned}
\mathcal{F}_n &= \sigma(\varepsilon_1, \dots, \varepsilon_n),\\
\mu_n &= b_0 + b_1 \varepsilon_{n-1} + \cdots + b_q \varepsilon_{n-q},\\
\sigma_n &\equiv \sigma = \text{const}.
\end{aligned}$$

Thus, in this model

$$h_n = b_0 + b_1 \varepsilon_{n-1} + \cdots + b_q \varepsilon_{n-q} + \sigma \varepsilon_n.$$

ARMA(p, q): AutoRegressive Moving Average model of order (p, q). Here the "initial values" $(\varepsilon_{1-q}, \dots, \varepsilon_{-1}, \varepsilon_0)$ and $(h_{1-p}, \dots, h_{-1}, h_0)$ are given, and

$$\begin{aligned}
\mathcal{F}_n &= \sigma(\varepsilon_1, \dots, \varepsilon_n),\\
\mu_n &= (a_0 + a_1 h_{n-1} + \cdots + a_p h_{n-p}) + (b_1 \varepsilon_{n-1} + \cdots + b_q \varepsilon_{n-q}),\\
\sigma_n &\equiv \sigma = \text{const}.
\end{aligned}$$

Hence, in the ARMA(p, q)-model

$$h_n = (a_0 + a_1 h_{n-1} + \cdots + a_p h_{n-p}) + (b_1 \varepsilon_{n-1} + \cdots + b_q \varepsilon_{n-q}) + \sigma \varepsilon_n.$$

5. In the linear models above it was assumed that the dispersion σ^2 is constant implying that the marginal one- and multidimensional distributions of the series $h_1, h_2, \ldots$ are Gaussian. The characteristic feature of the models introduced below (ARCH and GARCH) is that the dispersion is random, $\sigma_n = \sigma_n(\omega)$, which—because of the presence of the product $\sigma_n(\omega)\varepsilon_n(\omega)$ in (9.23)—makes these models *nonlinear* and non-Gaussian.

ARCH(p): AutoRegressive Conditional Heteroskedastic model of order p. Let again $\mathcal{F}_n = \sigma(\varepsilon_1, \ldots, \varepsilon_n)$. Assume that the sequence $(h_n)_{n\geq 1}$ obeys (9.23), where

$$\mu_n = \mathsf{E}(h_n \,|\, \mathcal{F}_{n-1}) = 0,$$

$$\sigma_n^2 = \mathsf{E}(h_n^2 \,|\, \mathcal{F}_{n-1}) = a_0 + \sum_{i=1}^{p} a_i h_{n-i}^2 \tag{9.25}$$

with $a_0 > 0$, $a_i \geq 0$, $\mathcal{F}_0 = \{\varnothing, \Omega\}$, and given "initial values" $h_{1-p}, \ldots, h_0$. In other words, we assume that σ_n^2 is a (linear) function of $h_{n-1}^2, \ldots, h_{n-p}^2$. This nonlinear model,

$$h_n(\omega) = \sigma_n(\omega)\varepsilon_n(\omega), \tag{9.26}$$

introduced by R. F. Engle ([71], 1982) permitted to explain many of observable properties of financial data, *e.g.*, the *cluster* property of the values of the variables h_n, $n \geq 1$.

If we replace the condition $\mu_n = \mathsf{E}(h_n \,|\, \mathcal{F}_{n-1}) = 0$ by

$$h_n = a_0 + a_1 h_{n-1} + \cdots + a_r h_{n-r},$$

we get the AR(r)/ARCH(p)-model

$$h_n = a_0 + a_1 h_{n-1} + \cdots + a_r h_{n-r} + \sigma_n \varepsilon_n,$$

where σ_n is defined in (9.25).

GARCH(p, q): Generalized AutoRegressive Conditional Heteroskedastic model of order (p, q). In this model again

$$\mu_n = \mathsf{E}(h_n \,|\, \mathcal{F}_{n-1}) = 0,$$

but instead of (9.25) one assumes that

$$\sigma_n^2 = \mathsf{E}(h_n^2 \,|\, \mathcal{F}_{n-1}) = a_0 + \sum_{i=1}^{p} a_i h_{n-i}^2 + \sum_{j=1}^{p} \beta_j \sigma_{n-j}^2, \tag{9.27}$$

with $a_0 > 0$, $a_i, \beta_j \geq 0$, and where $(\sigma^2_{1-q}, \dots, \sigma^2_0)$ are some "initial values". This model, introduced by T. Bollerslev ([35], 1986), proved to be useful in many respects, since the joint examination of parameters p and q in the model GARCH(p,q) allows one to diminish the value of p necessary for a good approximation by models ARCH(p) of empirical distributions of h_n.

There are numerous generalizations of the GARCH(p,q)- and ARCH(p)-models. Let us cite only a few names: EGARCH, TGARCH, HARCH,...; some information can be found in [161; Chap. II, § 3b].

In conclusion, let us underline that all models considered in this section assumed that there is only *one* source of randomness, $(\varepsilon_n)_{n\geq 1}$ (called 'white noise'), and the volatility σ_n is constant.

In the next section we will discuss models with *two* independent sources of randomness. One is the same sequence $(\varepsilon_n)_{n\geq 1}$, and the second is described by the *stochastic* volatility $\sigma = (\sigma_n(\omega))_{n\geq 1}$. This pair generates the sequence $h = (H_n)_{n\geq 0}$ by

$$h = \mu + \beta\sigma^2 + \sigma\varepsilon.$$

9.4 Conditionally Gaussian Models. II. IG- and $\mathbb{G}$IG-distributions for the Square of Stochastic Volatility and $\mathbb{G}$H-distributions for Returns

1. The abbreviations used in the title of this section have the following meaning:

IG stands for Inverse Gaussian (distribution),

$\mathbb{G}$IG stands for $\mathbb{G}$eneralized Inverse Gaussian (distribution) and

$\mathbb{G}$H stands for $\mathbb{G}$eneralized Hyperbolic (distribution).

The $\mathbb{G}$IG-distribution is determined by *three* parameters and is intended to be the distribution of the square of stochastic volatility, $\sigma^2 = \sigma^2(\omega)$.

The $\mathbb{G}$H-distribution is characterized by *five* parameters and describes the probability distribution of returns h given by

$$h = \mu + \beta\sigma^2 + \sigma\varepsilon, \tag{9.28}$$

where μ and β are constants, the square of stochastic volatility $\sigma^2 = \sigma^2(\omega)$ follows the $\mathbb{G}$IG-distribution (Law(σ^2) = $\mathbb{G}$IG), and the random variable $\varepsilon = \varepsilon(\omega)$ has normal distribution, $\mathcal{N}(0,1)$, and does not depend on σ.

These assumptions imply that the returns in the model (9.28) are generated by *two* sources of randomness, namely, by the pair (σ, ε).

The assumptions on independence of σ and ε and on normal distribution for ε imply that the probability distribution of h is a normal variance-mean mixture. Symbolically, this can be written as

$$\mathrm{Law}(h) = \mathsf{E}_{\sigma^2}\mathcal{N}(\mu + \beta\sigma^2, \sigma^2), \tag{9.29}$$

where E_{σ^2} stands for the averaging upon the probability distribution of σ^2, or as

$$\mathrm{Law}(h) = \mathbb{N} \circ \mathrm{Law}(\sigma^2). \tag{9.30}$$

Remark 9.1. Up to here we have used "$\circ$" when considering the processes obtained by the change of time from the "simple" processes (see Chap. 1, the end of Sec. 1.1). The product $\sigma\varepsilon$ has the distribution $B_{T(1)}$, where $B = (B_t)_{t\geq 0}$ is a Brownian motion and $T = (T(t))_{t\geq 0}$ is a random change of time, which does not depend on B and is such that $T(1) = \sigma^2$ (see Sec. 12.2).

The natural notation $\mathbb{N} \circ \mathrm{Law}(\sigma^2)$ for the distribution $\mathrm{Law}(B \circ T)$ of the process $B \circ T = (B_{T(t)})_{t\geq 0}$ makes clear the formula (9.30), since $\mathrm{Law}(h) = \mathrm{Law}(B_{T(1)})$.

In the sequel we shall see that many of either famous or less known distributions belong to the classes $\mathbb{GIG}$ or $\mathbb{GH}$. On the other hand, statistical research shows that the distributions of the class $\mathbb{GH}$ approximate fairly well the empirical distributions of various underlying financial instruments. (For more detail see, *e.g.*, [23], [26], [58], [59].)

Remark 9.2. Just as in the probability theory the random variables which have Gaussian, Poissonian, *etc.*, distributions are often called just Gaussian, Poissonian, *etc.*, random variables, here the random variables which follow $\mathbb{GH}$-, $\mathbb{GIG}$- or $\mathbb{N} \circ \mathbb{GIG}$-distributions will be referred to as $\mathbb{GH}$-, $\mathbb{GIG}$- or $\mathbb{N} \circ \mathbb{GIG}$-random variables.

2. To introduce the notion of $\mathbb{G}$eneralized Inverse Gaussian distribution we naturally begin with the definition of Inverse Gaussian (IG) distribution.

The latter is similar in its idea to the negative binomial distribution which is well known in discrete probability theory and mathematical statistics. Below is the exact definition.

Consider the independent Bernoulli random variables $\xi_1, \dots, \xi_N$ with $\mathsf{P}(\xi_i = 1) = p$, $\mathsf{P}(\xi_i = 0) = q$, where $p + q = 1$, $0 < p \leq 1$.

Let $S_0 = 0$ and define $S_n = \xi_1 + \cdots + \xi_n$, $n \leq N$. It is clear that $0 \leq S_N \leq N$.

Fix $0 \le r \le N$, and define $\tau_N(r)$ as the first time where the walk $(S_n)_{0\le n\le N}$ reaches the level r:

$$\tau_N(r) = \min\{0 \le n \le N \colon S_n = r\}. \tag{9.31}$$

(If the set $\{\cdot\}$ in (9.31) is empty, then we assume, *e.g.*, that $\tau_N(r) = \infty$.) Thus $\tau_N(r)$ takes values from the set $\{0, 1, 2, \dots, N, \infty\}$.

It is well known (see, *e.g.*, [160; Chap. II, § 3]) that the distribution of $\tau_N(r)$ is given by

$$\mathsf{P}(\tau_N(r) = k) = C_{k-1}^{r-1} p^r q^{k-r}. \tag{9.32}$$

The sequence $\tau_N = (\tau_N(r))_{0\le r\le N}$ has all the properties of a random change of time, considered in Chap. 1 for the continuous-time case. Thus, one can say that the negative binomial distribution is the distribution of a random change of measure $\tau_N = (\tau_N(r))_{0\le r\le N}$ determined by a random walk $(S_n)_{0\le n\le N}$.

Inverse (or Negative) Gaussian distribution IG is obtained in a similar way: instead of a random walk $(S_n)_{0\le n\le N}$ one should take a Wiener process (Brownian motion) $W^A = (W_s^A)_{s\ge 0}$ with drift,

$$W_s^A = As + W_s, \qquad s \ge 0,$$

where $A \ge 0$ and $W = (W_s)_{s\ge 0}$ is a standard Wiener process (Brownian motion).

By analogy with (9.31), define for $A \ge 0$ and $B > 0$

$$T^A(B) = \inf\{s \ge 0 \colon W_s^A = B\}, \tag{9.33}$$

i.e., let

$$T^A(B) = \inf\{s \ge 0 \colon As + W_s = B\} \tag{9.34}$$

be the first time when the process W^A reaches the level B, or, equivalently, the first time when the Wiener process W reaches the sloping line $B - As$.

With probability one the time $T^A(B)$ is finite for $A \ge 0$, $B > 0$, and its probability distribution has a density $p_{T^A(B)}(s)$ given by

$$p_{T^A(B)}(s) = \frac{B}{s}\,\varphi_s(B - As), \qquad s > 0, \tag{9.35}$$

where

$$\varphi_s(x) = \frac{1}{\sqrt{2\pi s}}\, e^{-x^2/(2s)} \tag{9.36}$$

is the density of the distribution of W_s.

Along with (9.35), cite also the formulae

$$\mathsf{E}T^A(B) = \frac{A}{B} \tag{9.37}$$

and, for $\lambda \geq 0$,

$$\mathsf{E}e^{-\lambda T^A(B)} = \exp\left\{AB\left(1 - \sqrt{1 + \frac{2\lambda}{A^2}}\right)\right\}. \tag{9.38}$$

Although these formulae are well known, for the completeness we prefer to trace their proofs.

It is easy to deduce (9.35) from (9.38). Indeed, straightforward calculations show that for any $B \geq 0$ and $\Lambda \geq 0$

$$\int_0^\infty \frac{B}{\sqrt{2\pi}\, t^{3/2}} e^{-B^2/(2t)} e^{-\Lambda t}\, dt = e^{-B\sqrt{2\Lambda}}.$$

Taking here $\Lambda = \lambda + A^2/2$, we see that

$$\int_0^\infty \frac{B}{t} \varphi_t(B - At) e^{-\lambda t}\, dt = \exp\left\{AB\left(1 - \sqrt{1 + \frac{2\lambda}{A^2}}\right)\right\}.$$

The formula (9.37) being an evident consequence of (9.38), it remains to establish (9.38).

To prove that the Laplace transformation $\mathsf{E}\exp\{-\lambda T^A(B)\}$ is really given by (9.38), we make use of a martingale method.

Consider the martingale

$$M_t = \exp\left\{\theta B_t - \frac{\theta}{2} t\right\}, \qquad t \geq 0,$$

where $\theta = -A + \sqrt{A^2 + 2\lambda}$, $\lambda \geq 0$.

By the optional sampling theorem,

$$\mathsf{E}M_{t\wedge T^A(B)} = 1.$$

Since $B_{t\wedge T^A(B)} \leq B - A(t \wedge T^A(B))$, we find that

$$0 \leq M_{t\wedge T^A(B)} \leq \exp\left\{B\left(-A + \sqrt{A^2 + 2\lambda}\right)\right\}.$$

Then, by the dominated convergence theorem,

$$\begin{aligned}
1 &= \lim_{t\to\infty} \mathsf{E}M_{t\wedge T^A(B)} \\
&= \lim_{t\to\infty} \mathsf{E}\Big[I(T^A(B) < \infty)\exp\left\{\theta B_{t\wedge T^A(B)} - \frac{\theta}{2}(t \wedge T^A(B))\right\} \\
&\qquad + I(T^A(B) = \infty)\exp\left\{\theta B_t - \frac{\theta}{2} t\right\}\Big] \\
&= \mathsf{E}I(T^A(B) < \infty)\exp\left\{\theta B_{T^A(B)} - \frac{\theta}{2} T^A(B)\right\} \\
&= \mathsf{E}\exp\left\{\theta B_{T^A(B)} - \frac{\theta}{2} T^A(B)\right\},
\end{aligned}$$

because $\mathsf{P}(T^A(B) < \infty) = 1$ for $A \geq 0$, $B \geq 0$.

Thus,

$$\mathsf{E}\exp\Big\{\theta[B - AT^A(B)] - \frac{\theta}{2}T^A(B)\Big\} = 1;$$

taking here $\theta = -A + \sqrt{A^2 + 2\lambda}$, one obtains

$$\mathsf{E}e^{-\lambda T^A(B)} = e^{-B\theta} = \exp\Big\{AB\Big(1 - \sqrt{1 + \frac{2\lambda}{A^2}}\Big)\Big\},$$

which is the required formula (9.38).

3. Rewrite (9.35) using (9.36):

$$p_{T^A(B)}(s) = \frac{B}{\sqrt{2\pi}}\, e^{AB}\, s^{-3/2}\, e^{-(A^2 s + B^2/s)/2}. \tag{9.39}$$

It is convenient to put $B^2 = b > 0$ and $A^2 = a > 0$. Then for the density

$$p(s; a, b) = p_{T^{\sqrt{a}}(\sqrt{b})}(s)$$

we get from (9.39):

$$p(s; a, b) = c_1(a, b)\, s^{-3/2}\, e^{-(as + b/s)/2}, \tag{9.40}$$

where the normalizing constant $c_1(a, b)$ is given by

$$c_1(a, b) = \sqrt{\frac{b}{2\pi}}\, e^{\sqrt{ab}}. \tag{9.41}$$

The class of densities $p(s; a, b)$ on $[0, \infty)$ defined by (9.40), with parameters $a \geq 0$, $b > 0$, is called *Inverse* (or *Negative*) *Gaussian distributions*, in abbreviated form

$$\mathrm{IG} = \mathrm{IG}(a, b).$$

The distributions of this class IG depend on two parameters a and b. The form itself of densities $p(s; a, b)$ given by (9.40) suggests how to get (at first glance, in a somewhat artificial way) distributions determined already by *three* parameters but containing as a subset the class IG.

Namely, on $[0, \infty)$ define *ad hoc* the functions

$$p(s; a, b, \nu) = c_2(a, b, \nu)\, s^{\nu - 1}\, e^{-(as + b/s)/2}, \tag{9.42}$$

where parameters a, b, and $\nu \in \mathbb{R}$ as well as the normalizing constant $c_2(a, b, \nu)$ are chosen such that the $p(s; a, b, \nu)$ are densities of some probability distributions on $[0, \infty)$.

It is easy to see that to satisfy this condition (of integrability) one should require that

$$\begin{aligned} &a \geq 0, \;\; b > 0 \;\;\text{ if }\;\; \nu < 0, \\ &a > 0, \;\; b > 0 \;\;\text{ if }\;\; \nu = 0, \\ &a > 0, \;\; b \geq 0 \;\;\text{ if }\;\; \nu > 0. \end{aligned} \tag{9.43}$$

Under these assumptions

$$\int_0^\infty s^{\nu-1}\, e^{-(as+b/s)/2}\, ds < \infty \tag{9.44}$$

and the normalizing constant $c_2(a,b,\nu)$ should be defined by

$$c_2(a,b,\nu) = \left[\int_0^\infty s^{\nu-1}\, e^{-(as+b/s)/2}\, ds\right]^{-1}. \tag{9.45}$$

It is known (see, *e.g.*, [1], [87; 8.432], [133]) that

$$K_\nu(y) \equiv \frac{1}{2}\int_0^\infty s^{\nu-1}\, e^{-y(s+1/s)/2}\, ds, \qquad y > 0, \tag{9.46}$$

is so-called *modified Bessel function of the third kind and index* ν, which is one of solutions of the following ordinary differential equation (for $y > 0$):

$$y^2 f''(y) + y f'(y) - (y^2 + \nu^2) f(y) = 0. \tag{9.47}$$

(The other solution is the function $I_\nu(y)$; for its definition, for the properties of the functions $K_\nu(y)$ and $I_\nu(y)$ as well as for the definitions and properties of the Bessel functions of the first and second kind, $J_\nu(y)$ and $N_\nu(y)$, which will appear in the formula (9.52) for the density of the Lévy measure see Subsec. 9.)

From (9.45) and (9.46) we find the representation for the normalizing constant:

$$c_2(a,b,\nu) = \frac{(a/b)^{\nu/2}}{2K_\nu(\sqrt{ab})}. \tag{9.48}$$

The probability distributions whose densities $p(s;a,b,\nu)$, $s > 0$, are given by (9.42) and characterized by the parameters (a,b,ν) which satisfy the conditions (9.43) are said to be

$\mathbb{G}$*eneralized Inverse Gaussian*

and denoted by

$$\mathbb{G}\mathrm{IG} = \mathbb{G}\mathrm{IG}(a,b,\nu).$$

If $\nu = -1/2$, $a \geq 0$, and $b > 0$, then we come back to IG(a, b)-distribution, which was the starting point for definition of $\mathbb{G}$IG-distributions (by passing from (9.40) to (9.42)):

$$\mathbb{G}\mathrm{IG}(a, b, -\tfrac{1}{2}) = \mathrm{IG}(a, b). \tag{9.49}$$

Remark 9.3. See [29; § 1.2] for the history of $\mathbb{G}$IG-distributions and many of their properties.

4. Let us discuss some general properties of the distributions of the class $\mathbb{G}$IG and several particular cases of these distributions.

A. $\mathbb{G}$IG-distributions are infinitely divisible [14], [90].

B. The density $f(y)$, $y > 0$, of Lévy's measure of the $\mathbb{G}$IG-distribution is given by

$$f(y) = \frac{e^{-ay/2}}{y}\left[\frac{1}{2}\int_0^\infty e^{-uy/(2b)}\, g_\nu(u)\, du + \max(0, \nu)\right], \tag{9.50}$$

or, equivalently,

$$f(y) = \frac{e^{-ay/2}}{y}\left[b\int_0^\infty e^{-uy}\, g_\nu(2bu)\, du + \max(0, \nu)\right], \tag{9.51}$$

where

$$g_\nu(u) = \frac{2}{\pi^2 u}\left[J^2_{|\nu|}(\sqrt{u}) + N^2_{|\nu|}(\sqrt{u})\right]^{-1}, \qquad u > 0. \tag{9.52}$$

Here the functions $J_\nu(u)$ and $N_\nu(u)$ are order ν Bessel functions of the first and second kind, respectively (see [1], [87], [133]). For the proof of the formula (9.50) see, *e.g.*, [104].

Remark 9.4. One can obtain the formula (9.50) from the formula

$$L(\lambda) = \exp\left\{-\int_0^\infty (1 - e^{-\lambda y}) f(y)\, dy\right\}$$

(see (5.37)), since the Laplace transform $L(\lambda) = \int_0^\infty e^{-\lambda s} p(s; a, b, \nu)\, ds$ can be found explicitly (see (9.58) below).

In two particular cases ($\nu = -\frac{1}{2}$, $a \geq 0$, $b > 0$ and $\nu > 0$, $a > 0$, $b = 0$) the formula (9.50) becomes significantly simpler.

Indeed, if $\nu = -\frac{1}{2}$, then

$$J_{1/2}(u) = \sqrt{\frac{2}{\pi u}}\sin u, \quad N_{1/2}(u) = -\sqrt{\frac{2}{\pi u}}\cos u.$$

Therefore

$$J_{1/2}^2(\sqrt{u}) + N_{1/2}^2(\sqrt{u}) = \frac{2}{\pi\sqrt{u}}. \tag{9.53}$$

This identity allows one to calculate the integral in (9.50); thus we find that

$$f(y) = \sqrt{\frac{b}{2\pi}}\,\frac{e^{-ay/2}}{y^{3/2}}. \tag{9.54}$$

Thus, for the IG-distribution, which is infinitely divisible, the density $f(y)$ of Lévy's measure is given by (9.54).

If $b = 0$, then directly from (9.50) we find that for $\nu > 0$, $a > 0$

$$f(y) = \frac{\nu e^{-ay/2}}{y}. \tag{9.55}$$

The case $a > 0$, $b = 0$, $\nu > 0$ corresponds to the so-called Gamma distribution (see (9.66) below).

C. The density of the $\mathbb{GIG}(a, b, \nu)$-distribution is unimodal with mode

$$m = \begin{cases} \dfrac{b}{2(1-\nu)}, & \text{if } a = 0, \\[2ex] \dfrac{(\nu - 1) + \sqrt{ab + (\nu-1)^2}}{a}, & \text{if } a > 0. \end{cases} \tag{9.56}$$

D. The Laplace transformation

$$L(\lambda) = \int_0^\infty e^{-\lambda s} p(s; a, b, \nu)\, ds, \qquad \lambda \geq 0, \tag{9.57}$$

is given by

$$L(\lambda) = \Big(1 + \frac{2\lambda}{a}\Big)^{-\nu/2} \frac{K_\nu\big(\sqrt{ab(1 + 2\lambda/a)}\big)}{K_\nu(\sqrt{ab})}. \tag{9.58}$$

For the proof it is enough to observe that

$$\begin{aligned} \int_0^\infty e^{-\lambda s} p(s; a, b, \nu)\, ds &= c_2(a, b, \nu) \int_0^\infty s^{\nu-1}\, e^{-\lambda s}\, e^{-(as + b/s)/2}\, ds \\ &= c_2(a, b, \nu) \int_0^\infty s^{\nu-1}\, e^{-(\tilde{a}s + b/s)/2}\, ds, \end{aligned}$$

where $\tilde{a} = 2\lambda + a$.

By (9.45),

$$\int_0^\infty s^{\nu-1}\, e^{-(\tilde{a}s + b/s)/2}\, ds = \frac{1}{c_2(\tilde{a}, b, \nu)}.$$

Consequently,

$$L(\lambda) = \frac{c_2(a,b,\nu)}{c_2(a+2\lambda,b,\nu)},$$

which together with (9.48) implies the required representation (9.58).

In the two particular cases $\nu = -\frac{1}{2}$, $a \geq 0$, $b > 0$ and $\nu > 0$, $a > 0$, $b = 0$ discussed in B, the Laplace transform $L(\lambda)$, $\lambda \geq 0$, has the form

$$L(\lambda) = e^{\sqrt{ab}(1-\sqrt{1+2\lambda/a})} \tag{9.59}$$

and

$$L(\lambda) = (1 + 2\lambda/a)^{-\nu}, \tag{9.60}$$

respectively. To prove (9.59) one should observe that for $\nu = -1/2$

$$K_{-1/2}(y) = \sqrt{\frac{\pi}{2}}\, y^{-1/2} e^{-y}$$

(see Subsec. 9 below). The formula (9.59) can be obtained also from (9.38) if one puts there $A = \sqrt{a}$, $B = \sqrt{b}$.

For the proof of (9.60) it is enough to pass to the limit $b \downarrow 0$ in (9.58) and take into account that for $\nu > 0$

$$K_\nu(y) \sim \Gamma(\nu)\, 2^{\nu-1} y^{-\nu}, \qquad y \to 0. \tag{9.61}$$

The Laplace transform (9.60) is well known—it is the Laplace transform of the Gamma distribution

$$\Gamma(a/2,\nu) \equiv \mathbb{GIG}(a,0,\nu), \tag{9.62}$$

whose density is given by

$$p(s;a,0,\nu) = \frac{(a/2)^\nu}{\Gamma(\nu)}\, s^{\nu-1}\, e^{-as/2}, \qquad a > 0, \quad \nu > 0. \tag{9.63}$$

E. If $a > 0$, $b > 0$, then the moments $\mathsf{E}\xi^n$, $n = 1, 2, \ldots$, of the $\mathbb{GIG}(a,b,\nu)$-random variable ξ are given by the formula

$$\mathsf{E}\xi^n = \left(\frac{b}{a}\right)^{n/2} \frac{K_{\nu+n}(\sqrt{ab})}{K_\nu(\sqrt{ab})}, \tag{9.64}$$

which can be obtained from the Laplace transform (9.58). For the cases $a = 0$, $b > 0$, $\nu < 0$ and $a > 0$, $b = 0$, $\nu > 0$ the corresponding formulae follows from (9.64) (one should pass to the limit as $a \downarrow 0$ and $b \downarrow 0$ and take into account the asymptotics (9.61)).

F. Above we mentioned two important particular cases of $\mathbb{GIG}$-distributions:

The case $a \geq 0$, $b > 0$, $\nu = -1/2$ which leads to the IG-distributions:

$$\mathbb{G}\mathrm{IG}(a, b, -1/2) = \mathrm{IG}(a, b); \tag{9.65}$$

the case $a > 0$, $b = 0$, $\nu > 0$ which leads to the Gamma distributions:

$$\mathbb{G}\mathrm{IG}(a, 0, \nu) = \Gamma(a/2, \nu). \tag{9.66}$$

The third case which is worth mentioning here is *the case* $a > 0$, $b = 0$, $\nu = 1$, where the density has the form

$$p(s; a, b, 1) = c_2(a, b, 1)\, e^{-(as+b/s)/2} \tag{9.67}$$

with the constant $c_2(a, b, 1)$ given by

$$c_2(a, b, 1) = \frac{\sqrt{a/b}}{2K_1(\sqrt{ab})}$$

(see (9.48)).

The distribution with the density (9.67) is commonly called *Positive Hyperbolic distribution* and denoted by $\mathrm{H}^+ = \mathrm{H}^+(a, b)$ (sometimes by PH or $\mathrm{PH}(a, b)$).

Thus, in addition to the cases (9.61) and (9.62) we get the third important case:

$$\mathbb{G}\mathrm{IG}(a, b, 1) = \mathrm{H}^+(a, b). \tag{9.68}$$

Directly from (9.58) we get that the Laplace transform of this distribution is given by

$$L(\lambda) = \Big(1 + \frac{2\lambda}{a}\Big)^{-1/2} \frac{K_1(\sqrt{ab(1 + 2\lambda/a)})}{K_1(\sqrt{ab})}. \tag{9.69}$$

Remark 9.5. We restricted ourselves by the discussion of three important representatives of the class of $\mathbb{G}$IG-distributions. In [26] one can find many other representatives of this class, *e.g.*:

$\mathbb{R}$IG ($\mathbb{R}$eciprocal Inverse Gaussian),

$\mathbb{R}$Gamma ($\mathbb{R}$eciprocal Gamma),

$\mathbb{R}\mathrm{H}^+$ ($\mathbb{R}$eciprocal Positive Hyperbolic),

see also Subsec. 4.

G. The basic facts on distributions of the class $\mathbb{G}$IG are compiled into Table 9.1.

H. Figures of the densities $p(s; a, b, \nu)$ for different values of the parameters a, b and ν are presented, for example, in [26], [30].

Table 9.1

$\mathbb{G}$IG-distributions (for σ^2)				
Parameters	Name	Density	Laplace transform	Density of Lévy measure
a, b, ν under (9.43)	$\mathbb{G}$IG Generalized Inverse Gaussian distribution	(9.42)	(9.59)	(9.50)

Important particular cases of $\mathbb{G}$IG-distributions				
Parameters	Name	Density	Laplace transform	Density of Lévy measure
$a \geq 0$ $b > 0$ $\nu = -1/2$	IG Inverse Gaussian	(9.40)	(9.59)	(9.54)
$a > 0$ $b > 0$ $\nu = 1$	PH or H$^+$ Positive Hyperbolic	(9.67)	(9.69)	(9.50)
$a > 0$ $b = 0$ $\nu > 1$	Gamma	(9.63)	(9.60)	(9.55)

5. In Table 9.1 we epitomized three particular cases of $\mathbb{G}$IG-distributions: Inverse Gaussian ($\nu = -1/2$), Positive Hyperbolic ($\nu = 1$) and Gamma ($\nu > 0$), in view of their wide applications in financial econometrics. Below we discuss two other interesting and important particular cases ($\nu = 0$ and $\nu = 1/2$).

$\mathbb{G}$IG-distribution with $\nu = 0$, *i.e.*, with density

$$p(x; a, b, 0) = \frac{\sqrt{a/b}}{2K_0(\sqrt{ab})x} e^{-(ax+b/x)/2}, \tag{9.70}$$

is named *Halphen's law* after the French hydrologist E. Halphen who came to this law when looking for distributions which would fit well the monthly flow X of river water. His observations showed that such distribution for X should decrease exponentially for very small and very large levels x and coincide (after appropriate scaling) with that of X^{-1}.

In the class of two-parameter distributions, the required distribution is the distribution with density (9.70), *i.e.*, $\mathbb{G}$IG-distribution with $\nu = 0$. Halphen himself called it *harmonic law*. For more general distributions with densities

$$f(x) = c\exp\{\theta_1 x + \theta_0 \log x + \theta_{-1} x^{-1}\}$$

he used the name *generalized harmonic laws.*

Cases $\nu = 1/2$ and $\nu = -1/2$ are, in a certain sense, *reciprocal.* Indeed, let $R^A(B)$ be the last time when the process $W^A = (At + W_t)_{t\geq 0}$ still stays below the threshold B:

$$R^A(B) = \sup\{t \geq 0 : At + W_t \leq B\}.$$

Then straightforward calculations yield:

$$\begin{aligned}
\mathsf{P}\big(R^A(B) \geq t\big) &= \mathsf{P}\big(As + W_s \geq B \;\text{ for all }\; s \geq t\big) \\
&= \mathsf{P}\Big(A\frac{1}{u} + W_{1/u} \geq B \;\text{ for all }\; \frac{1}{u} \geq t\Big) \\
&= \mathsf{P}\Big(A + uW_{1/u} \geq Bu \;\text{ for all }\; u \leq \frac{1}{t}\Big) \\
&= \mathsf{P}\Big(A - \widetilde{W}_u \geq Bu \;\text{ for all }\; u \leq \frac{1}{t}\Big) \\
&= \mathsf{P}\Big(-Bu - \widetilde{W}_u \geq -A \;\text{ for all }\; u \leq \frac{1}{t}\Big) \\
&= \mathsf{P}\Big(Bu + \widetilde{W}_u \leq A \;\text{ for all }\; u \leq \frac{1}{t}\Big) \\
&= \mathsf{P}\Big(T^B(A) \geq \frac{1}{t}\Big),
\end{aligned}$$

where the process

$$\widetilde{W}_u = -\begin{cases} uW_{1/u}, & u > 0, \\ 0, & u = 0, \end{cases}$$

is a Brownian motion and

$$T^B(A) = \inf\{t \geq 0 : Bu + \widetilde{W}_u \geq A\}.$$

Thus, the following reciprocality hold:

$$\mathsf{P}(R^A(B) \geq t) = \mathsf{P}\Big(T^B(A) \geq \frac{1}{t}\Big),$$

and corresponding densities $p_{R^A(B)}(t)$ and $p_{T^B(A)}(t)$ are interrelated by

$$p_{R^A(B)}(t) = p_{T^B(A)}\Big(\frac{1}{t}\Big)\frac{1}{t^2}. \tag{9.71}$$

By (9.39),

$$p_{T^B(A)}(s) = \frac{A}{\sqrt{2\pi}} e^{AB} s^{-3/2} e^{-(B^2 s + A^2/s)/2}.$$

Therefore from (9.71) we find that

$$p_{R^A(B)}(t) = \frac{A}{\sqrt{2\pi}}\, e^{AB}\, \frac{t^{3/2}}{t^2}\, e^{-(A^2t+B^2/t)/2}$$
$$= \frac{A}{\sqrt{2\pi}}\, e^{AB}\, \frac{1}{t^{1/2}}\, e^{-(A^2t+B^2/t)/2}. \tag{9.72}$$

Comparing this with (9.42), we see that $R^A(B)$ has $\mathbb{G}$IG-distribution with the parameter $\nu = -1/2$.

6. Now we turn to the definition of $\mathbb{G}$*eneralized Hyperbolic distributions.*

As already mentioned, this name is used for the distributions of returns h given by (9.1), where σ and ε are independent and σ^2 have $\mathbb{G}$IG-distribution. Thus, by definition,

$$\mathbb{G}\mathrm{H} = \mathbb{N} \circ \mathbb{G}\mathrm{IG}. \tag{9.73}$$

In other words, the $\mathbb{G}$H-distribution is a Normal variance-mean mixture

$$\mathbb{G}\mathrm{H} = \mathsf{E}_{\sigma^2} \mathcal{N}(\mu + \beta\sigma^2, \sigma^2), \tag{9.74}$$

where σ^2 has $\mathbb{G}$IG-distribution with the density

$$p_{\mathbb{G}\mathrm{H}}(x) = \int_0^\infty \exp\Big\{ -\frac{(x-(\mu+\beta u))^2}{2u} \Big\} \frac{1}{\sqrt{2\pi u}}\, p_{\mathbb{G}\mathrm{IG}}(u; a, b, \nu)\, du.$$

So, the $\mathbb{G}$H-distribution is characterized by *five* parameters (a, b, μ, β, ν); one often emphasizes this by writing

$$\mathbb{G}\mathrm{H} = \mathbb{G}\mathrm{H}(a, b, \mu, \beta, \nu) \tag{9.75}$$

(for both distributions and random variables with such distribution).

If $\nu = 1$, the distribution $\mathbb{G}\mathrm{H}(a, b, \mu, \beta, \nu)$ is commonly named *Hyperbolic* distribution and denoted by $\mathbb{H} = \mathbb{H}(a, b, \mu, \beta)$:

$$\mathbb{H} = \mathbb{G}\mathrm{H}(a, b, \mu, \beta, 1). \tag{9.76}$$

The graph of the logarithm of the density of $\mathbb{H}$-distribution is a *hyperbola* (see details in Subsec. 8)—this justifies the name 'hyperbolic' for this distribution and the term 'Generalized Hyperbolic distributions' (briefly $\mathbb{G}$H-class) for the whole class $\mathbb{N} \circ \mathbb{G}$IG.

7. Let $p^*(x; a, b, \mu, \beta, \nu)$ denote the density of the $\mathbb{G}\mathrm{H}(a, b, \mu, \beta, \nu)$-distribution. After averaging in (9.74) with respect to the $\mathbb{G}\mathrm{IG}(a, b, \nu)$-distribution (*i.e.*, with respect to the $\mathbb{G}$IG-distribution of the square σ^2 of stochastic volatility) we find that for all $x \in \mathbb{R}$

$$p^*(x; a, b, \mu, \beta, \nu) = c_3(a, b, \beta, \nu)\, \frac{K_{\nu-1/2}(\alpha\sqrt{b+[x-\mu]^2})}{(\sqrt{b+[x-\mu]^2})^{1/2-\nu}}\, e^{\beta(x-\mu)}, \tag{9.77}$$

where

$$\alpha = \sqrt{a + \beta^2}$$

and the normalizing constant $c_3(a, b, \beta, \nu)$ is given by

$$c_3(a, b, \beta, \nu) = \frac{(a/b)^{\nu/2}\, \alpha^{1/2-\nu}}{\sqrt{2\pi}\, K_\nu(\sqrt{ab})}. \tag{9.78}$$

Just as in (9.43), to guarantee that the function $p^*(x; a, b, \mu, \beta, \nu)$ is indeed the density of a probability distribution on $\mathbb{R}$, natural conditions of integrability must be fulfilled, which imply that the parameters (a, b, ν) satisfy (9.43).

In the case $a > 0$, $b = 0$, $\nu > 0$, one can determine the constant $c_3(a, b, \beta, \nu)$ by passing to the limit $b \downarrow 0$ on the right-hand side of (9.78). Taking into account that for $\nu > 0$

$$K_\nu(y) \sim 2^{\nu-1}\Gamma(\nu) y^{-\nu}, \qquad y \downarrow 0$$

(see Subsec. 9), we find that

$$c_3(a, 0, \beta, \nu) = \lim_{b \downarrow 0} c_3(a, b, \beta, \nu) = \frac{a^\nu}{\sqrt{\pi}\,\Gamma(\nu)(2\alpha)^{\nu-1/2}}, \tag{9.79}$$

where $\alpha = \sqrt{a + \beta^2}$.

Recall that distributions in Pearson's system are determined by *four* parameters (a, b_0, b_1, b_2) appearing in (9.21), whereas distributions of the class $\mathbb{GH}$ are determined by *five* parameters, which provides more possibilities (in comparison with Pearson's system) to construct models of distributions fitting well the empirical distributions of returns. The availability of a—although complicated but nevertheless explicit—formula (9.77) for the densities $p^*(x; a, b, \mu, \beta, \nu)$ also testifies in favor of the class $\mathbb{GH}$.

8. Now we list a number of properties of $\mathbb{GH}$-distributions, analogous to the properties of $\mathbb{GIG}$-distributions discussed above.

A*. $\mathbb{GH}$-distributions are infinitely divisible [14].

B*. The density $f^*(y)$, $y \in \mathbb{R} \setminus \{0\}$, of the Lévy measure $F^* = F^*(y)$ of the $\mathbb{GH}$-distribution is given by

$$f^*(y) = \frac{e^{\beta y}}{|y|} \left[b \int_0^\infty \exp\{-\sqrt{2u + \alpha^2}\, |y|\,\}\, g_\nu(2bu)\, du \right.$$
$$\left. + \exp\{-\alpha |y|\} \max(0, \nu) \right], \tag{9.80}$$

where the function $g_\nu(u)$ is defined in (9.52).

C*. If $\beta = 0$, then $\mathbb{G}$H-densities are unimodal with mode $m = \mu$. In the general case the mode m is determined as solution of the equation

$$\frac{m-\mu}{\sqrt{b+(m-\mu)^2}} \frac{K_{\nu-3/2}(\alpha\sqrt{b+(m-\mu)^2})}{K_{\nu-1/2}(\alpha\sqrt{b+(m-\mu)^2})} = \frac{\beta}{\alpha}, \tag{9.81}$$

where $\alpha = \sqrt{a+\beta^2}$.

This result is a consequence of the identity

$$\left.\frac{\partial p^*}{\partial x}\right|_{x=m} = 0$$

as well as of the fact that the derivatives $K'_{\nu-1/2}(y)$, which appear after differentiation of the right-hand side of (9.77), can be reduced to the Bessel functions, since

$$K'_\lambda(x) = -\frac{\lambda}{x} K_\lambda(x) - K_{\lambda-1}(x).$$

D*. The Laplace transform

$$L^*(\theta) = \int_0^\infty e^{\theta x} p^*(x; a, b, \mu, \beta, \nu)\, dx \tag{9.82}$$

is given (for complex θ such that $|\beta+\theta| < \alpha$) by the formula

$$L^*(\theta) = e^{\theta\mu} \left[\frac{a}{\alpha^2 - (\beta+\theta)^2}\right]^{\nu/2} \frac{K_\nu(\sqrt{b[\alpha^2 - (\beta+\theta)^2]})}{K_\nu(\sqrt{ab})}, \tag{9.83}$$

where $\alpha = \sqrt{a+b^2}$.

E*. If the random variable η has the $\mathbb{G}$H(a, b, μ, β, ν)-distribution, then

$$\mathsf{E}\eta = \mu + \beta\sqrt{\frac{b}{a}} \frac{K_{\nu+1}(\sqrt{ab})}{K_\nu(\sqrt{ab})} \tag{9.84}$$

and

$$\mathsf{D}\eta = \sqrt{\frac{b}{a}} \frac{K_{\nu+1}(\sqrt{ab})}{K_\nu(\sqrt{ab})} + \beta^2 \frac{b}{a}\left[\frac{K_{\nu+2}(\sqrt{ab})}{K_\nu(\sqrt{ab})} - \frac{K^2_{\nu+1}(\sqrt{ab})}{K^2_\nu(\sqrt{ab})}\right]. \tag{9.85}$$

F*. Many particular cases of $\mathbb{G}$H-distributions are worth special mentioning due to their numerous applications.

CASE $a > 0$, $b > 0$, $\nu = 1$. In this case $\mathbb{G}$IG$(a, b, 1)$ is H$^+(a, b)$ (positive hyperbolic distribution). This implies that the distribution

$$\mathbb{H}(a, b, \mu, \beta) \equiv \mathbb{G}\mathrm{H}(a, b, \mu, \beta, 1) = \mathbb{N} \circ \mathbb{G}\mathrm{IG}(a, b, 1)$$

can be rewritten in the form

$$\mathbb{H}(a, b, \mu, \beta) = \mathbb{N} \circ \mathrm{H}^+(a, b), \tag{9.86}$$

or, in a short form,

$$\mathbb{H} = \mathbb{N} \circ \mathrm{H}^{+}. \tag{9.87}$$

For $\mathbb{H} = \mathbb{H}(a, b, \mu, \beta)$, a special name, 'hyperbolic', is commonly used, which is justified as follows.

In the case $\nu = 1$ the density $p^*(x; a, b, \mu, \beta, 1)$ (because of the properties of Bessel's functions $K_{1/2}$ and K_1; see Subsec. 9 below) can be written in the form

$$p^*(x; a, b, \mu, \beta, 1) = c^*(a, b, \beta) \exp\Bigl\{-\alpha\sqrt{b + (x-\mu)^2} + \beta(x-\mu)\Bigr\}, \tag{9.88}$$

where

$$c^*(a, b, \beta) = \frac{a}{2b\alpha K_1(\sqrt{ab})}, \qquad \alpha = \sqrt{a + \beta^2}.$$

We see from (9.86) that the function

$$\log p^*(x; a, b, \mu, \beta, 1) = \log c^*(a, b, \beta) - \alpha\sqrt{b + (x-\mu)^2} + \beta(x-\mu)$$

is a *hyperbola* with asymptotes

$$-\alpha|x-\mu| + \beta(x-\mu).$$

In the considered case $\nu = 1$, the equation (9.77) (whose solution gives the value of the mode m) reduces to

$$\frac{m-\mu}{\sqrt{b + (m-\mu)^2}} = \frac{\beta}{\alpha},$$

which gives a unique mode

$$m = \mu + \sqrt{\frac{b}{a}}\,\beta.$$

Passing to the limit $b \downarrow 0$ in (9.88), we find that the density of the hyperbolic distribution with $a > 0$, $b = 0$, and $\nu = 1$ is given by

$$p^*(x; a, 0, \mu, \beta, 1) = \frac{\alpha^2 - \beta^2}{2\alpha} \exp\{\beta(x-\mu) - \alpha|x-\mu|\},$$

which coincides with the density of the *asymmetric Laplace distribution.*

Passing to the limit as $\alpha \to \infty$ and $b \to \infty$ so that $\sqrt{b}/\alpha \to \sigma^2$, one gets the *normal distribution* with the density

$$\frac{1}{\sqrt{2\pi\sigma^2}}\, e^{-(x-\mu)^2/(2\sigma^2)}.$$

CASE $a \geq 0$, $b > 0$, $\nu = -1/2$. In this case $\mathbb{G}\mathrm{IG}(a, b, -1/2) = \mathrm{IG}(a, b)$. According to (9.74), the corresponding $\mathbb{G}\mathrm{H}(a, b, \mu, \beta, -1/2)$-distribution is an $\mathbb{N} \circ$ IG-distribution. This clarifies why the $\mathbb{GH}(a, b, \mu, \nu, -1/2)$-distribution ($\nu = -1/2$) is called $\mathbb{N}$*ormal Inverse Gaussian* distribution with parameters (a, b, μ, β). The density of this distribution is given by

$$p^*(x; a, b, \mu, \beta, -1/2) = \frac{\alpha b}{\pi} e^{\sqrt{ab}} \frac{K_1(\alpha\sqrt{b + (x-\mu)^2})}{\sqrt{b + (x-\mu)^2}} e^{\beta(x-\mu)}, \qquad (9.89)$$

$x \in \mathbb{R}$, where $\alpha = \sqrt{a + \beta^2}$. The density $f^*(y)$ of the Lévy measure F^* is determined by

$$f^*(y) = e^{\beta y} \frac{\alpha\sqrt{b}}{\pi|y|} K_1(\alpha|y|), \qquad (9.90)$$

which follows from (9.80) for $\nu = -1/2$.

It is interesting to notice that the class of $\mathbb{N} \circ$ IG-distributions with fixed parameters a and β is closed with respect to *convolution*. More precisely, if independent random variables η_1 and η_2 have $\mathbb{N} \circ$ IG-distribution with parameters $(a, \gamma_1^2, \beta, \mu_1)$ and $(a, \gamma_2^2, \beta, \mu_2)$, then the sum $\eta_1 + \eta_2$ has $\mathbb{N} \circ$ IG-distribution with parameters $(a, (\gamma_1 + \gamma_2)^2, \beta, \mu_1 + \mu_2)$.

Indeed, if $a \geq 0$, $b > 0$, and $\nu = -1/2$, then, taking into account that

$$K_{-1/2}(y) = K_{1/2}(y) = \sqrt{\frac{\pi}{y}}\, y^{-1/2} e^{-y},$$

we get from (9.83) that (for complex θ such that $|\beta + \theta| < \alpha$)

$$L^*(\theta) = \exp\Big\{\theta\mu + \sqrt{b}\,\big(\sqrt{a} - \sqrt{a - 2\beta\theta - \theta^2}\,\big)\Big\}. \qquad (9.91)$$

Thus, if $b = \gamma^2$, then for fixed a and β the Laplace transform has the following property:

$$L^*(\theta; \gamma_1, \mu_1) \cdot L^*(\theta; \gamma_2, \mu_2) = L^*(\theta; \gamma_1 + \gamma_2, \mu_1 + \mu_2),$$

which implies the closedness with respect to convolution of $\mathbb{N} \circ$ IG-distributions.

Case $a > 0$, $b = 0$, $\nu > 0$. Another example of $\mathbb{GH}$-distribution with this closedness property is the Normal Gamma ($\mathbb{N} \circ$ Gamma) distribution, also called $\mathbb{V}$ariance Gamma ($\mathbb{VG}$), which is described in the following way.

As was mentioned in (9.54), the Gamma-distribution Gamma$(a/2, \nu)$ is a $\mathbb{GIG}(a, 0, \nu)$-distribution with the density $p(a, 0, \nu)$ given by (9.55). Therefore it is natural to name $\mathbb{N} \circ \mathbb{GIG}(a, 0, \nu)$ $(= \mathbb{GH}(a, 0, \mu, \beta, \nu))$ as Normal Gamma distribution and denote it by

$$\mathbb{N} \circ \text{Gamma} = \mathbb{N} \circ \text{Gamma}(a, \mu, \beta, \nu).$$

The density of this distribution is given by

$$p^*(x; a, 0, \mu, \beta, \nu) = \frac{a^\nu}{\sqrt{\pi}\,\Gamma(\nu)(2\alpha)^{\nu-1/2}}\,|x-\mu|^{\nu-1/2} \times K_{\nu-1/2}(\alpha|x-\mu|)\, e^{\beta(x-\mu)} \tag{9.92}$$

(to get this formula one should pass to the limit $b \downarrow 0$ in (9.78) and take into account (9.79)).

Passing to the same limit $b \downarrow 0$ in (9.83), one can find that the Laplace transform of the $\mathbb{N} \circ$ Gamma distribution with the density (9.92) is given by the formula

$$L^*(\theta) = e^{\mu\theta}\left(\frac{a}{a - 2\beta\theta - \theta^2}\right)^\nu. \tag{9.93}$$

This shows that if η_1 and η_2 are independent random variables with distributions $\mathbb{N} \circ \mathrm{Gamma}(a, \mu_1, \beta, \nu_1)$ and $\mathbb{N} \circ \mathrm{Gamma}(a, \mu_2, \beta, \nu_2)$, respectively, then their sum $\eta_1 + \eta_2$ has the $\mathbb{N} \circ \mathrm{Gamma}(a, \mu_1 + \mu_2, \beta, \nu_1 + \nu_2)$-distribution.

G*. The properties of $\mathbb{G}$H-distributions are summarized in Table 9.2.

9. As we have seen, to operate with distributions of classes $\mathbb{G}$IG and $\mathbb{G}$H one need

- the modified Bessel functions of the third kind and index $\nu \in \mathbb{R}$ (denoted by $K_\nu(y)$, $y > 0$);
- the Bessel functions of the first kind and index $\nu \in \mathbb{R}$ (denoted by $J_\nu(y)$, $y > 0$);
- the Bessel functions of the second kind and index $\nu \in \mathbb{R}$ (denoted by $N_\nu(y)$, $y > 0$).

According to standard guides on special functions (see, for example, [1], [87], [133], [173]), $K_\nu(y)$ and $N_\nu(y)$ are solutions to the ordinary differential equation (9.47). At that the function $K_\nu(y)$ admits the integral representation (9.46), which arises when one wants to find the normalizing constant $c_2(a, b, \nu)$ in (9.42).

The Bessel functions $J_\nu(y)$ and $N_\nu(y)$ appeared in our considerations when searching for the densities $f(y)$ and $f^*(y)$ of Lévy's measures of $\mathbb{G}$IG- and $\mathbb{G}$H-distributions (see (9.49) and (9.80)).

There are several different representations for $J_\nu(y)$, for example:

(a) series expansion

$$J_\nu(y) = \left(\frac{y}{2}\right)^\nu \sum_{k=0}^{\infty} \frac{(-1)^k}{k!\,\Gamma(\nu + k + 1)}\left(\frac{y}{2}\right)^{2k}; \tag{9.94}$$

Table 9.2

GH-distributions (for h)				
Parameters	Name	Density	Laplace transform	Density of Lévy's measure
a, b, ν under (9.43) and $\beta \in \mathbb{R}$, $\mu \in \mathbb{R}$	$\mathbb{GH}$ ($= \mathbb{N} \circ \mathbb{GIG}$) Generalized Hyperbolic or Normal Variance-mean mixture or Normal Generalized Inverse Gaussian	(9.78)	(9.83)	(9.80)

Important particular cases of $\mathbb{GH}$-distributions				
Parameters	Name	Density	Laplace transform	Density of Lévy's measure
$a \geq 0$ $b > 0$ $\nu = -1/2$ $\beta \in \mathbb{R}$ $\mu \in \mathbb{R}$	$\mathbb{N} \circ \mathrm{IG}$ Normal Inverse Gaussian	(9.89)	(9.91)	(9.90)
$a > 0$ $b > 0$ $\nu = 1$ $\beta \in \mathbb{R}$ $\mu \in \mathbb{R}$	$\mathbb{H} = \mathbb{N} \circ \mathrm{H}^+$ Hyperbolic or Normal Positive Hyperbolic	(9.88)	(9.83)	(9.80)
$a > 0$ $b = 0$ $\nu > 0$ $\beta \in \mathbb{R}$ $\mu \in \mathbb{R}$	$\mathbb{N} \circ$ Gamma Normal Gamma or Variance Gamma	(9.92)	(9.93)	(9.80)

(b) integral representation

$$J_\nu(y) = \frac{1}{\pi} \int_0^\pi \cos(y \sin u - \nu u)\, du - \frac{\sin \nu\pi}{\pi} \int_0^\infty e^{-y \sinh u - \nu u}\, du; \qquad (9.95)$$

in particular, for $\nu > -1/2$

$$J_\nu(y) = \frac{2(y/2)^\nu}{\sqrt{\pi}\Gamma(\nu+1/2)} \int_0^1 (1-u^2)^{\nu-1/2} \cos(yu)\, du, \qquad y \in \mathbb{R}. \tag{9.96}$$

The Bessel function $N_\nu(y)$ (sometimes referred to as *Weber function*) is defined as follows:

$$N_\nu(y) = \frac{J_\nu(y)\cos\nu\pi - J_{-\nu}(y)}{\sin\nu\pi}. \tag{9.97}$$

In the case $\nu = 1/2$

$$J_{1/2}(y) = \sqrt{\frac{2}{\pi y}}\, \sin y, \quad N_{1/2}(y) = -\sqrt{\frac{2}{\pi y}}\, \cos y. \tag{9.98}$$

These representations allow one to simplify significantly the formulae for the Lévy densities $f(y)$ for $\mathbb{GIG}$-distributions (see (9.49)) and $f^*(y)$ for $\mathbb{GH}$-distributions (see (9.80)).

For $K_\nu(y)$ the following asymptotical representations are well known (see, *e.g.*, [1]): as $y \to 0$

$$K_\nu(y) \sim \begin{cases} \Gamma(\nu)\, 2^{\nu-1} y^{-\nu}, & \nu > 0, \\ -\log y, & \nu = 0, \end{cases} \tag{9.99}$$

and as $y \to \infty$

$$K_\nu(y) = \sqrt{\frac{\pi}{2y}} \left(1 + \frac{4\nu^2 - 1}{8y} + O(y^{-2})\right) e^{-y}. \tag{9.100}$$

In the case $\nu = 1/2$ or $\nu = -1/2$

$$K_{1/2}(y) = K_{-1/2}(y) = \sqrt{\frac{\pi}{2}}\, y^{-1/2} e^{-y} \tag{9.101}$$

($K_\nu(y) = K_{-\nu}(y)$ for any ν).

Among other useful properties of the functions $K_\nu(y)$ are:

$$K_{\nu+1}(y) = \frac{2\nu}{y}\, K_\nu(y) + K_{\nu-1}(y); \tag{9.102}$$

$$K'_\nu(y) = -\frac{\nu}{y}\, K_\nu(y) - K_{\nu-1}(y); \tag{9.103}$$

$$K_{n+1/2}(y) = \sqrt{\frac{\pi}{2y}}\, e^{-y} \left[1 + \sum_{i=1}^{n} \frac{(n+i)!}{(n-i)!\, i!}\, (2y)^{-i}\right]. \tag{9.104}$$

Chapter 10

Martingale Measures in the Stochastic Theory of Arbitrage

10.1 Basic Notions and Summary of Results of the Theory of Arbitrage. I. Discrete Time Models

1. From the economics point of view the financial markets, many of which are described by the stochastic models considered above (linear, exponential, with change of time, *etc.*), are structures, dealing in which is subject to RISK. Therefore it is no wonder that the desire to master the very notion of "risk" gave rise to such conception and theories as "effective market", CAPM (Capital Asset Pricing Model), APT (Arbitrage Pricing Theory), theory of arbitrage, axiomatic theory of risk, *etc.*

In turn, the development of risk-*reduction* methods led to the appearance of such notions as diversification, hedging, financial derivatives, insurance, fair prices and so on. (For details see, for example, monographs [161], [74], [132].)

It should be emphasized that the progress in mathematical finance is essentially due to the comprehension of the fact that the *economical* notion of arbitrage on markets with stochastic structure is closely related to the *mathematical* notions of a martingale (as well as its relatives) and of a martingale measure.

In this chapter we review the essential concepts of the arbitrage theory and illustrate its applications to a number of financial markets models (driven by a Brownian motion or Lévy processes), which were studied in the previous chapters. These models relate to the continuous-time case, where the construction of the arbitrage theory needs overcoming the significant theoretical difficulties, caused in a great part by difficulty of dealing with rather large classes of strategies (and their "compactification") and respective functions of gain, income, capital, whose description calls for quite complicated constructions of stochastic integrals.

2. In the discrete-time case the stochastic theory of arbitrage is quite transparent, and it is natural to begin with this case, which will display the essential concepts and results.

We suppose given a filtered probability space $(\Omega, \mathcal{F}, (\mathcal{F}_n)_{n\geq 0}, \mathsf{P})$, which describes the "stochastics" of the considered financial markets. We shall assume that $\mathcal{F}_0 = \{\varnothing, \Omega\}$.

It is useful to notice that if the choice of elements of the space $(\Omega, \mathcal{F}, (\mathcal{F}_n)_{n\geq 0})$ usually does not cause serious difficulties, the choice of "physical" measure (probability) P is far from being so evident. In many cases it is advisable to consider instead of one concrete measure P a certain *family* $\mathcal{P}$ of such measures P. In [161; Chap. V, § 1c] a "chaotic" model is studied, when $(\Omega, \mathcal{F}, (\mathcal{F}_n)_{n\geq 0})$ is equipped with no probability measure and all considerations are taken not with respect to one or another probability measure but for *each* ω from Ω.

3. Having in view the further applications to the operations on financial markets, we assume to be given a certain set $\mathcal{A}$ of financial *assets* (in other words, primary securities), which consists of a *bank account*

$$B = (B_n)_{n\geq 0}$$

and of a finite number d of *stocks*

$$S = (S_n)_{n\geq 0},$$

where $S_n = (S_n^1, \dots, S_n^d)$.

A financial market with such assets will be called *(B, S)-market.*

The difference between a bank account and a stock consists in the following: the variables B_n are $\mathcal{F}_{n-1}$-measurable ($\mathcal{F}_{-1} = \mathcal{F}_0$), whereas the variables S_n^i are $\mathcal{F}_n$-measurable. In other words, the variables B_n are predictable and S_n^i are optional.

If one follows carefully the lines of the proofs of the "fundamental theorems" stated below (for the exact formulations see below), one can observe that, in fact, not the "proper" values S_n^i of stock prices are important but their "relative" (discounted) values S_n^i/B_n. (We assume everywhere that $B_n > 0$, $S_n^i \geq 0$.) This fact justifies the often used assumption $B_n \equiv 1$, $n \geq 0$.

The operations on a securities market consist in that some traders buy securities, others sell them. At that, *ask* prices S^a and *bid* prices S^b, generally speaking, do not coincide: the *spread* $S^a - S^b \geq 0$. (The statistics of prices shows that the spread and volatility are positively correlated. Thus, the growth of volatility, increasing the risk because of a lesser accuracy of a price-movement forecast, induces traders to augment spread as a compensation for greater risk.)

Trading with stocks is always subject to transaction costs, that should be kept in mind when constructing strategies of behavior on securities markets. (See further Sec. 10.3, devoted to the arbitrage questions in the presence of transaction costs.)

In the first approximation, for simplicity's sake, one can think that the bid price and the ask price coincide ($S^a = S^b$) and thus on the market there is only one price, which we denote be S. (A method of reduction of two prices S^a and S^b to one is to set, for example, $S = \sqrt{S^a \cdot S^b}$.)

4. Now consider a trader acting on a (B, S)-market with a bank account $B = (B_n)_{n \geq 0}$ and asset prices $S = (S_n)_{n \geq 0}$, where $S_n = (S_n^1, \dots, S_n^d)$.

One says that a stochastic sequence $\pi = (\beta, \gamma)$, where the sequences $\beta = (\beta_n)_{n \geq 0}$ and $\gamma = (\gamma_n)_{n \geq 0}$ with $\gamma_n = (\gamma_n^1, \dots, \gamma_n^d)$ are predictable (i.e., β_n and γ_n^i are $\mathcal{F}_{n-1}$-measurable), forms a *portfolio* of assets of a trader.

The quantity

$$X_n^\pi = \beta_n B_n + \sum_{i=1}^{d} \gamma_n^i S_n^i \tag{10.1}$$

is called the *capital of the portfolio* π at time n.

For simplicity of notation we write the scalar product $(\gamma_n, S_n) = \sum_{i=1}^{n} \gamma_n^i S_n^i$ as $\gamma_n S_n$. Then (10.1) takes the form

$$X_n^\pi = \beta_n B_n + \gamma_n S_n. \tag{10.2}$$

Observe that for any two sequences $a = (a_n)_{n \geq 0}$ and $b = (b_n)_{n \geq 0}$

$$\Delta(a_n b_n) = a_n \Delta b_n + b_{n-1} \Delta a_n, \qquad n \geq 1, \tag{10.3}$$

with the evident notation $\Delta a_n = a_n - a_{n-1}$.

Then (10.2) implies that

$$\Delta X_n^\pi = \big[\beta_n \Delta B_n + \gamma_n \Delta S_n\big] + \big[B_{n-1} \Delta \beta_n + S_{n-1} \Delta \gamma_n\big]. \tag{10.4}$$

Since the real capital alterations come from the changes in bank account and in asset values only, the quantity in the last square brackets in (10.4) is null, and therefore

$$\Delta X_n^\pi = \beta_n \Delta B_n + \gamma_n \Delta S_n. \tag{10.5}$$

Clearly this quantity determines the one-step income. The aggregated income over the time n (denoted by G_n^π) will be equal to

$$G_n^\pi = \sum_{k=1}^{n} \Delta X_k^\pi, \tag{10.6}$$

and (the starting capital being X_0^π) the capital at time n will be

$$X_n^\pi = X_0^\pi + G_n^\pi. \tag{10.7}$$

A portfolio $\pi = (\beta, \gamma)$ satisfying (10.5) (its capital is then determined by (10.7)) is called *self-financing* (notation: $\pi \in \mathrm{SF}$), and the condition $B_{n-1}\Delta\beta_n + S_{n-1}\Delta\gamma_n = 0$ is called the *condition of self-financing.*

By straightforward verification one can see that a self-financing portfolio π has the following important feature:

$$\Delta\left(\frac{X_n^\pi}{B_n}\right) = \gamma_n \Delta\left(\frac{S_n}{B_n}\right). \tag{10.5*}$$

5. Recall some definitions from the arbitrage theory. The "classical" definitions of arbitrage and no arbitrage on (B, S)-market are the following.

Definition 10.1. One says that a (B, S)-market with a *finite* number $d < \infty$ of assets and a *finite* time horizon $N < \infty$ provides an *arbitrage opportunity* (or allows *arbitrage*), if there exists a self-financing portfolio π such that

(a) $X_0^\pi = 0$;
(b) $X_N^\pi \geq 0$ (P-a.s.) [equivalently, $G_N^\pi \geq 0$ (P-a.s.)], and
(c) $X_N^\pi > 0$ [equivalently, $G_N^\pi > 0$] with a positive P-probability.

Remark 10.1. In the case of a finite time horizon, $N < \infty$, it is assumed that we deal with a filtered probability space $(\Omega, \mathcal{F}_N, (\mathcal{F}_n)_{0\leq n\leq N}, \mathsf{P})$, where P is a probability measure on $\mathcal{F}_N$. (In many cases, for the sake of clarity, it is expedient to write P_N instead of P.)

Definition 10.2. A (B, S)-market (with $d < \infty$ and a *finite* time horizon $N < \infty$) is said to be *arbitrage free* [or to meet the *NA* (*no arbitrage*) *condition*] if it provides no arbitrage opportunity.

With the notation

$$\mathbb{G} = \{G_N^\pi : \pi \in \mathrm{SF}\},$$

the set of income from self-financing strategies, one can reformulate the NA condition in the following widely used form:

$$\mathbb{G} \cap \mathbb{L}_{(\geq 0)} = \{0\}, \tag{10.8}$$

where $\mathbb{L}_{(\geq 0)}$ is the set of nonnegative random variables. [Equality (10.8) is understood in the sense of identity of classes of P-equivalent random variables, *i.e.*, if $\xi \in G \cap \mathbb{L}_{(\geq 0)}$, then ξ is P-equivalent to the "null" random variable ($\mathsf{P}(\xi = 0) = 1$).]

To formulate the "first fundamental theorem of the arbitrage theory", introduce the set

$$\mathcal{M}(\mathsf{P}) = \left\{ \widetilde{\mathsf{P}} : \ \widetilde{\mathsf{P}} \sim \mathsf{P} \ \text{and} \ \frac{S}{B} = \left(\frac{S^1}{B}, \dots, \frac{S^d}{B} \right) \ \text{is a } \widetilde{\mathsf{P}}\text{-martingale} \right\},$$

consisting of the measures $\widetilde{\mathsf{P}}$ on $\mathcal{F}_N$ which are equivalent to the measure P and for which the d-variate process S/B is a martingale, *i.e.*,

$$\mathsf{E}_{\widetilde{\mathsf{P}}} \left| \frac{S_n^i}{B_n} \right| < \infty$$

for all $n = 0, 1, \dots, N$ and $i = 1, \dots, d$ and

$$\mathsf{E}_{\widetilde{\mathsf{P}}} \left(\frac{S_n^i}{B_n} \,\Big|\, \mathcal{F}_{n-1} \right) = \frac{S_{n-1}^i}{B_{n-1}}$$

for all $n = 1, \dots, N$ and $i = 1, \dots, d$, where $\mathsf{E}_{\widetilde{\mathsf{P}}}$ denotes averaging with respect to the measure $\widetilde{\mathsf{P}}$.

It should be emphasized that, generally speaking, we do not assume that the expectations $\mathsf{E}[S_n^i / B_n]$ with respect to the initial "physical" measure P are finite.

Theorem 10.1. (The first fundamental theorem of the arbitrage theory; discrete time.) *In the above-formulated model of (B, S)-market, for No Arbitrage (NA) it is necessary and sufficient that the set $\mathcal{M}(\mathsf{P})$ of martingale measures is nonempty*:

$$\text{NA} \iff \mathcal{M}(\mathsf{P}) \neq \varnothing. \tag{10.9}$$

This remarkable theorem—proved first for the case of finite Ω in [91] and then for the general case in [53]—was the key result which established the connection between the economics notion of arbitrage and the martingale theory.

In addition to [91] and [53], where the proof of this theorem is given, let us cite the papers [105], [99], [147], [167], where different proofs were proposed and other equivalent formulations were given. For the sake of completeness let us outline the key steps of the proof of this theorem (see details in [161; Chap. V, §§ 2c, 2d]).

Sufficiency: $\mathcal{M}(\mathsf{P}) \neq \varnothing \Longrightarrow \text{NA}$.

We have to show that if for some self-financing strategy π its income G_N^π satisfies

$$G_N^\pi = \sum_{k=1}^{N} (\beta_k \Delta B_k + \gamma_k \Delta S_k) \geq 0 \quad (\mathsf{P}\text{-a.s.}),$$

then $G_N^\pi = 0$ (P-a.s.).

Consider the discounted incomes $(G_n^\pi / B_n)_{0 \leq n \leq N}$ with $G_0^\pi = 0$. In view of (10.5*), assuming that $X_0^\pi = 0$, we find that

$$\Delta\left(\frac{G_n^\pi}{B_n}\right) = \gamma_n \Delta\left(\frac{S_n}{B_n}\right) \quad \left(= \sum_{i=1}^{d} \gamma_n^i \Delta\left(\frac{S_n^i}{B_n}\right)\right).$$

Since $\mathcal{M}(\mathsf{P}) \neq \varnothing$, the sequences $(S_n^i / B_n)_{0 \leq n \leq N}$ are martingales for all $i = 1, \ldots, d$. The sequences $(\sum_{i=1}^d \gamma_n^i \Delta(S_n^i / B_n))_{0 \leq n \leq N}$ with predictable $(\gamma_n^i)_{0 \leq n \leq N}$, being martingale transforms, are local martingales (see Remark 3.2 on page 44 and, for more detail, [161; Chap. II, § 1c]).

Since

$$\frac{G_N^\pi}{B_N} \geq 0 \quad (\mathsf{P}\text{- and } \widetilde{\mathsf{P}}\text{-a.s.}), \tag{10.10}$$

the local martingale $(G_N^\pi / B_n)_{0 \leq n \leq N}$ is just a martingale with respect to the measure $\widetilde{\mathsf{P}}$ (see [161; Chap. II, lemma in § 1c]). Therefore

$$\mathsf{E}_{\widetilde{\mathsf{P}}}\left(\frac{G_N^\pi}{B_N}\right) = \frac{G_0^\pi}{B_0} = 0$$

and thus, in view of (10.10), $G_N^\pi = 0$ ($\widetilde{\mathsf{P}}$- and P-a.s.).

Necessity: NA $\Longrightarrow \mathcal{M}(\mathsf{P}) \neq \varnothing$.

The greater part of known proofs appeal somehow or other to the ideas and results of functional analysis and have the character of "proof of existence" (of martingale measures). However, from the point of view of probability theory, it would be preferable to construct *explicitly* at least one martingale measure. Such a proof can be found in [161; Chap. V, § 2d]. The very idea of this proof is suggested by the following purely probabilistic proposition (see [147] and [161; Chap. V, § 2d]).

Lemma 10.1. *Let* P *be a probability measure on* $(\mathbb{R}, \mathcal{B}(\mathbb{R}))$ *and let* $\xi(x) = x$ *be a coordinate random variable,* $x \in \mathbb{R}$, *such that*

$$\mathsf{P}(\xi > 0) > 0 \quad \textit{and} \quad \mathsf{P}(\xi < 0) > 0. \tag{10.11}$$

Then there exists a probability measure $\widetilde{\mathsf{P}}$ *such that*

$$\mathsf{E}_{\widetilde{\mathsf{P}}} e^{a\xi} < \infty; \tag{10.12}$$

in particular,

$$\mathsf{E}_{\widetilde{\mathsf{P}}}|\xi| < \infty, \tag{10.13}$$

for all $a \in \mathbb{R}$, *and*

$$\mathsf{E}_{\widetilde{\mathsf{P}}}\xi = 0. \tag{10.14}$$

Notice that one can interpret condition (10.11) as the absence of arbitrage in the *one-step* (B, S)-model ($S_0 = 0$, $S_1 = \xi$, $B_0 = B_1 = 1$), and condition (10.14) as the martingale property: $\mathsf{E}(\xi \mid \mathcal{F}_0) = \mathsf{E}(S_1 \mid \mathcal{F}_0) = S_0$, where $\mathcal{F}_0 = \{\varnothing, \Omega\}$, $S_0 = 0$.

6. In the class of arbitrage-free financial markets an important role is played by the so-called *complete* markets, for which the set of martingale measures $\mathcal{M}(\mathsf{P})$ consists of exactly one measure ($|\mathcal{M}(\mathsf{P})| = 1$).

The following definition of such markets has a transparent economical sense.

Definition 10.3. The arbitrage-free (B, S)-market (with $d < \infty$ and finite time horizon N) is said to be *complete* or N-*complete* if for any $\mathcal{F}_N$-measurable random variable $F_N = F_N(\omega)$ there exist a self-financing portfolio π and $x \in \mathbb{R}$ such that $X_0^\pi = x$ and

$$x + G_N^\pi = F_N \quad (\mathsf{P}\text{-a.s.}). \tag{10.15}$$

Theorem 10.2. (The second fundamental theorem of the arbitrage theory; discrete time.) *Let a* $(B, S) = (B_n, S_n)_{n \le N}$*-market be arbitrage free. Then the following conditions are equivalent:*

(i) *the* (B, S)*-market is complete;*
(ii) *the set* $\mathcal{M}(\mathsf{P})$ *of martingale measures consists of a single element* (i.e., $|\mathcal{M}(\mathsf{P})| = 1$);
(iii) *in* $\mathcal{M}(\mathsf{P})$ *one can find a martingale measure* $\widetilde{\mathsf{P}}$ *such that any bounded or nonnegative martingale* $M = (M_n, \mathcal{F}_n, \widetilde{\mathsf{P}})_{n \le N}$ *admits a* $\frac{S}{B}$*-representation,* i.e., *there is a predictable process* $\alpha = (\alpha_n)_{n \le N}$ *such that* $\widetilde{\mathsf{P}}$*-a.s.*

$$M_n = M_0 + \sum_{k=1}^{n} \alpha_k \Delta\left(\frac{S_k}{B_k}\right), \qquad n \le N.$$

For the proof and other characterizations see [99] and [161; Chap. V, § 4].

7. The classical example of an arbitrage-free complete (B, S)-market is provided by CRR-model (Cox–Ross–Rubinstein), in which

$$\begin{aligned} \Delta B_n &= r B_{n-1}, \\ \Delta S_n &= \rho_n S_{n-1}, \end{aligned} \tag{10.16}$$

where $n \leq N$, B_0 and S_0 are positive constants. The variables $\rho_1, \ldots, \rho_N$ are assumed independent and identically distributed, each one taking two values a and b such that $-1 < a < r < b$ and

$$\mathsf{P}(\rho_n = b) = p, \qquad \mathsf{P}(\rho_n = a) = q,$$

where $0 < p < 1$ and $p + q = 1$.

In this CRR-model there exists a unique martingale measure $\widetilde{\mathsf{P}}$, which is such that $\rho_1, \ldots, \rho_N$ are again $\widetilde{\mathsf{P}}$-independent and identically distributed, and

$$\widetilde{\mathsf{P}}(\rho_n = b) = \tilde{p}, \qquad \widetilde{\mathsf{P}}(\rho_n = a) = \tilde{q},$$

where

$$\tilde{p} = \frac{r-a}{b-a}, \qquad \tilde{q} = \frac{b-r}{b-a}$$

(see details in [161; Chap. V, § 3f]).

8. The first and second fundamental theorems of the arbitrage theory are widely applied in practice and, first of all, in problems of determining the set of the so-called *mutually acceptable* prices of one or another contingent claim $F_N = F_N(\omega)$.

Let us give a pictorial interpretation of these prices.

Suppose that on the market considered there is a "seller" and a "buyer". The latter buys from the seller at a price x some derivative (for example, an option) and the seller at time N pays to the buyer a ($\mathcal{F}_N$-measurable) random variable $F_N = F_N(\omega)$ whose values are determined by the evolution of stochastic sequences $B = (B_n)_{0 \leq n \leq N}$ and $S = (S_n)_{0 \leq n \leq N}$.

So, in the case of a standard call option and for $d = 1$

$$F_N(\omega) = (S_N(\omega) - K)^+, \tag{10.17}$$

and in the case of put option

$$F_N(\omega) = (K - S_N(\omega))^+, \tag{10.18}$$

where the quantity K—called *strike* or *exercise* price—is stipulated in advance when concluding the contract.

These cases relate to the so-called European options whose distinctive feature is that the buyer of the option is passive, *i.e.*, once having bought the option, he/she is simply waiting for the time N, when he/she get the payment equal to $F_N = F_N(\omega)$. In practice, for example in the case of a call option with F_N given by (10.17), it works in the following way. If $S_N < K$, then $F_N = 0$, *i.e.* the buyer gets nothing and since he/she had paid the amount x for the option, he/she loses. However if $S_N > K$, then, on the contract conditions, the buyer may purchase the stock at the price K, which is less than the real price S_N. After doing so, he/she can sell the stock at the market price S_N. As a result his/her gain will be equal to

$$F_N - x = (S_N - K)^+ - x. \tag{10.19}$$

The question arises as to what is a *mutually acceptable* (acceptable both for the seller and buyer) price of the contract with the contingent claim $F_N = F_N(\omega)$.

Such a price, suitable for both buyer and seller, is naturally understood as a price when neither buyer nor seller has a *riskless* gain. In other words, the amount x should be chosen to provide arbitrage opportunities neither for the seller nor for the buyer.

It is rather clear that the complement to the set of mutually acceptable prices decomposes into two domains, the first containing the prices favorable for the buyer (evidently, it is the domain of small x), the second containing the prices which the seller would prefer (of course, this it a domain of large x).

After these explanatory remarks, let us go to definitions assuming without loss of generality that $B_n \equiv 1$ (and so $G_N^\pi = \sum_{k=1}^N \gamma_k \Delta S_k$).

From (10.19) we see that the gain of the buyer is determined by $F_N - x$. If he/she has bought γ_0 stocks ($\gamma_0 \in \mathbb{R}$, the case $\gamma_0 < 0$ corresponds to the short-selling), then his/her gain will be equal to $\gamma_0(F_N - x)$. This observation makes natural the following definition.

Definition 10.4. The *interval of mutually acceptable prices* of a contingent claim F_N is the set

$$\mathbb{I}(F_N) = \big[\mathbb{C}_*(F_N), \mathbb{C}^*(F_N)\big], \tag{10.20}$$

where $\mathbb{C}_*(F_N)$ (*lower price*) and $\mathbb{C}^*(F_N)$ (*upper price*) are specified as follows:

$$\begin{aligned} \mathbb{C}_*(F_N) &= \sup\big\{x : \exists\pi \text{ with } X_0^\pi = x \text{ and } x + G_N^\pi \le F_N \ (\mathsf{P}\text{-a.s.})\big\}, \\ \mathbb{C}^*(F_N) &= \inf\big\{x : \exists\pi \text{ with } X_0^\pi = x \text{ and } x + G_N^\pi \ge F_N \ (\mathsf{P}\text{-a.s.})\big\}. \end{aligned} \tag{10.21}$$

In [161; Chap. V, § 1b] and [132; Chap. 2, § 2.5.2], one can find detailed motivation why the two prices, the lower price $\mathbb{C}_*(F_N)$ (called also a buyer price) and the upper price $\mathbb{C}^*(F_N)$ (called also a seller price), should be defined just as in (10.21). The essence is that

- if the contract price x is in $[0, \mathbb{C}_*(F_N))$, then the buyer can get a riskless gain (by short selling);

- if the contract price x is in $(\mathbb{C}^*(F_N), \infty)$, then the seller can get a riskless gain (by choosing an adequate portfolio);
- however, if the contract price x is in $[\mathbb{C}_*(F_N), \mathbb{C}^*(F_N)]$, then neither seller nor buyer can get a riskless gain (each of them can either gain or loose—depending on the randomness of stock prices).

Since in the first case the seller will not agree to a low price (from the interval $[0, \mathbb{C}_*(F_N)))$ and in the second case the buyer will not buy the contract at a high price (from $(\mathbb{C}^*(F_N), \infty))$, the set of *admissible* prices both for seller and for buyer is $[\mathbb{C}_*(F_N), \mathbb{C}^*(F_N)]$. If the contract price is from this interval, then both seller and buyer understand that none of them will have riskless gain and, due to "randomness", the losses cannot be reduced to zero but the gain is possible (with a positive probability).

Theorem 10.3. (Basic formulae for lower and upper prices.) *Let the (B, S)-market be arbitrage free and let $F_N = F_N(\omega)$ be a nonnegative $\mathcal{F}_N$-measurable contingent claim. Then*

$$\mathbb{C}_*(F_N) = B_0 \inf_{\widetilde{\mathsf{P}} \in \mathcal{M}(\mathsf{P})} \mathsf{E}_{\widetilde{\mathsf{P}}} \frac{F_N}{B_N} \tag{10.22}$$

and

$$\mathbb{C}^*(F_N) = B_0 \sup_{\widetilde{\mathsf{P}} \in \mathcal{M}(\mathsf{P})} \mathsf{E}_{\widetilde{\mathsf{P}}} \frac{F_N}{B_N}, \tag{10.23}$$

where $\mathcal{M}(\mathsf{P})$ is the set of martingale measures $\widetilde{\mathsf{P}}$ which are equivalent to the initial measure P.

Corollary 10.1. *In the case of complete arbitrage-free markets the set $\mathcal{M}(\mathsf{P})$ consists of exactly one measure $\widetilde{\mathsf{P}}$. The theorem above implies that $\mathbb{C}_*(F_N) = \mathbb{C}^*(F_N)$.* (This common quantity denoted by $\mathbb{C}(F_N)$, is called the *rational* (*fair*) price. For the concrete calculations of these prices in the CRR-model see [161; Chap. V, § 1c].)

9. The proof of Theorem 10.3 will be given only for the complete arbitrage-free (B, S)-market which consist of a bank account $(B_n)_{n \le N}$ and a stock $S = (S_n)_{n \le N}$. (In the case $S = (S^1, \dots, S^d)$, where $d < \infty$, the proof is analogous.)

Since we assumed that the considered market is arbitrage free and complete,

(a) there exists a martingale measure $\widetilde{\mathsf{P}}$ equivalent to the martingale measure P ($\widetilde{\mathsf{P}} \sim \mathsf{P}$) such that the sequence $S/B = (S_n/B_n)_{n \le N}$ is a martingale (Theorem 10.1);

(b) this martingale measure $\widetilde{\mathsf{P}}$ is a unique martingale measure in the class $\mathcal{P}(\mathsf{P})$ of measure which are equivalent to the measure P (Theorem 10.2).

The statements (a) and (b) imply that if a portfolio $\pi^* = (\beta^*, \gamma^*)$ is such that $X_0^{\pi^*} = x^*$ and $X_N^{\pi^*} = F_N$ (P-a.s.) then, in view of (10.5*),

$$\Delta\left(\frac{X_n^{\pi^*}}{B_n}\right) = \gamma_n^* \Delta\left(\frac{S_n}{B_n}\right),$$

and thus

$$\frac{X_N^{\pi^*}}{B_N} = \frac{X_0^{\pi^*}}{B_0} + \sum_{n=1}^{N} \gamma_n^* \Delta\left(\frac{S_n}{B_n}\right). \tag{10.24}$$

Taking expectation $\mathsf{E}_{\widetilde{\mathsf{P}}}$ of both sides, we find that

$$\mathsf{E}_{\widetilde{\mathsf{P}}} \frac{X_N^{\pi^*}}{B_N} = \frac{X_0^{\pi^*}}{B_0}.$$

Since $X_N^{\pi^*} = F_N$ and $X_0^{\pi^*} = x^*$, we have

$$B_0 \mathsf{E}_{\widetilde{\mathsf{P}}} \frac{F_N}{B_N} = x^*. \tag{10.25}$$

Observe that the left-hand side of (10.25) does not depend on the structure of the hedge (x^*, π^*). In other words, if π° is another portfolio such that $X_0^{\pi^\circ} = x^\circ$ and $X_N^{\pi^\circ} = F_N$ (P-a.s.) then again

$$B_0 \mathsf{E}_{\widetilde{\mathsf{P}}} \frac{F_N}{B_N} = x^\circ$$

and thus $x^\circ = x^*$.

Consequently, the price of the so-called *perfect hedge* defined by

$$\mathbb{C}(F_N) = \inf\{x : \text{there exists } \pi \text{ such that} X_0^\pi = x \text{ and } X_N^\pi = F_N \ (\mathsf{P}\text{-a.s.})\}$$

(it is clear that $\mathbb{C}(F_N) = \mathbb{C}_*(F_N) = \mathbb{C}^*(F_N)$) is determined by the formula

$$\boxed{\mathbb{C}(F_N) = \mathsf{E}_{\widetilde{\mathsf{P}}} F_N}, \tag{10.26}$$

where $F_N = F_N(\omega)$ is a nonnegative $\mathcal{F}_N$-measurable claim.

10. The formula (10.26)— which is called a *basic formula* for the price of perfect hedging of European options on the arbitrage-free complete markets— answers the question about the fair price of an option with the pay-off function $F_N = F_N(\omega)$. But this does not cover all problems here, since one need also to know how to find a portfolio π^* of perfect hedging for which $X_0^{\pi^*} = \mathbb{C}(F_N)$ and $X_N^\pi = F_N$ (P-a.s.).

Now we describe a standard trick which enables one to construct this hedge.

Consider a martingale $M = (M_n, \mathcal{F}_n, \widetilde{\mathsf{P}}_n)_{n \le N}$ with $M_n = \mathsf{E}_{\widetilde{\mathsf{P}}}(F_N/B_N \,|\, \mathcal{F}_n)$. The market (B, S) is assumed to be arbitrage free and complete. Being so, by the second fundamental theorem 10.2 for M the following $\frac{S}{B}$*-representation* holds:

$$M_n = M_0 + \sum_{k=1}^{n} \gamma_k \Delta\left(\frac{S_k}{B_k}\right), \tag{10.27}$$

where the γ_k are $\mathcal{F}_{k-1}$-measurable.

Compose a portfolio $\pi^* = (\beta^*, \gamma^*)$ with $\gamma^* = \gamma$, where γ can be found from (10.27), and with $\beta^* = (\beta_n^*)_{n \le N}$, where

$$\beta_n^* = M_n - \frac{\gamma_n S_n}{B_n}.$$

It is easy to see that the portfolio $\pi^* = (\beta^*, \gamma^*)$ is self-financing.

For this portfolio

$$\frac{X_0^{\pi^*}}{B_0} = M_0$$

and

$$\Delta\left(\frac{X_n^{\pi^*}}{B_n}\right) = \gamma_n^* \Delta\left(\frac{S_n}{B_n}\right) = \gamma_n \Delta\left(\frac{S_n}{B_n}\right) = \Delta M_n.$$

Thus for all $0 \le n \le N$

$$\frac{X_n^{\pi^*}}{B_n} = M_n = \mathsf{E}_{\widetilde{\mathsf{P}}}\left(\frac{F_N}{B_N} \,\Big|\, \mathcal{F}_n\right)$$

and therefore ($\widetilde{\mathsf{P}}$-a.s. and P-a.s.)

$$X_N^{\pi^*} = F_N,$$

i.e., the portfolio π^* is a perfect hedge.

Summarize the above results in the following theorem which gives main formulae for both a perfect hedge and its capital.

Theorem 10.4. *Let (B, S) be an arbitrage free complete market and let $F_N = F_N(\omega)$ be a nonnegative pay-off function. For the European option with this pay-off function,*

(a) *the fair price $\mathbb{C} = \mathbb{C}(F_N)$ is given by*

$$\mathbb{C} = B_0 \mathsf{E}_{\widetilde{\mathsf{P}}} \frac{F_N}{B_N}; \tag{10.28}$$

(b) *the portfolio* $\pi^* = (\beta^*, \gamma^*)$ *described above reproduces* F_N *perfectly:*

$$X_N^{\pi^*} = F_N \quad (\mathsf{P}\text{-}a.s.); \tag{10.29}$$

(c) *the values* γ_k^*, $k \le N$, *can be found from the* $\frac{S}{B}$*-representation:*

$$\mathsf{E}_{\widetilde{\mathsf{P}}}\left(\frac{F_N}{B_N}\,\Big|\,\mathcal{F}_n\right) = \mathsf{E}_{\widetilde{\mathsf{P}}}\frac{F_N}{B_N} + \sum_{k=1}^{n} \gamma_k^* \Delta\left(\frac{S_k}{B_k}\right); \tag{10.30}$$

(d) *the values* β_k^*, $k \le N$, *are determined by the formulae*

$$\beta_k^* = \frac{X_k^{\pi^*} - \gamma_j^* S_k}{B_k}; \tag{10.31}$$

(e) *the capital*

$$X_n^{\pi^*} = B_n \mathsf{E}_{\widetilde{\mathsf{P}}}\left(\frac{F_N}{B_N}\,\Big|\,\mathcal{F}_n\right) \qquad (\mathsf{P}\text{- } and\ \widetilde{\mathsf{P}}\text{-}a.s.).$$

10.2 Basic Notions and Summary of Results of the Theory of Arbitrage. II. Continuous-Time Models

1. By analogy with the discrete-time case we assume that we are given a filtered probability space $(\Omega, \mathcal{F}, (\mathcal{F}_t)_{t\ge 0}, \mathsf{P})$ which satisfies the "standard" conditions (see Sec. 3.1).

On this filtered probability space consider a set $\mathcal{A}$ of financial assets which consists (just as in the discrete-time case) of a bank account $B = (B_t)_{t\ge 0}$ and a finite number d of stocks $S = (S_t^1, \dots, S_t^d)_{t\ge 0}$.

If we choose the bank account as a "unit of measurement" then it is clear that one is interested not in values S in themselves but in their relative values, *i.e.*, in discounted prices

$$\frac{S}{B} = \left(\frac{S_t^1}{B_t}, \dots, \frac{S_t^d}{B_t}\right)_{t\ge 0}.$$

We denote these discounted prices by $X = (X_t^1, \dots, X_t^d)_{t\ge 0}$.

One of the key topics in both theory of mathematical finance and practice of financial trading ("technical analysis") consists in realizing which processes simulate the prices X. We have already cited different models (such as CRR, $\mathbb{N} \circ$ IG, $\mathbb{N} \circ \mathbb{G}$IG (= $\mathbb{G}$H), $\mathbb{N} \circ$ H$^+$ ($\mathbb{H}$), $\mathbb{N} \circ$ Gamma ($\mathbb{V}$G), CGMY, CIR, ...) which are aimed to "catch" one or another property exhibited by real prices (see, *e.g.*, Sec. 9.1 or Sec. 12.1 and Sec. 12.4).

When choosing a model for X, one should, of course, be aware of for which models the theoretical apparatus is developed that will allow the effective studying of the financial problems.

In Sec. 10.1 we considered a model of an arbitrary process $X = (X_n)_{n\geq 0}$ with discrete time $n \geq 0$, which can naturally be embedded in the continuous-time model where the process $\widetilde{X} = (\widetilde{X}_t)_{t\geq 0}$ with piecewise-constant trajectories is such that

$$\widetilde{X}_t = X_n \quad \text{for} \quad n \leq t < n+1. \tag{10.32}$$

A natural next step in constructing models of the processes $X = (X_t)_{t\geq 0}$ is to address the so-called processes with *discrete intervention of chance* which, by definition, have the following structure:

$$X_t = \sum_{k\geq 1} x_k I(\tau_k \leq t), \tag{10.33}$$

where $\tau_1, \tau_2, \dots$ $(\tau_1 < \tau_2 < \cdots)$ are times when the process X jumps and the x_k are the sizes of its jumps $(\Delta X_{\tau_k} \equiv X_{\tau_k} - X_{\tau_k -} = x_k)$.

Processes of this type are very attractive from both theoretical and practical points of view. Among them the first to cite are, for example, $X_t = \alpha N_t^{(1)} - \beta N_t^{(2)}$, where $N^{(1)} = (N^{(1)})_{t\geq 0}$ and $N^{(2)} = (N^{(2)})_{t\geq 0}$ are two independent Poisson processes, considered in Sec. 7.1 (Example 7.1), or two independent "telegraph" signals (see [125; § 15.4]). The "arbitrage theory" and "pricing theory" for such processes are very similar to those discussed for the discrete time in Sec. 10.1 and Sec. 11.1.

2. Below the main attention will be paid to processes X which are semimartingales (see Chap. 3 for the underlying theoretical results).

The extensive use of semimartingales in mathematical finance is justified by a variety of reasons. The main among them are that, firstly, this class is wide enough and, secondly, for it the theory of stochastic integration is well developed which suits fine for the construction of arbitrage theory.

Let us pass to the necessary definitions restricting ourselves to the case $d = 1$ (*i.e.*, the case of a *single* stock).

Remark 10.2. Notice that when $d = 1$, the theorem on stochastic integration stated in Chap. 3 suffices to formulate basic results of the arbitrage theory, whereas in the case $d > 1$ we have to engage the theory of vector stochastic integration (see the paper [44] devoted to such theory and its applications to the theory of arbitrage).

Slightly modifying the notation used in Sec. 10.1 we shall say that the pair $\pi = (x, H)$, where $H = (H_t)_{t\geq 0}$ is a predictable process (see Sec. 3.1) and $x \in \mathbb{R}$, forms the *strategy* (of trader or investor), if the stochastic integrals

$$(H \cdot X)_t = \int_0^t H_s \, dX_s, \qquad t \geq 0, \tag{10.34}$$

are well defined (see Sec. 3.3).

The variable

$$G_t^\pi = \int_0^t H_s \, dX_s \tag{10.35}$$

will be referred to as the *return* of the strategy π over time t and the variable

$$X_t^\pi = x + G_t^\pi \tag{10.36}$$

will be referred to as the *capital* at the time t.

The following definition is closely related to Definition 10.1 given for the discrete-time case (see Sec. 10.1). In the definition below some technical conditions appear as an effect of the continuity of time (which provides "too many possibilities" for trading that can result in strategies with economically unacceptable features).

Definition 10.5. One says that the strategy $\pi = (x, H)$ *realizes the arbitrage* if

(a) $x = 0$;
(b) there exists a constant $b \geq 0$ such that $\mathsf{P}\big(X_t^\pi \geq -b$ for all $t \geq 0\big) = 1$;
(c) the limit $X_\infty^\pi \equiv \lim_{t\to\infty} X_t^\pi$ exists P-a.s.;
(d) $X_\infty^\pi \geq 0$ (P-a.s.);
(e) $\mathsf{P}(X_\infty^\pi > 0) > 0$.

The situation when such a strategy does not exist is identified as "absence of arbitrage" (NA, No Arbitrage).

Note that the admissibility condition (*i.e.*, $\mathsf{P}\big(X_t^\pi \geq -b$ for all $t \geq 0\big) = 1$ for some constant b) introduced in (b) is essential. Without it, examples are easy to construct when arbitrage exists. For instance, if X is a Brownian motion $B = (B_t)_{t\geq 0}$ with $B_0 = x = 0$, and $H_t = I(t \leq \tau_a)$, where $\tau_a = \inf\{t : B_t = a\}$, $a > 0$, then $\mathsf{P}(\tau_a < \infty) = 1$ and $X_\infty = B_{\tau_a} = a$ (P-a.s.). Consequently, there exist arbitrage opportunities. (Observe that if, from the very beginning, we study the arbitrage problems on a finite interval $[0, T]$ (with X_∞^π replaced by X_T^π), then for any finite stopping time τ ($\tau \leq T$) and a Brownian motion B we have $\mathsf{E}B_\tau = 0$.

3. In the discrete-time case the first fundamental theorem (Theorem 10.1 in Sec. 10.1) stated that on a finite time interval for a finite number of stocks

$$\boxed{\text{NA} \Leftrightarrow \mathcal{M}(\mathsf{P}) \neq \varnothing}, \tag{10.37}$$

where $\mathcal{M}(\mathsf{P}) = \{\mathsf{Q} \sim \mathsf{P} : X \text{ is a } \mathsf{Q}\text{-martingale}\}$.

In other words, the necessary and sufficient condition for No Arbitrage is the existence of an equivalent martingale measure. Unfortunately, in the continuous-time case the statement of type (10.37) is generally not valid. The main difficulty is that the NA condition does not implies existence of a martingale (or even local martingale) measure. Therefore to get the equivalence of type (10.37) we have to replace the NA condition by stronger ones.

Owing to works by Delbaen and Schachermayer (see their book [56] which both provides "a guided tour to arbitrage theory" and reviews their original papers) it became clear what is the analog of (10.37) in the continuous-time case.

To formulate their results we need the following definition.

Definition 10.6. A sequence $\pi_k = (x_k, H_k)$, $k \geq 1$, is said to realize *free lunch with vanishing risk* (FLVR) if for all $k \geq 1$

(a) $x_k = 0$;
(b) there exists a constant $b_k > 0$ such that $\mathsf{P}(X_t^{\pi_k} \geq -b_k$ for all $t \geq 0) = 1$;
(c) the limit $X_\infty^{\pi_k} = \lim_{t\to\infty} X_t^{\pi_k}$ exists P-a.s.;
(d) $X_\infty^{\pi_k} \geq -\frac{1}{k}$ and there exist constants $\delta_1 > 0$, $\delta_2 > 0$ (which do not depend on k) such that

$$\mathsf{P}(X_\infty^{\pi_k} > \delta_1) > \delta_2.$$

If such a sequence π_k, $k \geq 1$, does not exist one says that the considered model meets the NFLVR (*no free lunch with vanishing risk*) condition.

4. Delbaen and Schachermayer consider one more condition, namely, FLBR (*free lunch with bounded risk*) but we restrict ourselves to the formulation of the first fundamental theorem in continuous-time case under NA and NFLVR.

Theorem 10.5. (The first fundamental theorem of the arbitrage theory; continuous time; case of locally bounded processes.) *Assume that the process $X = (X_t)_{t\geq 0}$ is locally bounded* (i.e., *there exists a* *sequence of stopping times* $(\tau_n)_{n\geq 1}$ *such that* $\tau_n \uparrow \infty$ *a.s. and* $|X_{t\wedge\tau_n}| \leq n$ *for all* $n \geq 1$).

Let

$$\mathrm{L}\mathcal{M}(\mathsf{P}) = \{\mathsf{Q} \sim \mathsf{P} : X \text{ is a } \mathsf{Q}\text{-local martingale}\}.$$

The necessary and sufficient condition for the absence of arbitrage in the NFLVR *sense is the nonemptiness of the set* $\mathrm{L}\mathcal{M}(\mathsf{P})$:

$$\boxed{\text{NFLVR} \iff \mathrm{L}\mathcal{M}(\mathsf{P}) \neq \varnothing}, \tag{10.38}$$

and the following implications hold:

$$\text{NA} \impliedby \text{NFLVR} \iff \mathrm{L}\mathcal{M}(\mathsf{P}) \neq \varnothing \impliedby \mathcal{M}(\mathsf{P}) \neq \varnothing. \tag{10.39}$$

Remark 10.3. The assumption of local boundedness should not be thought of as a condition of technical character. As found by Delbaen and Schachermayer, the NFLVR condition by itself (without this local boundedness assumption) does not guarantee the existence of an equivalent local martingale measure.

To get free of this assumption we have to replace the class of local martingale by a larger class of so-called σ-martingales which are defined as follows.

Definition 10.7. A semimartingale $X = (X_t)_{t\geq 0}$ given on a stochastic basis $(\Omega, \mathcal{F}, (\mathcal{F}_t)_{t\geq 0}, \mathsf{P})$ is called a *σ-martingale* if there exists a sequence of predictable sets $D_n \subseteq \Omega \times \mathbb{R}_+$ such that

$$D_n \subseteq D_{n+1}, \quad \bigcup D_n = \Omega \times \mathbb{R}_+,$$

and for every $n \geq 1$ the process

$$X_t^{D_n}(\omega) = \int_0^t I_{D_n}(\omega, s)\, dX_s(\omega) \tag{10.40}$$

is a uniformly integrable martingale.

This definition, proposed by Goll and Kallsen (see [86], [85]), is equivalent to the following one.

Definition 10.8. A semimartingale $X = (X_t)_{t\geq 0}$ given on a stochastic basis $(\Omega, \mathcal{F}, (\mathcal{F}_t)_{t\geq 0}, \mathsf{P})$ is called a *σ-martingale* if there exist a local martingale $M = (M_t)_{t\geq 0}$ and an integrable (with respect to M) predictable process $H = (H_t)_{t\geq 0}$ such that

$$X_t = X_0 + (H \cdot M)_t.$$

Let

$$\sigma\mathcal{M}(\mathsf{P}) = \{\mathsf{Q} \sim \mathsf{P} : X \text{ is a } \sigma\text{-martingale with respect to } \mathsf{Q}\}.$$

Theorem 10.6. (The first fundamental theorem of the arbitrage theory; continuous time; general case.) 1) *The necessary and sufficient condition for the absence of arbitrage in the* NFLVR *sense is the nonemptiness of the set* $\sigma\mathcal{M}(\mathsf{P})$:

$$\text{NFLVR} \iff \sigma\mathcal{M}(\mathsf{P}) \neq \varnothing. \tag{10.41}$$

2) *Let the process* X *be nonnegative* ($X \geq 0$). *Then the necessary and sufficient condition for the absence of arbitrage in the* NFLVR *sense is the nonemptiness of the set* $\mathrm{L}\mathcal{M}(\mathsf{P})$:

$$\text{NFLVR} \iff \mathrm{L}\mathcal{M}(\mathsf{P}) \neq \varnothing. \tag{10.42}$$

Notice that evidently

$$\mathrm{L}\mathcal{M}(\mathsf{P}) \neq \varnothing \implies \sigma\mathcal{M}(\mathsf{P}) \neq \varnothing.$$

One can construct an example (see [44; Example 5.3]) that the reverse is not true, *i.e.*,

$$\sigma\mathcal{M}(\mathsf{P}) \neq \varnothing \not\Rightarrow \mathrm{L}\mathcal{M}(\mathsf{P}) \neq \varnothing.$$

5. Similarly to the discrete-time case consider the problem of completeness of a market.

Definition 10.9. A model of a market described by the semimartingale prices $X = (X_t)_{t\geq 0}$ is *complete* if for any bounded $\mathcal{F}$-measurable random variable $F = F(\omega)$ one can find a strategy π such that

1) for some constants a and b

$$\mathsf{P}\big(a \leq X_t^\pi \leq b \text{ for all } t \geq 0\big) = 1;$$

2) the limit $X_\infty^\pi = \lim_{t\to\infty} X_t^\pi$ exists P-a.s.;
3) $X_\infty^\pi = F$ P-a.s.

To formulate the main theorem ("second fundamental theorem") we need the following definition.

Definition 10.10. Let $X = (X_t)_{t\geq 0}$ be a semimartingale given on a filtered probability space $(\Omega, \mathcal{F}, (\mathcal{F}_t)_{t\geq 0}, \mathsf{Q})$. One says that a local martingale $M = (M_t)_{t\geq 0}$ given on this probability space *admits* X*-representation* if one can find an integrable predictable process $H = (H_t)_{t\geq 0}$ such that for all $t \geq 0$

$$M_t = M_0 + \int_0^t H_s\, dX_s \qquad (\mathsf{P}\text{-a.s.}) \tag{10.43}$$

(in a short form: $M = M_0 + H \cdot X$).

Theorem 10.7. (The second fundamental theorem of the arbitrage theory; continuous time; general case.)

(a) *Assume that the family of σ-martingale measures is nonempty* (i.e., $\sigma\mathcal{M}(\mathsf{P}) \neq \varnothing$). *Then the following conditions are equivalent*:

1) *the model described by the semimartingale* $X = (X_t)_{t\geq 0}$ *is complete*;
2) *the set* $\sigma\mathcal{M}(\mathsf{P})$ *consists of a single measure* ($|\sigma\mathcal{M}(\mathsf{P})| = 1$);
3) *in the set* $\sigma\mathcal{M}(\mathsf{P})$ *there exists a measure* Q *such that any local martingale* $M = (M_t)_{t\geq 0}$ *admits (with respect to this measure) the* X*-representation.*

(b) *Assume the semimartingale* X *is nonnegative and* ($\mathrm{L}\mathcal{M}(\mathsf{P}) \neq \varnothing$). *Then the following conditions are equivalent*:

1) *the model described by the semimartingale* $X = (X_t)_{t\geq 0}$ *is complete*;
2) *the set* $\mathrm{L}\mathcal{M}(\mathsf{P})$ *consists of a single measure* ($|\mathrm{L}\mathcal{M}(\mathsf{P})| = 1$);
3) *in the set* $\mathrm{L}\mathcal{M}(\mathsf{P})$ *there exists a measure* Q *such that any local martingale* $M = (M_t)_{t\geq 0}$ *admits (with respect to this measure) the* X*-representation.*

The item (a) can be stated schematically as

$$\text{completeness} \iff |\sigma\mathcal{M}(\mathsf{P})| = 1 \iff X\text{-representation holds.} \quad (10.44)$$

The item (b) can be formulated as

$$\text{completeness} \iff |\mathrm{L}\mathcal{M}(\mathsf{P})| = 1 \iff X\text{-representation holds.} \quad (10.45)$$

For the proof of the statements of the theorem see [44]. The part "completeness $\Leftarrow$ $|\sigma\mathcal{M}(\mathsf{P})| = 1$" is proved in [56]. Notice also that the implication "completeness $\Leftarrow$ $|\mathrm{L}\mathcal{M}(\mathsf{P})| = 1$" follows from [98; (11.2)] and [3].

6. From the first and second fundamental theorems stated above one can draw some conclusions which are important from the point of view of their applications to the real assets markets. (Note here that we dealt only with one-dimensional processes X. The case of multidimensional processes $X = (X^1, \dots, X^d)$ requires no changes except that the stochastic integral $H \cdot X$ should be understood as a *vector* stochastic integral, see details in [44].)

(i) If the process X is *continuous* (and thus locally bounded) then

$$\mathrm{NA} \Longleftarrow \mathrm{NFLVR} \iff \mathrm{L}\mathcal{M}(\mathsf{P}) \neq \varnothing.$$

(ii) If the process X is *nonnegative* then

$$\text{NA} \Longleftarrow \text{NFLVR} \Longleftrightarrow \sigma\mathcal{M}(\mathsf{P}) \neq \varnothing \Longleftarrow \text{L}\mathcal{M}(\mathsf{P}) \neq \varnothing.$$

Thus in the case of continuous or nonnegative semimartingales X, the existence of at least one local martingale measure ensures NFLVR and NA.

(iii) In the case of nonnegative semimartingales we have

$$|\,\text{L}\mathcal{M}(\mathsf{P})| = 1 \Longrightarrow \text{completeness.}$$

In other words, the existence of exactly one local martingale measure ensures completeness of the model (in the sense of Definition 10.9).

7. Similarly to the discrete-time case the first and second fundamental theorems are important for specifying both the set of mutually acceptable prices, say in option contracts, and the structure of hedging portfolios.

We formulate an analog of Theorem 10.4, assuming that, firstly, the considered (B, S)-market, where $B = (B_t)_{t\le T}$ and $S = (S_t)_{t\le T}$, is arbitrage free (meets the NA-condition) and, secondly, the class $\text{L}\mathcal{M}(\mathsf{P})$ consists of exactly one measure (which is a martingale measure) and thus the (B, S)-market is complete. [For both models (Bachelier's and Black–Scholes–Merton's) considered below in Chap. 11 these conditions are satisfied.]

Theorem 10.8. *Under the assumptions above*:

(a) *the rational (fair, mutually acceptable) price $\mathbb{C}_T$ of an option with nonnegative and $\mathcal{F}_T$-measurable pay-off function $F_T = F_T(\omega)$ is determined by the formula*

$$\mathbb{C}_T = B_0 \mathsf{E}_{\widetilde{\mathsf{P}}} \frac{F_T}{B_T};$$

(b) *there exists a portfolio $\pi^* = (\beta^*, \gamma^*)$ for which $X_T^{\pi^*} = F_T$ (a perfect hedge, replication)*;

(c) *the values γ_t^*, $0 \le t \le T$, can be found from the S/B-representation*

$$\mathsf{E}_{\widetilde{\mathsf{P}}}\left(\frac{F_T}{B_T}\,\Big|\,\mathcal{F}_t\right) = \mathsf{E}_{\widetilde{\mathsf{P}}}\frac{F_T}{B_T} + \int_0^t \gamma_u^*\, d\Big(\frac{S_u}{B_u}\Big), \qquad t \le T;$$

(d) *the values β_t^*, $0 \le t \le T$, are determined by the formula*

$$\beta_t^* = \frac{X_t^{\pi^*} - \gamma_t^* S_t}{B_t};$$

(e) *the capital*

$$X_t^{\pi^*} = B_t \mathsf{E}_{\widetilde{\mathsf{P}}}\left(\frac{F_T}{B_T}\,\Big|\,\mathcal{F}_t\right) \qquad (\mathsf{P}\text{- } \textit{and } \widetilde{\mathsf{P}}\text{-}\textit{a.s.}).$$

10.3 Arbitrage in a Model of Buying/Selling Assets with Transaction Costs

1. Imagine that one has a bank account which allows him/her to deposit and withdraw money. Let B_0 be the state of this account at time $n = 0$. Assume there exists a stock free to buy or to sell. Let $S^a = (S^a_n)_{0\le n\le N}$ and $S^b = (S^b_n)_{0\le n\le N}$ be the prices of buying (ask price) and selling (bid price) of a single stock. Naturally, the market reality is that $S^a_n \ge S^b_n$. It is often convenient to think that there exists a "basic" price $S = (S_n)_{0\le n\le N}$ such that

$$S^a_n = (1+\lambda)S_n, \qquad \lambda \ge 0, \tag{10.46}$$

and

$$S^b_n = (1-\mu)S_n, \qquad \mu \ge 0. \tag{10.47}$$

Imagine that at time $n = 0$, being aware of prices S^a_0 and S^b_0 one decided to buy or sell H_0 stocks. When buying, one just takes $H_0 S^a_0$ units from the bank account leaving it equal to $B_1 = B_0 - H_0 S^a_0$. If one sells stocks (one can own or borrow them—the second situation is named *short selling*), the bank account becomes equal to $B_1 = B_0 + H_0 S^b_0$. At time $n = 1$ one can both buy and sell stocks (at prices S^a_1 and S^b_1, respectively) and so on.

Thus, at time N, on conditions that one has no more stocks, the amount on the bank account will be

$$\mathbb{G}_N = \sum_{n=0}^{N} \Big[-H_n I(H_n > 0) S^a_n - H_n I(H_n < 0) S^b_n\Big], \tag{10.48}$$

where H_n's are $\mathcal{F}_n$-measurable and $\sum_{n=0}^N H_n = 0$. [The latter condition $\sum_{n=0}^N H_n = 0$ means that at the terminal time N one has no stocks. The condition of $\mathcal{F}_n$-measurability for H_n implies that at time n one knows the prices S^a_n and S^b_n (as well as all preceding prices) and takes decision relying on this information.]

Observe that if $S^a_n = S^b_n = S_n$ for all $0 \le n \le N$, then $\mathbb{G}_N$ admits the following usual representation:

$$\mathbb{G}_N = \sum_{n=1}^{N} \gamma_n (S_n - S_{n-1}), \tag{10.49}$$

where

$$\begin{aligned} \gamma_1 &= H_0, \\ \gamma_2 &= \gamma_1 + H_1, \\ &\dots\dots\dots\dots\dots\dots \\ \gamma_{N-1} &= \gamma_{N-2} + H_{N-2}, \end{aligned} \tag{10.50}$$

and, in view of condition $\sum_{n=0}^{N} H_n = 0$,

$$\gamma_n = H_0 + H_1 + \cdots + H_{N-1}.$$

Since we assumed H_n's to be $\mathcal{F}_n$-measurable, the variables γ_n are $\mathcal{F}_{n-1}$-measurable (just as in the previous sections when we constructed the portfolio).

2. Let's give the following definition.

Definition 10.11. The model of buying/selling stocks considered above is said to be *arbitrage free* if the set

$$\mathcal{G}_N = \Big\{ \mathbb{G}_N : \ \mathbb{G}_N = \sum_{n=0}^{N} \big[-H_n I(H_n > 0) S_n^a - H_n I(H_n < 0) S_n^b \big], \quad H_n\text{'s are } \mathcal{F}_n\text{-measurable and } \sum_{n=0}^{N} H_n = 0 \Big\} \tag{10.51}$$

is such that

$$\mathcal{G}_N \cap L_+^0 = \{0\}, \tag{10.52}$$

where L_+^0 is the set of nonnegative random variables given on the initial probability space $(\Omega, \mathcal{F}, \mathsf{P})$.

Theorem 10.9. *The above-described buying/selling model driven by the strategy $\pi = (H_0, H_1, \ldots, H_N)$ with the income specified by (10.48) is arbitrage-free if and only if there exist a measure Q equivalent to the measure P ($\mathsf{Q} \sim \mathsf{P}$) and a Q-martingale $M = (M_n, \mathcal{F}_n, \mathsf{Q})_{0 \le n \le N}$ such that*

$$S^a \le M \le S^b \qquad \mathsf{Q}\text{-}a.s. \tag{10.53}$$

Remark 10.4. To a variable degree of generalization, this theorem in such a form and its proof can be found in [89].

10.4 Asymptotic Arbitrage: Some Problems

1. Our previous considerations of the arbitrage theory relied on the assumption that there is a single (B, S)-market given on a filtered probability space. In the discrete-time case it was $(\Omega, \mathcal{F}, (\mathcal{F}_k)_{0 \le k \le N}, \mathsf{P})$, where $B = (B_k)_{0 \le k \le N}$ and $S = (S_k)_{0 \le k \le N}$ are bank account and stocks, respectively (more exactly, their monetary values). In the continuous-time case it was $(\Omega, \mathcal{F}, (\mathcal{F}_t)_{0 \le t \le T}, \mathsf{P})$ with the bank account $B = (B_t)_{0 \le t \le T}$ and stocks $S = (S_t)_{0 \le t \le T}$.

When formulating the fundamental theorems (Sec. 10.1 and Sec. 10.2) we have seen the importance to consider the probability measures Q which are absolutely continuous with respect or equivalent to the initial measure P. It is in terms of such measures that the first and second fundamental theorems were formulated and the formulae for rational prices were deduced.

2. There are many reasons for extending the notion of (B, S)-market introduced above, namely for considering the triangular array $(\mathbb{B}, \mathbb{S})$ ('scheme of series') of such markets.

More exactly, we assume now that for each $n \geq 1$ "its own" (B^n, S^n)-market is specified,

$$(B^n, S^n) = \big((B^k, S^k)\big)_{0 \leq k \leq k*(n)}, \quad \text{where} \quad S_k^n = \big(S_k^{n,1}, \dots, S_k^{n,d(n)}\big),$$

$$0 \leq k(n) < \infty, \quad 1 \leq d(n) < \infty,$$

on "its own" probability space

$$\big(\Omega^n, \mathcal{F}^n, (\mathcal{F}_k^n)_{0 \leq k \leq k(n)}, \mathsf{P}^n\big).$$

As usual, we assume that $\mathcal{F}_0^n = \{\varnothing, \Omega^n\}$ and $\mathcal{F}^n = \mathcal{F}_{k(n)}^n$.

The case $d(n) \uparrow \infty$ fits the situation when the number of traded stocks increases; the condition $k(n) \uparrow \infty$ would reflect the fact that the markets exist for a longer and longer period of time (these latter markets are referred to as 'large').

The main problem which we will consider now is about conditions for existence of the so-called asymptotic arbitrage whose definition is as follows.

Let $X^{\pi(n)} = \big(X_k^{\pi(n)}\big)_{k \leq k(n)}$ be the capital of a self-financing portfolio $\pi(n)$ on a (B^n, S^n)-market (see [161; Chap. VI, § 3]). As explained earlier, without loss of generality we can assume that $B_k^n \equiv 1$. Under this assumption

$$X_k^{\pi(n)} = X_0^{\pi(n)} + \sum_{l=1}^{k} (\gamma_l^n, \Delta S_l^n), \tag{10.54}$$

where $(\gamma_l^n, \Delta S_l^n) = \sum_{i=1}^{d(n)} \gamma_l^{n,i} \Delta S_l^{n,i}$ with evident notation γ_l^n and ΔS_l^n (see Sec. 10.1).

Definition 10.12. One says that the sequence of strategies $\pi = (\pi(n))_{n \geq 1}$ realizes the *asymptotic arbitrage in the scheme of series* $(\mathbb{B}, \mathbb{S}) = \{(B^n, S^n),\ n \geq 1\}$ if

(a) $\lim_n X_n^{\pi(n)} = 0$;

(b) $X^{\pi(n)}_{k(n)} \geq -c(n)$ (P^n-a.s.) for all $n \geq 1$, where $0 \leq c(n) \downarrow 0$ as $n \to \infty$;

(c) $\lim_{\varepsilon \downarrow 0} \limsup_n \mathsf{P}^n\big(X^{\pi(n)}_{k(n)} \geq \varepsilon\big) > 0$.

It is easy to catch the analogy between the notion of asymptotic arbitrage and that of NFLVR (see Definition 10.6 in Sec. 10.2). However there are some distinctions, which results from that we now deal with a scheme of series of markets.

3. Our aim is to establish "martingale" conditions for the absence of asymptotic arbitrage.

To this end it is useful to introduce the notion of contiguity of probability measures, which is a natural generalization of a notion of absolute continuity of probability measures.

Definition 10.13. Let $(E^n, \mathcal{E}^n)$, $n \geq 1$, be a sequence of measurable spaces endowed with measures Q^n and $\widetilde{\mathsf{Q}}^n$, $n \geq 0$. The sequence $(\widetilde{\mathsf{Q}}^n)_{n\geq 1}$ is said to be *contiguous* with respect to $(\mathsf{Q}^n)_{n\geq 1}$ (notation: $(\widetilde{\mathsf{Q}}^n) \triangleright (\mathsf{Q}^n)$) if for any sequence of sets A^n, $n \geq 1$, such that $\mathsf{Q}^n(A^n) \to 0$ as $n \to \infty$ we have that $\widetilde{\mathsf{Q}}^n(A^n) \to 0$ as $n \to \infty$.

In the asymptotical mathematical statistics this notion is well known, and there are various criteria which allow one to check whether the given sequences are contiguous. We cite several of them.

Let

$$\overline{\mathsf{Q}}^n = \frac{1}{2}(\mathsf{Q}^n + \widetilde{\mathsf{Q}}^n) \quad \text{and} \quad \zeta^n = \frac{d\mathsf{Q}^n}{d\overline{\mathsf{Q}}^n}, \quad \widetilde{\zeta}^n = \frac{d\widetilde{\mathsf{Q}}^n}{d\overline{\mathsf{Q}}^n}.$$

Let also

$$Z^n = \frac{\widetilde{\zeta}^n}{\zeta^n}.$$

The variable Z^n is just the Radon–Nikodým derivative of the absolutely continuous component of $\widetilde{\mathsf{Q}}^n$ with respect to the measure Q^n, or, equivalently, Z^n is the *Lebesgue derivative* of the measure $\widetilde{\mathsf{Q}}^n$ with respect to the measure Q^n, which we denote by $\big[d\widetilde{\mathsf{Q}}^n/d\mathsf{Q}^n\big]_{\mathrm{Leb}}$. (Notice that $\frac{d\widetilde{\mathsf{Q}}}{d\mathsf{Q}}$ is the standard notation for the Radon–Nikodým derivative of the measure $\widetilde{\mathsf{Q}}$ with respect to the measure Q when $\widetilde{\mathsf{Q}} \ll \mathsf{Q}$.)

For studying the contiguity problems the following notion of the *Hellinger integral* $H(\alpha; \mathsf{Q}, \widetilde{\mathsf{Q}})$ of order $\alpha \in (0,1)$ can prove to be useful:

$$H(\alpha; \mathsf{Q}, \widetilde{\mathsf{Q}}) = \mathsf{E}_{\overline{\mathsf{Q}}}\zeta^{\alpha}\widetilde{\zeta}^{1-\alpha}, \quad \text{where } \overline{\mathsf{Q}} = \tfrac{1}{2}(\mathsf{Q} + \widetilde{\mathsf{Q}}), \tag{10.55}$$

or (in symbolic form)

$$H(\alpha; \mathrm{Q}, \widetilde{\mathrm{Q}}) = \int_E (d\mathrm{Q})^\alpha (d\widetilde{\mathrm{Q}})^{1-\alpha}. \tag{10.56}$$

The case $\alpha = 1/2$ is of a particular importance. Putting $H(\mathrm{Q}, \widetilde{\mathrm{Q}}) = H(\frac{1}{2}; \mathrm{Q}, \widetilde{\mathrm{Q}})$ we see that

$$H(\mathrm{Q}, \widetilde{\mathrm{Q}}) = \int_E \sqrt{d\mathrm{Q}\, d\widetilde{\mathrm{Q}}}. \tag{10.57}$$

The quantity $\rho = \rho(\mathrm{Q}, \widetilde{\mathrm{Q}})$ such that

$$\rho^2(\mathrm{Q}, \widetilde{\mathrm{Q}}) = \frac{1}{2} \int_E \left(\sqrt{d\mathrm{Q}} - \sqrt{d\widetilde{\mathrm{Q}}}\right)^2 \tag{10.58}$$

bears the name of the *Hellinger distance* between Q and $\widetilde{\mathrm{Q}}$. It is clear that $\rho(\mathrm{Q}, \widetilde{\mathrm{Q}}) = \sqrt{1 - H(\mathrm{Q}, \widetilde{\mathrm{Q}})}$.

Example 10.1. Let

$$\mathrm{Q} = \mathrm{Q}_1 \times \mathrm{Q}_2 \times \cdots, \qquad \widetilde{\mathrm{Q}} = \widetilde{\mathrm{Q}}_1 \times \widetilde{\mathrm{Q}}_2 \times \cdots,$$

where Q_k and $\widetilde{\mathrm{Q}}_k$ are the Gaussian measures on $(\mathbb{R}, \mathcal{B}(\mathbb{R}))$ with densities

$$q_k(x) = \frac{1}{\sqrt{2\pi}\sigma_k} \exp\left\{-\frac{(x-\mu_k)^2}{2\sigma_k^2}\right\}$$

and

$$\tilde{q}_k(x) = \frac{1}{\sqrt{2\pi}\sigma_k} \exp\left\{-\frac{(x-\tilde{\mu}_k)^2}{2\sigma_k^2}\right\}.$$

Then

$$H(\mathrm{Q}, \widetilde{\mathrm{Q}}) = \prod_{k=1}^{\infty} H(\mathrm{Q}_k, \widetilde{\mathrm{Q}}_k),$$

where

$$H(\mathrm{Q}_k, \widetilde{\mathrm{Q}}_k) = \int_{\mathbb{R}} q_k^\alpha(x) \tilde{q}_k^{1-\alpha}(x)\, dx = \exp\left\{-\frac{\alpha(1-\alpha)}{2} \left(\frac{\mu_k - \tilde{\mu}_k}{\sigma_k}\right)^2\right\}.$$

Therefore

$$H(\alpha; \mathrm{Q}, \widetilde{\mathrm{Q}}) = \exp\left\{-\frac{\alpha(1-\alpha)}{2} \sum_{k=1}^{\infty} \left(\frac{\mu_k - \tilde{\mu}_k}{\sigma_k}\right)^2\right\} \tag{10.59}$$

and

$$H(\mathrm{Q}, \widetilde{\mathrm{Q}}) = \exp\left\{-\frac{1}{8} \sum_{k=1}^{\infty} \left(\frac{\mu_k - \tilde{\mu}_k}{\sigma_k}\right)^2\right\}. \tag{10.60}$$

For the proof of the following lemma see [100; Chap. V, Lemma 1.6].

Lemma 10.2. *The following conditions are equivalent:*

(a) $(\widetilde{\mathsf{Q}}^n) \lhd (\mathsf{Q}^n)$;
(b) $\lim_{\varepsilon\downarrow 0} \limsup_n \widetilde{\mathsf{Q}}^n(\zeta^n < \varepsilon) = 0$;
(c) $\lim_{N\uparrow\infty} \limsup_n \widetilde{\mathsf{Q}}^n(Z^n > N) = 0$;
(d) $\lim_{\alpha\downarrow 0} \liminf_n H(\alpha; \mathsf{Q}^n, \widetilde{\mathsf{Q}}^n) = 1$.

4. Before we formulate conditions for absence of asymptotic arbitrage in the general case (see Theorem 10.11 on page 222) let us study the so-called stationary case, where all (B^n, S^n)-markets have the following structure:

$$(B^n, S^n) = ((B_k, S_k))_{k\le k(n)}, \qquad k(n) = n, \quad d(n) = d.$$

These two conditions on $k(n)$ and $d(n)$ imply that the market (B^{n+1}, S^{n+1}) acting right up to the time $n+1$ is the extension of the market (B^n, S^n) acting right up to the time n and the number d of stocks does not vary in time.

In the considered stationary case it is convenient to assume that a filtered probability space $(\Omega, \mathcal{F}, (\mathcal{F}_k)_{k\ge 1}, \mathsf{P})$ is given and for every $n \ge 1$ the (B^n, S^n)-market is defined on "its own" space $(\Omega, \mathcal{F}, (\mathcal{F}_k)_{k\le n}, \mathsf{P}_n)$, where $\mathsf{P}_n = \mathsf{P}|\mathcal{F}_n$.

Suppose that for each $n \ge 1$ the (B^n, S^n)-market is *arbitrage free*. The question which we are interested in is under what conditions there is no arbitrage as $n \to \infty$.

To formulate such conditions introduce the following notation:

$\mathcal{P}(\mathsf{P}_n)$ is the family of martingale measures for the (B^n, S^n)-market;

$$\mathbb{Z}_n = \left\{ Z_n : Z_n = \frac{d\widetilde{\mathsf{P}}_n}{d\mathsf{P}_n},\ \widetilde{\mathsf{P}}_n \in \mathcal{P}(\mathsf{P}_n) \right\};$$

$$\mathbb{Z}_\infty = \left\{ Z_\infty : Z_\infty = \limsup_n \frac{d\widetilde{\mathsf{P}}_n}{d\mathsf{P}_n},\ \widetilde{\mathsf{P}}_n \in \mathcal{P}(\mathsf{P}_n) \right\}.$$

Observe that for every sequence of consistent martingale measures $(\widetilde{\mathsf{P}}_n)_{n\ge 1}$ the sequence $(Z_n, \mathcal{F}_n)_{n\ge 1}$ is a nonnegative P-martingale and therefore with respect to the measure P the limit $\lim Z_n$ $(= Z_\infty)$ exists such that $0 \le \mathsf{E} Z_\infty \le 1$.

Theorem 10.10. *In the stationary case for the arbitrage-free (B^n, S^n)-markets, where $B^n = (B_k)_{k\le n}$, $S^n = (S_k)_{k\le n}$, $n \ge 1$, the condition*

$$\lim_{\varepsilon\downarrow 0} \limsup_n \inf_{Z_n \in \mathbb{Z}_n} \mathsf{P}(Z_n < \varepsilon) = 0 \tag{10.61}$$

is necessary and sufficient for the absence of asymptotic arbitrage in the sense of Definition 10.12 (*see page* 217).

The condition

$$\lim_{\varepsilon \downarrow 0} \inf_{Z_\infty \in \mathbb{Z}_\infty} \mathsf{P}(Z_\infty < \varepsilon) = 0 \tag{10.62}$$

is sufficient for the absence of asymptotic arbitrage.

We shall prove only the sufficiency part. For the necessity part see [161; Chap. VI, § 3c].

Let $(\widetilde{\mathsf{P}}_n)_{n\geq 1}$ be a sequence of martingale measures and let $\pi = (\pi(n))_{n\geq 1}$ be a sequence of strategies on the (B^n, S^n)-markets, $n \geq 1$. The sequence $(X_k^{\pi(n)})_{k\leq n}$ (because of its definition $X_k^{\pi(n)} = X_0^{\pi(n)} + \sum_{l=1}^k (\gamma_l^n, \Delta S_l^n)$) is a martingale transform and thus a local martingale; for the details see [161; Chap. II, § 1c, p. 98]. Then, by the lemma in the same § 1c of Chap. II [161] and Definition 10.12 (b), we find that

$$X_0^{\pi(n)} = \mathsf{E} Z_n X_n^{\pi(n)}.$$

Consequently, for all $\varepsilon > 0$

$$\begin{aligned}
\mathsf{E} Z_n X_n^{\pi(n)} &= \mathsf{E} Z_n X_n^{\pi(n)} \big[I\big(-c(n) \leq X_n^{\pi(n)} < 0\big) \\
&\qquad\qquad + I\big(0 \leq X_n^{\pi(n)} < \varepsilon\big) + I\big(X_n^{\pi(n)} \geq \varepsilon\big) \big] \\
&\geq -c(n) + \mathsf{E} Z_n X_n^{\pi(n)} I(Z_n \geq \varepsilon) I\big(X_n^{\pi(n)} \geq \varepsilon\big) \\
&\geq -c(n) + \varepsilon^2 \mathsf{P}\big(X_n^{\pi(n)} \geq \varepsilon,\ Z_n \geq \varepsilon\big) \\
&\geq -c(n) + \varepsilon^2 \big[\mathsf{P}\big(X_n^{\pi(n)} \geq \varepsilon\big) - \mathsf{P}(Z_n < \varepsilon)\big].
\end{aligned} \tag{10.63}$$

Therefore

$$X_0^{\pi(n)} + c(n) + \varepsilon^2 \mathsf{P}(Z_n < \varepsilon) \geq \varepsilon^2 \mathsf{P}\big(X_n^{\pi(n)} \geq \varepsilon\big). \tag{10.64}$$

If conditions (a) and (b) in the definition of the sequence $(\pi(n))_{n\geq 1}$ of strategies which realize the arbitrage (see Definition 10.12 on page 217) are fulfilled, then from (10.64) we obtain that

$$\lim_{\varepsilon \downarrow 0} \limsup_n \inf_{Z_n \in \mathbb{Z}_n} \mathsf{P}(Z_n < \varepsilon) \geq \lim_{\varepsilon \downarrow 0} \limsup_n \mathsf{P}\big(X_n^{\pi(n)} \geq \varepsilon\big). \tag{10.65}$$

This leads, in view of (10.61), to

$$\lim_{\varepsilon \downarrow 0} \limsup_n \mathsf{P}\big(X_n^{\pi(n)} \geq \varepsilon\big) = 0.$$

Hence under the assumption (10.61) condition (c) of Definition 10.12 necessarily fails.

Thus the condition (10.61) guarantees the absence of asymptotic arbitrage.

To establish the sufficiency of (10.62) for the absence of arbitrage, it is enough to observe that—again from (10.64)—it follows that for any sequence of consistent martingale measures $(\widetilde{\mathsf{P}}_n)_{n\geq 1}$

$$\mathsf{P}(Z_\infty > \varepsilon) \geq \limsup_n \mathsf{P}(Z_n < \varepsilon) \geq \limsup_n \mathsf{P}\big(X_n^{\pi(n)} \geq \varepsilon\big)$$

and therefore

$$\lim_{\varepsilon\downarrow 0} \inf_{Z_\infty \in \mathbb{Z}_\infty} \mathsf{P}(Z_\infty < \varepsilon) \geq \lim_{\varepsilon\downarrow 0} \limsup_n \mathsf{P}\big(X_n^{\pi(n)} \geq \varepsilon\big). \tag{10.66}$$

Thus, if (10.62) holds, then condition (c) of Definition 10.12 cannot be fulfilled, and, consequently, (10.62) ensures the absence of asymptotic arbitrage.

5. Now proceed to study of the general case (without the assumption of stationarity).

We shall consider the sequence $(\mathbb{B},\mathbb{S}) = \{(B^n,S^n),\ n\geq 1\}$ of large markets which for all $n\geq 1$ are arbitrage free. For every $n\geq 1$ the (B^n,S^n)-market is defined on the filtered probability space $(\Omega^n,\mathcal{F}^n,(\mathcal{F}_k^n)_{k\leq k(n)},\mathsf{P}^n)$, where $\mathcal{F}_{k(n)}^n = \mathcal{F}^n$.

Let $\widetilde{\mathsf{P}}_{k(n)}^n$ be a martingale measure and let $\widetilde{\mathsf{P}}_{k(n)}^n \sim \mathsf{P}_{k(n)}^n$ and $Z_{k(n)}^n = \frac{d\widetilde{\mathsf{P}}_{k(n)}^n}{d\mathsf{P}_{k(n)}^n}$. By analogy with (10.64) we find that

$$X_0^{\pi(n)} + c(n) + \varepsilon^2\mathsf{P}(Z_{k(n)}^n < \varepsilon) \geq \varepsilon^2\mathsf{P}\big(X_{k(n)}^{\pi(n)} \geq \varepsilon\big). \tag{10.67}$$

From this, just as in Theorem 10.10, we obtain the following theorem.

Theorem 10.11. *The condition*

$$\lim_{\varepsilon\downarrow 0}\limsup_n \inf_{Z_{k(n)}^n \in \mathbb{Z}_{k(n)}^n} \mathsf{P}^n(Z_{k(n)}^n < \varepsilon) = 0 \tag{10.68}$$

is sufficient (as well as necessary) for the absence of asymptotic arbitrage.

6. Let return to the stationary case. Let $(\widetilde{\mathsf{P}}_n)_{n\geq 1}$ be a sequence of martingale measures. According to Lemma 10.2,

$$\mathsf{P}(Z_\infty > 0) = 1 \iff (\mathsf{P}_n) \lhd (\widetilde{\mathsf{P}}_n), \tag{10.69}$$

where $Z_\infty = \lim \frac{d\widetilde{\mathsf{P}}_n}{d\mathsf{P}_n}$ (the limit exists P-a.s.) and $(\mathsf{P}_n) \lhd (\widetilde{\mathsf{P}}_n)$ means that the sequence of the measures $(\mathsf{P}_n)_{n\geq 1}$ is contiguous with respect to the sequence of the measures $(\widetilde{\mathsf{P}}_n)_{n\geq 1}$.

From Theorem 10.10 we conclude that if $\mathsf{P}(Z_\infty > 0) = 1$ then there is no asymptotic arbitrage. Together with (10.64) this leads to the following result.

Theorem 10.12. *In the stationary case the contiguity condition* $(\mathsf{P}_n) \triangleleft (\widetilde{\mathsf{P}}_n)$ *implies the absence of asymptotic arbitrage.*

7. Let us give an example of application of the last theorem.

Let $(\Omega, \mathcal{F}, \mathsf{P})$ be a probability space, where $\Omega = \{-1, 1\}$ is the space of binary sequences $x = (x_1, x_2, \dots)$ such that $x_i = \pm 1$ and $\mathsf{P}\{x : (x_1, x_2, \dots, x_n)\} = 2^{-n}$. If we put $\varepsilon_i(x) = x_i$ then, evidently, $\varepsilon = (\varepsilon_1, \varepsilon_2, \dots)$ is a Bernoulli sequence of independent and identically distributed random variables, $\mathsf{P}(\varepsilon_i = \pm 1) = 1/2$.

Now define a stationary sequence of (B^n, S^n)-markets.

Assume that for every $n \geq 1$ the (B^n, S^n)-market is defined on $(\Omega, \mathcal{F}^n, \mathsf{P}^n)$, where $\mathcal{F}^n = \mathcal{F}_n = \sigma(\varepsilon_1, \dots, \varepsilon_n)$ and $\mathsf{P}^n = \mathsf{P}|\mathcal{F}^n$. At that $B^n = (B^n_k)_{k \leq n}$, where $B^n_k \equiv 1$ and $S^n = (S_1, \dots, S_n)$ is defined recursively,

$$S_k = S_{k-1}(1 + \rho), \qquad S_0 = 1, \tag{10.70}$$

where $\rho_k = \mu_k + \sigma_k \varepsilon_k$, $\sigma_k > 0$, and $\max(-\sigma_k, \sigma_{k-1}) < \mu_k < \sigma_k$.

Rewrite (10.70) in the form

$$S_k = S_{k-1}(1 + \sigma_k(\varepsilon_k - b_k)), \tag{10.71}$$

where $b_k = -\mu_k / \sigma_k$. (Observe that $|b_k| < 1$.)

In the considered model for every $n \geq 1$ there exists a unique martingale measure $\widetilde{\mathsf{P}}^n$ (see [161; Chap. V, § 3f]), which is the direct product $\widetilde{\mathsf{P}}^n = \widetilde{\mathsf{P}}^n_1 \times \dots \times \widetilde{\mathsf{P}}^n_n$, and with respect to this measure the random variables $\varepsilon_1, \dots, \varepsilon_n$ are independent and

$$\widetilde{\mathsf{P}}^n(\varepsilon_k = 1) = \frac{1}{2}(1 + b_k), \qquad \widetilde{\mathsf{P}}^n(\varepsilon_k = -1) = \frac{1}{2}(1 - b_k).$$

It is easy to calculate that

$$H(\alpha; \widetilde{\mathsf{P}}, \mathsf{P}) = \prod_{k=1}^{n} \left[\frac{(1 + b_k)^\alpha + (1 - b_k)^\alpha}{2} \right].$$

Since by Lemma 10.2

$$(\mathsf{P}^n) \triangleleft (\widetilde{\mathsf{P}}^n) \quad \Longleftrightarrow \quad \lim_{\alpha \downarrow 0} \liminf_{n} H(\alpha; \widetilde{\mathsf{P}}^n, \mathsf{P}^n) = 1,$$

we find that in the considered case

$$(\mathsf{P}^n) \lhd (\widetilde{\mathsf{P}}^n) \iff \lim_{\alpha\downarrow 0} \liminf_n \prod_{k=1}^n \left[\frac{(1+b_k)^\alpha + (1-b_k)^\alpha}{2}\right] = 1$$

$$\iff \lim_{\alpha\downarrow 0} \limsup_n \prod_{k=1}^n \left[1 - \frac{(1+b_k)^\alpha + (1-b_k)^\alpha}{2}\right] = 0.$$

From this we deduce that

$$(\mathsf{P}^n) \lhd (\widetilde{\mathsf{P}}^n) \iff \sum_{k=1}^\infty b_k^2 < \infty.$$

Since $b_k = -\mu_k/\sigma_k$, we conclude by Theorem 10.12 that in the considered case the condition

$$\sum_{k=1}^\infty \left(\frac{\mu_k}{\sigma_k}\right)^2 < \infty$$

is sufficient (and necessary; see [161; Chap. VI, § 3c, Theorem 1]) for the absence of asymptotic arbitrage.

Chapter 11

Change of Measure in Option Pricing

11.1 Overview of the Pricing Formulae for European Options

1. The general theory of arbitrage, in particular the notions of hedging and rational price, constitute a solid mathematical and economical base for a detailed analysis of one of the most widely known and used derivative financial instruments, which are options.

Now we shall consider in greater detail two main types of options, namely European and American.

Generally, an *option* is a contract (derivative security) issued by a firm, a bank or another financial organization and giving its buyer the right to buy or sell, for example, shares (stocks), bonds, currency, *etc.*, on specified condition at a fixed instant (*European options*) or during a certain period of time in the future with the selection of time of exercise at an arbitrary (*e.g.*, random) instant (*American options*).

In this section we collected some classical results on European options, restricting ourselves to the so-called *standard call options* and *standard put options* (for other possibilities see, *e.g.*, [161], [132]).

2. As already explained in Chap. 10 (see Sec. 10.1, Subsec. 8 therein) a *standard call option of European type* with maturity time N (now we consider the discrete-time case) is characterized by the *strike* or *exercise price* K (fixed at the instant of writing the contract) at which the buyer will be able to buy, say, shares (stocks), whose market price at time N is S_N.

The situation $S_N > K$ is good for the buyer because, by the contract terms (for a standard call option), he/she has the right to buy shares at the lower price K ($< S_N$). Once bought, they can be immediately sold at the market price S_N. So, his/her gains from this operation will be equal to $S_N - K$.

On the other hand, if $S_N < K$ then the buyer will not purchase shares at the price K because of the possibility to buy them at a lower price S_N.

As a results of these rules, the buyer's gain F_N at time N can be expressed by the formula

$$F_N = (S_N - K)^+, \tag{11.1}$$

where $a^+ = a \vee 0$ ($= \max(a, 0)$).

The buyer must pay for the purchase of the contract some premium $\mathbb{C}_N$ so that his/her net profit from the call option is

$$(S_N - K)^+ - \mathbb{C}_N. \tag{11.2}$$

Respectively the writer's (*i.e.*, seller's) gain is

$$\mathbb{C}_N - (S_N - K)^+. \tag{11.3}$$

In the case of a standard European option with maturity date N the buyer has right to sell stock (at time N) at price K.

If the price of the contract is $\mathbb{P}_N$, then the buyer's net profit will be equal to

$$(K - S_N)^+ - \mathbb{P}_N. \tag{11.4}$$

The seller's net profit will be

$$\mathbb{P}_N - (K - S_N)^+. \tag{11.5}$$

This description shows that the European call options are characterized by

(i) their maturity date N;

(ii) the pay-off function F_N.

For a standard call option

$$F_N = (S_N - K)^+,$$

For a standard put option

$$F_N = (K - S_N)^+.$$

Now it is the right time to list other popular options:

- standard call option with aftereffect:

$$F_N = (S_N - K_N)^+,$$

where $K_N = c\min(S_0, S_1, \ldots, S_N)$, $c = \text{const}$;

- arithmetic Asian call option:

$$F_N = (\overline{S}_N - K)^+,$$

where $\overline{S}_N = \frac{1}{N+1}\sum_{k=0}^N S_k$;
- standard put option with aftereffect:

$$F_N = (K_N - S_N)^+,$$

where $K_N = c\max(S_0, S_1, \dots, S_N)$, $c = \text{const}$;
- arithmetic Asian put option:

$$F_N = (K - \overline{S}_N)^+,$$

where $\overline{S}_N = \frac{1}{N+1}\sum_{k=0}^N S_k$.

3. **(a)** When the writer sells an option, say a standard call option, he/she gets from the buyer the premium $\mathbb{C}_N$. A basic care of the seller is to meet the contract engagements, *i.e.*, to ensure the possibility to pay the amount $F_N = (S_N - K)^+$.

To this end the seller of the option creates his/her own portfolio of securities $\pi = (\beta, \gamma)$, where (in accordance with Sec. 10.1) $\beta = (\beta_0, \beta_1, \dots, \beta_N)$ characterizes the state of his/her bank account and $\gamma = (\gamma_0, \gamma_1, \dots, \gamma_N)$ shows the balance of buying/selling stocks. (Now we consider the (B,S)-market described on page 196.)

Let X_n^π be the capital at time n corresponding to a portfolio π. According to Sec. 10.1, the fair price $\mathbb{C}_N$ of the standard call option with pay-off function F_N (under assumption that the (B,S)-market is *arbitrage free* and *complete*) is determined by

$$\mathbb{C}_N = \inf\{x \geq 0 : \text{ there exists } \pi \text{ such that } X_0^\pi = x \text{ and } X_N^\pi = F_N \text{ (P-a.s.)}\}. \quad (11.6)$$

Theorem 10.4 states that one can find $\mathbb{C}_N$ by the formula

$$\mathbb{C}_N = B_0 \mathsf{E}_{\widetilde{\mathsf{P}}_N} \frac{F_N}{B_N}, \quad (11.7)$$

where the expectation $\mathsf{E}_{\widetilde{\mathsf{P}}_N}$ is taken with respect to the martingale measure $\widetilde{\mathsf{P}}_N$, equivalent (on $\mathcal{F}_N$) to the measure $\mathsf{P}_N = \mathsf{P}|\mathcal{F}_N$ and such that $(S_n/B_n)_{0\leq n\leq N}$ is a $\widetilde{\mathsf{P}}_N$-martingale. The same theorem suggests how to determine the components β^* and γ^* of a hedging portfolio $\pi^* = (\beta^*, \gamma^*)$.

Let us cite the well-known result on the structure of $\mathbb{C}_N$, β^*, and γ^* for the CRR-model described in (10.16):

$$\Delta B_n = rB_{n-1}, \qquad \Delta S_n = \rho S_{n-1}.$$

Theorem 11.1 (CRR formulae). *In the CRR model:*

(a) *The fair (rational) price $\mathbb{C}_N$ of a standard European call option with pay-off function $F_N = (S_N - K)^+$ is given by*

$$\mathbb{C}_N = S_0 \mathbb{B}(K_0, N; p^*) - K(1+r)^{-N} \mathbb{B}(K_0, N; \tilde{p}), \tag{11.8}$$

where

$$K_0 = 1 + \left[\log \frac{K}{S_0(1+a)^N} \Big/ \log \frac{1+b}{1+a} \right],$$

$$\tilde{p} = \frac{r-a}{b-a}, \qquad p^* = \frac{1+b}{1+r}\tilde{p},$$

$$\mathbb{B}(K_0, N; p) = \sum_{k=K_0}^{N} C_N^k p^k (1-p)^{N-k}.$$

If $K_0 > N$ then $\mathbb{C}_N = 0$.

(b) *The fair (rational) price $\mathbb{P}_N$ of a standard European put option with pay-off function $F_N = (K - S_N)^+$ is given by*

$$\mathbb{P}_N = \mathbb{C}_N - S_0 + K(1+r)^{-N}. \tag{11.9}$$

(*The relations* (11.8) *and* (11.9) *imply that*

$$\mathbb{P}_N = -S_0\big[1 - \mathbb{B}(K_0, N; p^*0)\big] + K(1+r)^{-N}\big[1 - \mathbb{B}(K_0, N; \tilde{p})\big].) \tag{11.10}$$

Notice that (11.9), called the *call-put parity* formula, follows by very simple arguments: since

$$(K - S_N)^+ = (S_N - K)^+ - S_N + K, \tag{11.11}$$

taking expectation with respect to the martingale measure $\widetilde{\mathsf{P}}_N$ leads to

$$\begin{aligned}
\mathbb{P}_N &= \mathsf{E}_{\widetilde{\mathsf{P}}_N}(1+r)^{-N}(K - S_N)^+ \\
&= \mathbb{C}_N - \mathsf{E}_{\widetilde{\mathsf{P}}_N}(1+r)^{-N} S_N + K(1+r)^{-N} \\
&= \mathbb{C}_N - S_0 + K(1+r)^{-N},
\end{aligned}$$

which is nothing else that the required formula (11.9).

One can prove (11.8) by straightforward calculations, starting from the fact that, in view of (11.7),

$$\mathbb{C}_N = B_0 \mathsf{E}_{\mathsf{P}_N} \widetilde{Z}_N \frac{F_N}{B_N}, \tag{11.12}$$

where $\widetilde{Z}_N$ is the Radon–Nikodým derivative:

$$\widetilde{Z}_N = \frac{d\widetilde{\mathsf{P}}_N}{d\mathsf{P}_N}.$$

Since $\mathsf{P}_1(\rho_1 = b) = p$, $\mathsf{P}(\rho_a = a) = q$ and $\widetilde{\mathsf{P}}_1(\rho_1 = b) = \tilde{p}$, $\widetilde{\mathsf{P}}_1(\rho_1 = a) = \tilde{q}$, where

$$\tilde{p} = \frac{r-a}{b-a}, \qquad \tilde{q} = \frac{b-r}{b-a},$$

we have

$$\widetilde{Z}_1 = \begin{cases} \tilde{p}/p & \text{if } \rho_1 = b, \\ \tilde{q}/q & \text{if } \rho_1 = a. \end{cases}$$

With the notation $\mu = \mathsf{E}_1\rho_1$ and $\sigma^2 = \mathsf{D}_1\rho$ we find that $\mu = bp + aq$, $\sigma^2 = pq(b-a)^2$, and

$$\widetilde{Z}_1 = \begin{cases} 1 - \dfrac{\mu - r}{\sigma^2}\,(b-\mu) & \text{if } \rho_1 = b, \\ 1 - \dfrac{\mu - r}{\sigma^2}\,(a-\mu) & \text{if } \rho_1 = a. \end{cases}$$

Due to P_N- and $\widetilde{\mathsf{P}}_N$-independence of the random variables $\rho_1, \dots, \rho_n$,

$$\widetilde{Z}_N = \prod_{k=1}^{N} \Big(1 - \frac{\mu - r}{\sigma^2}\,(\rho_k - \mu)\Big). \tag{11.13}$$

Denoting

$$h_k \equiv -\frac{\mu - r}{\sigma^2}\,(\rho_k - \mu) \quad \text{and} \quad H_n = h_1 + \dots + h_n,$$

one can represent $\widetilde{Z}_N$ in the form

$$\widetilde{Z}_N = \mathcal{E}(H)_N, \tag{11.14}$$

where $\mathcal{E}(H)_N$ is the stochastic exponential (see (4.4)).

Therefore (11.12) takes the following form:

$$\mathbb{C}_N = B_0 \mathsf{E}_{\mathsf{P}_N} \mathcal{E}(H)_N \frac{F_N}{B_N}. \tag{11.15}$$

For a call option the pay-off function is $F_N = (S_N - K)^+$. Then

$$\begin{aligned} \mathbb{C}_N &= B_0 \mathsf{E}_{\mathsf{P}_N} \mathcal{E}(H)_N \frac{(S_N - K)^+}{B_N} = (1+r)^{-N} \mathsf{E}_{\mathsf{P}_N} \mathcal{E}(H)_N (S_N - K)^+ \\ &= (1+r)^{-N} \mathsf{E}_{\mathsf{P}_N} \mathcal{E}(H)_N (S_N - K)\, I(S_N > K). \end{aligned} \tag{11.16}$$

Further calculations, taking into account (11.14) and (11.13), lead to the required formula (11.8).

3. (b) To describe the structure of an optimal hedge $\pi^* = (\beta^*, \gamma^*)$ introduce the function

$$\widehat{F}_n(x;p) = \sum_{k=0}^{n} F\big(x(1+b)^k(1+a)^{n-k}\big)C_n^k p^k(1-p)^{n-k}, \tag{11.17}$$

where $F(x) = (x-K)^+$ for call options and $F(x) = (K-x)^+$ for put options.

Theorem 11.2. *If $F_N = F(S_N)$ then a perfect hedge $\pi^* = (\beta^*, \gamma^*)$ in the CRR model is described by the formulae*

$$\gamma_k^* = (1+r)^{-(N-k)} \frac{\widehat{F}_{N-k}(S_{k-1}(1+b);p^*) - \widehat{F}_{N-k}(S_{k-1}(1+a);p^*)}{S_{k-1}(b-a)}, \tag{11.18}$$

where $p^* = (r-a)/(b-a)$, and

$$\beta_k^* = \frac{1}{B_N}\bigg\{\widehat{F}_{N-k+1}(S_{k-1};p^*) - \frac{1+r}{b-a} \times \Big[\widehat{F}_{N-k}\big(S_{k-1}(1+b);p^*\big) - \widehat{F}_{N-k}\big(S_{k-1}(1+a);p^*\big)\Big]\bigg\}. \tag{11.19}$$

Proof. The quantities γ_k^* can be found from the $\frac{S}{B}$-representation (10.30) (with $\widetilde{\mathsf{P}} = \widetilde{\mathsf{P}}_N$) and the quantities β_k^* from (10.31). Thus we have to consider the martingale $(\mathsf{E}_{\widetilde{\mathsf{P}}_N}(F_N/B_N \mid \mathcal{F}_n))_{0\le n\le N}$ and to deduce for it a representation of the form (10.30) in order to "extract" from it the values of γ_k^*; the detailed calculations were traced in [161; Chap. VI, § 4] and then are omitted here (similar calculations are made below in Subsecs. 4(b) and 5(b) for the Bachelier and Samuelson–Black–Merton–Scholes models).

The quantities β_k^* can be found, according to (10.31), by the formulae

$$\beta_k^* = \frac{X_k^{\pi^*} - \gamma_k^* S_k}{B_k},$$

where $X_k^{\pi^*} = (1+r)^{-(N-k)}\widehat{F}_{N-k}(S_k;\tilde{p})$ (see also Theorem 1 in [161; Chap. VI, § 4]). □

4. (a) In the Bachelier (B,S)-model on the time interval $[0,T]$ the bank account is thought of as unvarying ($B_t \equiv 1$, $t \ge 0$) and stock prices are assumed to be driven by a *linear Brownian motion* with drift (cf. (1.22)):

$$S_t = S_0 + \mu t + \sigma W_t, \qquad t \le T, \tag{11.20}$$

where $W = (W_t)_{t\ge 0}$ is a standard Wiener process (a Brownian motion) on some probability space $(\Omega, \mathcal{F}, \mathsf{P})$, $\sigma > 0$, $\mu \in \mathbb{R}$.

Although the stock prices S_t in this model are allowed to take negative values, which is unsuitable from the financial and economic point of view ("the prices must be nonnegative"), but for $S_0 > 0$ and $\mu > 0$ the process $(S_t)_{t\geq 0}$ remains positive for quite a long time. Indeed, let

$$\tau_0 = \inf\{t \geq 0 : S_t = 0\}.$$

Then

$$\mathsf{P}(\tau_0 < \infty) = e^{-2\mu S_0/\sigma^2} \tag{11.21}$$

(see [161; p. 760]). In other words, for large $\mu S_0/\sigma^2$ the probability that the process $(S_t)_{t\geq 0}$ has everywhere-positive trajectories is rather close to one $(= 1 - e^{-2\mu S_0/\sigma^2})$.

Apparently this argument was also taken into account when L. Bachelier [4] used Brownian motion with drift to describe the stock prices.

The model (11.20) is of interest because it is both arbitrage free and complete. Indeed, let

$$\widetilde{Z}_T = \exp\Big\{-\frac{\mu}{\sigma} W_T - \frac{1}{2}\Big(\frac{\mu}{\sigma}\Big)^2 T\Big\} \tag{11.22}$$

and, in addition to the measure P_T on $(\Omega, \mathcal{F}_T, (\mathcal{F}_t)_{t\leq T})$, define the measure $\widetilde{\mathsf{P}}_T$ by

$$d\widetilde{\mathsf{P}}_T = \widetilde{Z}_T \, d\mathsf{P}_T.$$

By Girsanov's theorem (Sec. 6.3)

$$\mathrm{Law}\big(S_0 + \mu t + \sigma W_t;\ t \leq T \,|\, \widetilde{\mathsf{P}}_T\big) = \mathrm{Law}\big(S_0 + \sigma W_t;\ t \leq T \,|\, \mathsf{P}_T\big). \tag{11.23}$$

Hence, with respect to $\widetilde{\mathsf{P}}_T$ the process $(S_t)_{t\leq T}$ is a martingale and thus $\widetilde{\mathsf{P}}_T$ is a martingale measure. It is equivalent to the measure P_T ($\widetilde{\mathsf{P}}_T \sim \mathsf{P}$) and moreover is a unique martingale measure (see [161; Chap. VII, § 4a, Subsec. 5]).

Theorem 11.3 (Bachelier's formulae). (a) *In the model* (11.20) *the rational price $\mathbb{C}_T$ of the standard European call option with the pay-off function $F_T = (S_T - K)^+$ is given by the formula*

$$\mathbb{C}_T = (S_0 - K)\,\Phi\Big(\frac{S_0 - K}{\sigma\sqrt{T}}\Big) + \sigma\sqrt{T}\,\varphi\Big(\frac{S_0 - K}{\sigma\sqrt{T}}\Big), \tag{11.24}$$

where $\varphi(x) = (2\pi)^{-1/2}e^{-x^2/2}$, $\Phi(x) = \int_{-\infty}^{x} \varphi(y)\,dy$.

In particular, for $S_0 = K$ (in the so-called at-the-money case)

$$\mathbb{C}_T = \sigma\sqrt{\frac{T}{2\pi}}.$$

(b) *In the model* (11.20) *the rational price* $\mathbb{C}_T$ *of the standard European put option with the pay-off function* $F_T = (K - S_T)^+$ *is given by the formula*

$$\mathbb{P}_T = -(S_0 - K)\,\Phi\Big(\frac{K - S_0}{\sigma\sqrt{T}}\Big) + \sigma\sqrt{T}\,\varphi\Big(\frac{S_0 - K}{\sigma\sqrt{T}}\Big). \qquad (11.25)$$

In particular, for $S_0 = K$

$$\mathbb{P}_T = \mathbb{C}_T = \sigma\sqrt{\frac{T}{2\pi}}.$$

Proof. (See [161; Chap. VIII, § 1a].) In view of Theorem 10.8 and (11.22)

$$\begin{aligned}\mathbb{C}_T &= \mathsf{E}_{\widetilde{\mathsf{P}}_T} F_T = \mathsf{E}_{\widetilde{\mathsf{P}}_T}(S_0 + \mu T + \sigma W_T - K)^+ \\ &= \mathsf{E}_{\mathsf{P}_T}(S_0 - K + \sigma W_T)^+ = \mathsf{E}_{\mathsf{P}}(S_0 - K + \sigma\sqrt{T}W_1)^+, \end{aligned} \qquad (11.26)$$

where the measure P is such that $\mathrm{Law}(W_1 \,|\, \mathsf{P}) = \mathcal{N}(0,1)$.

Simple calculations show that

$$\mathsf{E}_{\mathsf{P}}(a + bW_1)^+ = a\,\Phi\Big(\frac{a}{b}\Big) + b\,\varphi\Big(\frac{a}{b}\Big) \qquad (11.27)$$

for all $a \in \mathbb{R}$ and $b \geq 0$.

Putting here $a = S_0 - K$, $b = \sigma\sqrt{T}$, we get the required formula (11.24).

The formula (11.25) follows from (11.24) and the call-put parity $\mathbb{P}_T = \mathbb{C}_T - S_0 + K$, which in turn is implied by the identities $(K - S_T)^+ = (S_T - K)^+ - S_T + K$ and $\mathsf{E}_{\widetilde{\mathsf{P}}_T} S_T = S_0$. □

4. (b) Let us investigate now the structure of a hedge performing the perfect replication, $X_T^{\pi^*} = F_T$ (P-a.s.).

According to Theorem 10.8(c), the quantities $(\gamma_t^*)_{t\leq T}$ can be "extracted" from the S/B-representation

$$\mathsf{E}_{\widetilde{\mathsf{P}}}\Big(\frac{F_T}{B_T}\,\Big|\,\mathcal{F}_t\Big) = \mathsf{E}_{\widetilde{\mathsf{P}}}\frac{F_T}{B_T} + \int_0^t \gamma_u^*\,d\Big(\frac{S_u}{B_u}\Big), \qquad t \leq T, \qquad (11.28)$$

or, because

$$X_t^{\pi^*} = B_t \mathsf{E}_{\widetilde{\mathsf{P}}}\Big(\frac{F_T}{B_T}\,\Big|\,\mathcal{F}_t\Big), \qquad (11.29)$$

from the representation

$$\frac{X_t^{\pi^*}}{B_t} = \mathsf{E}_{\widetilde{\mathsf{P}}}\frac{F_T}{B_T} + \int_0^t \gamma_u^*\,d\Big(\frac{S_u}{B_u}\Big). \qquad (11.30)$$

In the considered case $B \equiv 1$, therefore

$$X_t^{\pi^*} = \mathsf{E}_{\widetilde{\mathsf{P}}}(F_T \,|\, \mathcal{F}_t) = \mathsf{E}_{\widetilde{\mathsf{P}}}((S_T - K)^+ \,|\, \mathcal{F}_t).$$

By the generalized Bayes formula (see Lemma 6.2)

$$X_t^{\pi^*} = \mathsf{E}_{\widetilde{\mathsf{P}}}\big((S_T - K)^+ \,|\, \mathcal{F}_t\big) = \mathsf{E}_{\mathsf{P}}\bigg(\frac{\widetilde{Z}_T}{\widetilde{Z}_t}(S_T - K)^+ \,\bigg|\, \mathcal{F}_t\bigg),$$

where $\widetilde{Z}_T$ is defined in (11.22) and

$$\widetilde{Z}_t = \exp\Big\{-\frac{\mu}{\sigma}\,W_t - \frac{1}{2}\Big(\frac{\mu}{\sigma}\Big)^2 t\Big\}.$$

Since $S_t = S_0 + \mu t + \sigma W_t$, we have

$$X_t^{\pi^*} = \mathsf{E}_{\mathsf{P}}\bigg(\exp\Big\{-\frac{\mu}{\sigma}\,(W_T - W_t) - \frac{1}{2}\Big(\frac{\mu}{\sigma}\Big)^2 (T-t)\Big\} \times \big[(S_t - K) + \sigma(W_T - W_t) + \mu(T-t)\big]^+\bigg). \tag{11.31}$$

Letting $a = S_t - K$ and $b = \sigma\sqrt{T-t}$, by simple transformations we get from (11.31) that

$$X_t^{\pi^*} = \mathsf{E}(a + b\xi)^+, \tag{11.32}$$

where the expectation E is taken with respect to the law of the standard Gaussian random variable $\xi \sim \mathcal{N}(0,1)$.

On the other hand, by (11.27)

$$\mathsf{E}(a + b\xi)^+ = a\,\Phi\Big(\frac{a}{b}\Big) + b\,\varphi\Big(\frac{a}{b}\Big), \tag{11.33}$$

where, as usual, $\varphi(x) = (2\pi)^{-1/2}e^{-x^2/2}$, $\Phi(x) = \int_{-\infty}^{x}\varphi(y)\,dy$.

Putting

$$C(t,x) = (x - K)\,\Phi\bigg(\frac{x-K}{\sigma\sqrt{T-t}}\bigg) + \sigma\sqrt{T-t}\,\varphi\bigg(\frac{x-K}{\sigma\sqrt{T-t}}\bigg), \tag{11.34}$$

we find from (11.33) that

$$X_t^{\pi^*} = C(t, S_t).$$

By (11.30)

$$dX_t^{\pi^*} = \gamma_t^*\,dS_t. \tag{11.35}$$

Since $C(t,x) \in C^{1,2}$, the Itô formula can be applied to $C(t, S_t)$:

$$dC(t,S_t) = \frac{\partial C}{\partial x}\,dS_t + \bigg(\frac{\partial C}{\partial t} + \frac{1}{2}\sigma^2\frac{\partial^2 C}{\partial x^2}\bigg)\,dt. \tag{11.36}$$

Comparing this with (11.34), we find that

$$\gamma_t^* = \frac{\partial C}{\partial s}(t, S_t) \tag{11.37}$$

and

$$\frac{\partial C}{\partial t} + \frac{1}{2}\sigma^2 \frac{\partial^2 C}{\partial x^2} = 0. \tag{11.38}$$

Having the explicit formula (11.34) for $C(t, x)$, one can deduce, thanks to (11.37), the explicit formula for γ_t^*:

$$\gamma_t^* = \Phi\left(\frac{S_t - K}{\sigma\sqrt{T-t}}\right). \tag{11.39}$$

Finally, to find β_t^* use the relation

$$X_t^{\pi^*} = C(t, S_t) = \beta_t^* + \gamma_t^* S_t.$$

Hence, in view of (11.34) and (11.39),

$$\beta_t^* = -K\,\Phi\left(\frac{S_t - K}{\sigma\sqrt{T-t}}\right) + \sigma\sqrt{T-t}\,\varphi\left(\frac{S_t - K}{\sigma\sqrt{T-t}}\right). \tag{11.40}$$

Thus we have proved the following theorem ([161; Chap. VIII, § 1a]).

Theorem 11.4. *In Bachelier's model the components* $\beta^* = (\beta_t^*)_{t\le T}$ *and* $\gamma^* = (\gamma_t^*)_{t\le T}$ *of the hedging portfolio* $\pi^* = (\beta^*, \gamma^*)$ *are determined by the formulae* (11.40) *and* (11.39).

Corollary 11.1. *Suppose that* $\lim_{t\uparrow T} S_t > K$. *Then as* $t \uparrow T$

$$\gamma_t^* \to 1 \quad \textit{and} \quad \beta_t^* \to -K. \tag{11.41}$$

If $\lim_{t\uparrow T} S_t < K$, *then as* $t \uparrow T$

$$\gamma_t^* \to 0 \quad \textit{and} \quad \beta_t^* \to 0. \tag{11.42}$$

5. (a) As noticed above, a weakness the Bachelier model lies in that it allows the prices S_t to be negative. The Samuelson–Black–Merton–Scholes model (proposed by P. A. Samuelson [152]) is more realistic. Therein the prices S_t, $t \ge 0$, are driven by a *geometric* (in the terminology of [152] *economic*) Brownian motion:

$$S_t = S_0 e^{H_t}, \tag{11.43}$$

where

$$H_t = \left(\mu - \frac{\sigma^2}{2}\right)t + \sigma W_t, \tag{11.44}$$

with $W = (W_t)_{t\ge 0}$ a Wiener process (Brownian motion). By Itô's formula,

$$dS_t = S_t\big(\mu\,dt + \sigma\,dW_t\big). \tag{11.45}$$

According to (4.8)–(4.8),

$$e^{H_t} = \mathcal{E}(\widetilde{H})_t, \tag{11.46}$$

where

$$\widetilde{H}_t = H_t + \frac{1}{2}\langle H^c \rangle_t = \left(\mu - \frac{\sigma^2}{2}\right)t + \sigma W_t + \frac{\sigma^2}{2}t = \mu t + \sigma W_t. \tag{11.47}$$

The formula (11.45) is often written in a symbolic form:

$$\frac{dS_t}{S_t} = \mu\, dt + \sigma\, dW_t. \tag{11.48}$$

However one should fully realize that the expression (11.48) is not the result of the formal division of both sides of (11.45) by S_t. Notation (11.45) is merely a convenient and illustrative form of the integral relation

$$S_t = S_0 + \int_0^t S_u \mu\, du + \int_0^t S_u \sigma\, dW_u, \tag{11.49}$$

which does not imply the "divided-by-S_t" relation

$$\int_0^t \frac{dS_u}{S_u} = \int_0^t \mu\, du + \int_0^t \sigma\, dW_u, \qquad i.e., \quad \int_0^t \frac{dS_u}{S_u} = \mu t + \sigma W_t.$$

From (11.46), (11.47) we see that the well-grounded passage from (11.45) to (11.48) must use the stochastic exponential (11.46) or, equivalently, the stochastic logarithm

$$\widetilde{H} = \mathcal{L}(e^H)$$

(see details in Sec. 4.1).

The model (11.43)–(11.44), proposed by Samuelson, underlies the well-known Black–Merton–Scholes model of the (B, S)-market, where the bank account $B = (B_t)_{t \geq 0}$ is such that

$$dB_t = rB_t\, dt,$$

that is,

$$B_t = B_0 e^{rt}, \tag{11.50}$$

and the stock prices $S = (S_t)_{t \geq 0}$ are described by geometric Brownian motion:

$$dS_t = S_t(\mu\, dt + \sigma\, dW_t),$$

that is,

$$S_t = S_0 e^{(\mu - \sigma^2/2)t + \sigma W_t}. \tag{11.51}$$

Our main interest will be to establish the Black–Scholes formulae for the rational prices $\mathbb{C}_T$ and $\mathbb{P}_T$ of call and put options with pay-off functions $F_T = (S_T - K)^+$ and $F_T = (K - S_T)^+$, respectively.

By analogy with (11.22) introduce the variable

$$\widetilde{Z}_T = \exp\left\{-\frac{\mu - r}{\sigma} W_T - \frac{1}{2}\left(\frac{\mu - r}{\sigma}\right)^2 T\right\}. \tag{11.52}$$

(When $r = 0$ the formula (11.52) turns into (11.22).)

Define the measure $\widetilde{\mathsf{P}}_T$ by

$$d\widetilde{\mathsf{P}}_T = \widetilde{Z}_T \, d\mathsf{P}_T, \tag{11.53}$$

where P_T is the initial physical measure on the filtered measurable space $(\Omega, \mathcal{F}_T, (\mathcal{F}_t)_{t \le T})$.

We have

$$\mathrm{Law}\big(\mu t + \sigma W_t;\ t \le T \,|\, \widetilde{\mathsf{P}}_T\big) = \mathrm{Law}\big(rt + \sigma \widetilde{W}_t;\ t \le T \,|\, \widetilde{\mathsf{P}}_T\big), \tag{11.54}$$

where

$$\widetilde{W}_t = W_t + \frac{\mu - r}{\sigma} t.$$

The process $\widetilde{W} = (\widetilde{W}_t)_{t \le T}$ is a Wiener process with respect to the measure $\widetilde{\mathsf{P}}_T$. Therefore from (11.54) we find that

$$\mathrm{Law}\big(\mu t + \sigma W_t;\ t \le T \,|\, \widetilde{\mathsf{P}}_T\big) = \mathrm{Law}\big(rt + \sigma W_t;\ t \le T \,|\, \mathsf{P}_T\big). \tag{11.55}$$

Thus

$$\begin{aligned}
\mathrm{Law}\big(S_t;\ t \le T \,|\, \widetilde{\mathsf{P}}_T\big) &\equiv \mathrm{Law}\big(S_0 e^{(\mu - \sigma^2/2)t + \sigma W_t};\ t \le T \,|\, \widetilde{\mathsf{P}}_T\big) \\
&= \mathrm{Law}\big(S_0 e^{(r - \sigma^2/2)t + \sigma W_t};\ t \le T \,|\, \mathsf{P}_T\big)
\end{aligned}$$

and

$$\mathrm{Law}\left(\frac{S_t}{B_t};\ t \le T \,\Big|\, \widetilde{\mathsf{P}}_T\right) = \mathrm{Law}\left(\frac{S_0}{B_0} e^{-(\sigma^2/2)t + \sigma W_t};\ t \le T \,\Big|\, \mathsf{P}_T\right), \tag{11.56}$$

where $(e^{-(\sigma^2/2)t + \sigma W_t})_{t \le T}$ is evidently a P_T-martingale.

From (11.56) we deduce that the process $(S_t/B_t)_{t \le T}$ is a martingale with respect to the measure $\widetilde{\mathsf{P}}_T$. In other words, the measure $\widetilde{\mathsf{P}}_T$ is a martingale measure for the process $(S_t/B_t)_{t \le T}$. Is it important that this martingale measure is unique (for a proof see [161; Chap. VII, § 4a, Subsec. 5]).

Therefore the considered (B, S)-market is arbitrage free and complete. So, by Theorem 10.8 the rational price $\mathbb{C}_T$ of a standard call option with pay-off function $F_T = (S_T - K)^+$ is given by

$$\mathbb{C}_T = B_0 \mathsf{E}_{\widetilde{\mathsf{P}}_T} \frac{F_T}{B_T} = \frac{B_0}{B_T} \mathsf{E}_{\widetilde{\mathsf{P}}_T} F_T. \tag{11.57}$$

Form this and (11.55) we find that

$$\begin{aligned}\mathbb{C}_T &= e^{-rT}\mathsf{E}_{\widetilde{\mathsf{P}}_T}(S_T - K)^+ = e^{-rT}\mathsf{E}_{\widetilde{\mathsf{P}}_T}(S_0 e^{(\mu-\sigma^2/2)T+\sigma W_T} - K)^+ \\ &= e^{-rT}\mathsf{E}_{\mathsf{P}_T}(S_0 e^{(r-\sigma^2/2)T+\sigma W_T} - K)^+ \\ &= e^{-rT}\mathsf{E}_{\mathsf{P}_T}(S_0 e^{rT} e^{-(\sigma^2/2)T+\sigma\sqrt{T}W_1} - K)^+ \\ &= e^{-rT}\mathsf{E}(ae^{bW_1-b^2/2} - K)^+, \end{aligned} \tag{11.58}$$

where $a = S_0 e^{rT}$, $b = \sigma\sqrt{T}$, and E is the expectation with respect to the law of the standard normal, $\mathcal{N}(0,1)$, random variable W_1.

Straightforward calculations (cf. (11.27)) yield

$$\mathsf{E}(ae^{bW_1-b^2/2} - K)^+ = a\Phi\left(\frac{\log\frac{a}{K} + \frac{1}{2}b^2}{b}\right) - K\Phi\left(\frac{\log\frac{a}{K} - \frac{1}{2}b^2}{b}\right) \tag{11.59}$$

and therefore, in view of (11.58),

$$\mathbb{C}_T = S_0\Phi\left(\frac{\log\frac{a}{K} + \frac{1}{2}b^2}{b}\right) - Ke^{-rT}\Phi\left(\frac{\log\frac{a}{K} - \frac{1}{2}b^2}{b}\right) \tag{11.60}$$

(cf. (11.8) for the CRR-model).

For the rational price $\mathbb{P}_T$ of the put option, owing to the call-put duality

$$\mathbb{P}_T = \mathbb{C}_T - S_0 + Ke^{-rt} \tag{11.61}$$

(cf. (11.9)), we get from (11.60) that

$$\begin{aligned}\mathbb{P}_T = -S_0&\left[1 - \Phi\left(\frac{\log\frac{S_0}{K} + T(r+\frac{\sigma^2}{2})}{\sigma\sqrt{T}}\right)\right] \\ &+ Ke^{-rT}\Phi\left[1 - \left(\frac{\log\frac{S_0}{K} + T(r+\frac{\sigma^2}{2})}{\sigma\sqrt{T}}\right)\right]. \end{aligned} \tag{11.62}$$

The notation

$$d_{\pm}(T) = \frac{\log\frac{S_0}{K} + T(r+\frac{\sigma^2}{2})}{\sigma\sqrt{T}} \tag{11.63}$$

provides the more compact form for (11.61) and (11.62):

$$\mathbb{C}_T = S_0\Phi(d_+(T)) - Ke^{-rT}\Phi(d_-(T)) \tag{11.64}$$

and

$$\mathbb{P}_T = -S_0\Phi(-d_+(T)) + Ke^{-rT}\Phi(-d_-(T)). \tag{11.65}$$

Thus, we have proved the following theorem.

Theorem 11.5 (Black–Scholes' formulae). *In the Black–Merton–Scholes model of the* (B, S)*-market defined by* (11.50), (11.51), *the rational prices* $\mathbb{C}_T$ *and* $\mathbb{P}_T$ *of call and put options are determined by* (11.60) *and* (11.62), *respectively.*

5. (b) Let us turn to the description of the perfect-hedging portfolio $\pi^* = (\beta^*, \gamma^*)$ for the call options. We shall follow the same ideas as were used in the CRR- and Bachelier's models. (The case of put options can be studied in an analogous way.)

From (11.66), by the generalized Bayes' formula (see Lemma 6.2 and calculations above for Bachelier's model), we find that

$$\begin{aligned} X_t^{\pi^*} &= B_t \mathsf{E}_{\widetilde{\mathsf{P}}_T}(B_T F_T \,|\, \mathcal{F}_T) = e^{-r(T-t)} \mathsf{E}_{\widetilde{\mathsf{P}}_T}((S_T - K)^+ \,|\, \mathcal{F}_T) \\ &= e^{-r(T-t)} \mathsf{E}_{\mathsf{P}_T} \left\{ \frac{\widetilde{Z}_T}{\widetilde{Z}_t} (S_T - K)^+ \,\Big|\, \mathcal{F}_T \right\}, \end{aligned} \tag{11.66}$$

where $\widetilde{Z}_T$ is specified in (11.52) and

$$\widetilde{Z}_t = \exp\Big\{-\frac{\mu - r}{\sigma} W_t - \frac{1}{2}\Big(\frac{\mu - r}{\sigma}\Big)^2 t\Big\}.$$

Transformations of the right-hand side in (11.66), analogous to those which allowed to pass from (11.31) to (11.32) in Bachelier's model, yield that

$$X_t^{\pi^*} = e^{-r(T-t)} \mathsf{E}\big(a e^{b\xi - b^2/2} - K\big)^+, \tag{11.67}$$

where $a = S_t e^{r(T-t)}$, $b = \sigma\sqrt{T-t}$, and the expectation E is taken with respect to the law of a standard Gaussian random variable $\xi \sim \mathcal{N}(0,1)$.

Taking into account (11.59) and denoting

$$\begin{aligned} C(t,x) = -x\Phi\Big(&\frac{\log\frac{x}{K} + (T-t)(r + \frac{\sigma^2}{2})}{\sigma\sqrt{T-t}}\Big) \\ &+ K e^{-r(T-t)} \Phi\Big(\frac{\log\frac{x}{K} + (T-t)(r - \frac{\sigma^2}{2})}{\sigma\sqrt{T-t}}\Big), \end{aligned} \tag{11.68}$$

we find that

$$X_t^{\pi^*} = C(t, S_t).$$

In the same way as in Bachelier's model (see Subsec. 4(b)) we conclude that

$$\gamma_t^* = \frac{\partial C}{\partial s}(t, S_t) \tag{11.69}$$

and

$$\frac{\partial C}{\partial t} + r x \frac{\partial C}{\partial x} + \frac{1}{2}\sigma^2 x^2 \frac{\partial^2 C}{\partial x^2} = rC. \tag{11.70}$$

From (11.68) and (11.69) it follows that

$$\gamma_t^* = \Phi\left(\frac{\log\frac{S_t}{K} + (T-t)(r+\frac{\sigma^2}{2})}{\sigma\sqrt{T-t}}\right). \tag{11.71}$$

From $X_t^* = \beta_t^* B_t + \gamma_t^* S_t$ $(= C(t, S_t))$ and (11.68) we deduce that

$$\beta_t^* = -\frac{K}{B_0}\, e^{-rT}\, \Phi\left(\frac{\log\frac{S_t}{K} + (T-t)(r-\frac{\sigma^2}{2})}{\sigma\sqrt{T-t}}\right). \tag{11.72}$$

Thus, we have proved the following theorem.

Theorem 11.6. *In the Black–Merton–Scholes model the components of the hedging portfolio $\pi^* = (\beta^*, \gamma^*)$ for the call option are determined by the formulae* (11.72) *and* (11.71).

Corollary 11.2. *Using* (11.72) *and* (11.71), *one can immediately establish that if $t \uparrow T$ and $S_T > K$ then $\gamma_t^* \to 1$ and therefore $\gamma_t^* S_t \to S_T$ and $\beta_t^* B_t \to -K$. If $S_T < K$ then $\gamma_t^* S_t \to 0$ and $\beta_t^* B_t \to 0$.* (Cf. the corollary to Theorem 11.4 on page 234.)

5. (c) The function $C(t,x) \in C^{1,2}$ introduced above was shown to obey the equation

$$\frac{\partial C}{\partial t} + rx\frac{\partial C}{\partial x} + \frac{1}{2}\sigma^2 x^2 \frac{\partial^2 C}{\partial x^2} = rC \tag{11.73}$$

(with boundary condition $C(t,x) = (x-K)^+$).

It is interesting to reproduce the original 1973 proofs by Black and Scholes [32] and Merton [130] of formulae for the rational prices $\mathbb{C}_T$ and $\mathbb{P}_T$. Their method, in essence, started from the above equation for $C(t,x)$.

Namely, these authors proceeded from the following. Let the rational price $\mathbb{C}_{[t,T]}$ of an option with the pay-off function F_T on the interval $[t,T]$ be given by

$$\mathbb{C}_{[t,T]} = \inf\{x:\ \text{there exists } \pi \text{ such that } X_t^\pi = x \\ \text{and } X_T^\pi = F_T\ (\mathsf{P}\text{-a.s.})\}.$$

By arguments of "equilibrium" of the considered market (resulting in the equality $\mu = r$) and by Markov property of the process $S = (S_t)_{t \le T}$ they found, under assumption that $\mathbb{C}_{[t,T]}$ is of the from $Y(t, S_t)$ for some function $Y = Y(t,x)$, that the following equation must be satisfied:

$$\frac{\partial Y}{\partial t} + rx\frac{\partial C}{\partial x} + \frac{1}{2}\sigma^2 x^2 \frac{\partial^2 C}{\partial x^2} = rC \tag{11.74}$$

with boundary condition $Y(T,x) = (x-K)^+$.

This equation is of Feynman–Kac type (see [161; Chap. III, § 3f] or [165; Chap. 6, § 4]) and its solution can be represented as

$$Y(t,x) = \widetilde{\mathsf{E}}_{t,x}\Big\{e^{-r(T-t)}\big(\widetilde{S}_T - K\big)^+\Big\}, \tag{11.75}$$

where $\widetilde{\mathsf{E}}_{t,x}$ is the averaging with respect to the measure $\widetilde{\mathsf{P}}_{t,x}$ of the diffusion Markov process $(\widetilde{S}_u)_{t\le u\le T}$ which obeys the equation

$$d\widetilde{S}_u = \widetilde{S}_u\big(r\,du + \sigma\,d\widetilde{W}_u\big) \tag{11.76}$$

with some Wiener process $\widetilde{W}$ and with the initial condition $\widetilde{S}_t = x$.

Taking into account that $\mathrm{Law}\big(\widetilde{S}_u,\ t\le u\le T \mid \widetilde{S}_t = x,\ \widetilde{\mathsf{P}}_{t,x}\big) = \mathrm{Law}\big(S_u,\ t\le u\le T \mid S_t = x,\ \widetilde{\mathsf{P}}_{t,x}\big)$, from (11.75) and (11.66) we conclude that in fact $Y(t,x) = C(t,x)$.

Thus the original approach of Black–Scholes and Merton leads to the same function $C(t,x)$ which was used above to deduce both the formulae for rational prices $\mathbb{C}_T$, $\mathbb{P}_T$ and the formulae for the hedging portfolio $\pi^* = (\beta^*,\gamma^*)$ in the case of call options.

11.2 Overview of the Pricing Formulae for American Options

1. As was mentioned in Sec. 11.1, in contrast to European options with maturity time N, in the case of American options the buyer may execute it at *any* time $\tau \le N$. It is natural that this time must be chosen using only the information already obtained—without anticipation. The exact description is the following.

Let $(\Omega, \mathcal{F}, (\mathcal{F}_n)_{n\le N}, \mathsf{P})$ be a filtered probability space ($\mathcal{F}_N = \mathcal{F}$) which describes "stochastics" of financial markets. $\mathcal{F}_n$ is the σ-algebra of events observable up to time n and is sometimes treated as "information" available in the market (both for the buyer and the seller) up to time n.

Taking in mind the above property of nonanticipation, we will assume that the exercise time $\tau = \tau(\omega)$ is a *Markov time*, or *stopping time*, *i.e.*, a random variable taking values in the set $\{0,1,\dots,N\}$ and such that for each n from this set the event

$$\{\tau = n\} \in \mathcal{F}_n$$

(we consider now the discrete-time case; $\mathcal{F}_0$ is chosen to be the trivial σ-algebra $\{\varnothing, \Omega\}$).

To each time $0 \leq n \leq N$, we relate the pay-off (nonnegative, $\mathcal{F}_n$-measurable) function $F_n = F_n(\omega)$: if the buyer chooses a time $\tau = \tau(\omega)$ to execute the option, then the gain will be equal to $F_\tau = F_{\tau(\omega)}(\omega)$.

For example, for the standard call option we have

$$F_\tau = (S_\tau - K)^+,$$

where $S = (S_n)_{0 \leq n \leq N}$ is the price of a certain, say, asset on the financial market.

In the case of the standard put option

$$F_\tau = (K - S_\tau)^+.$$

2. The main problems related to American options consist in determining the rational (fair) price of the option, in describing the "optimal" exercise time τ^* of the option and in finding the "optimal" hedging strategy of the seller.

Just as in the case of European options we will assume (cf. Sec. 10.1) that all our considerations develop on the (B, S)-market which consists of a bank account $B = (B_n)_{0 \leq n \leq N}$ and one stock $S = (S_n)_{0 \leq n \leq N}$; a (self-financing) portfolio of the seller is denoted by $\pi = (\beta, \gamma)$ and $X_n^\pi = \beta_n B_n + \gamma_n S_n$ is the capital of the portfolio π at time n.

Now we give the definition of lower and upper prices, $\widetilde{\mathbb{C}}_*(F)$ and $\widetilde{\mathbb{C}}^*(F)$, of American options which is analogous to Definition 10.4 for European option.

We assume, for simplicity, that $B_n \equiv 1$ and use the notation (cf. (10.6))

$$G_n^\pi = \sum_{k=1}^{n} \gamma_k \Delta S_k.$$

Definition 11.1. The *lower price* of an American option with the system of (nonnegative) pay-off functions $F = (F_n)_{0 \leq n \leq N}$ is

$$\widetilde{\mathbb{C}}_*(F) = \sup\{x : \exists(\pi, \tau) \text{ with } X_0^\pi = x \text{ and } x + G_\tau^\pi \leq F_\tau \ (\mathsf{P}\text{-a.s.})\}. \tag{11.77}$$

The *upper price* is

$$\widetilde{\mathbb{C}}^*(F) = \inf\{x : \exists(\pi, \tau)) \text{ with } X_0^\pi = x \text{ and } x + G_\tau^\pi \geq F_\tau \ (\mathsf{P}\text{-a.s.})\}. \tag{11.78}$$

Just as in the case of a European option (cf. Sec. 10.1), the prices $\widetilde{\mathbb{C}}_*(F)$ and $\widetilde{\mathbb{C}}^*(F)$ are also called a buyer's price and a seller's price, respectively.

3. In this subsection we will formulate the main result on how to calculate the upper price $\widetilde{\mathbb{C}}^*(F)$ (*i.e.*, the maximal price which the buyer is willing to pay) for the case of no-arbitrage markets.

As in the case of European options, the buyer knows that if he/she is proposed to purchase an option at a price higher than $\widetilde{\mathbb{C}}^*(F)$, then the seller will certainly have a riskless gain ("free lunch"), thus the buyer will not accept this price. (We restrict ourselves to studying the upper price, taking the position of the buyer.)

Denote by $\mathcal{M}(\mathsf{P})$ the class of all martingale measures (see Sec. 10.1) and by $\mathfrak{M}_N$ the class of all stopping times $\tau = \tau(\omega)$ taking values in $\{0, 1, \ldots, N\}$.

Theorem 11.7. *Let the set $\mathcal{M}(\mathsf{P})$ be nonempty (which is equivalent to the absence of arbitrage possibilities on the market) and such that $\sup_{\widetilde{\mathsf{P}} \in \mathcal{M}(\mathsf{P})} \mathsf{E}_{\widetilde{\mathsf{P}}} F_n < \infty$. Then*

$$\widetilde{\mathbb{C}}^*(F) = \sup_{\tau \in \mathfrak{M}_N, \, \widetilde{\mathsf{P}} \in \mathcal{M}(\mathsf{P})} \mathsf{E}_{\widetilde{\mathsf{P}}} F_\tau. \tag{11.79}$$

If the measure $\widetilde{\mathsf{P}}$ is a unique martingale measure (equivalently, if the market is arbitrage free and complete), then

$$\widetilde{\mathbb{C}}^*(F) = \widetilde{\mathbb{C}}_*(F) \tag{11.80}$$

and

$$\widetilde{\mathbb{C}}^*(F) = \sup_{\tau \in \mathfrak{M}_N} \mathsf{E}_{\widetilde{\mathsf{P}}} F_\tau. \tag{11.81}$$

Remark 11.1. If we reject the above assumption $B_n \equiv 1$, then the formulae of Theorem 11.7 remain valid up to replacing F_n by $F_n B_0 / B_n$. The martingality of the measure $\widetilde{\mathsf{P}}$ from $\mathcal{M}(\mathsf{P})$ will then imply that the sequence $(S_n/B_n)_{0 \le n \le N}$ is a $\widetilde{\mathsf{P}}$-martingale and $\widetilde{\mathsf{P}} \ll \mathsf{P}$.

For the proof of this theorem and of its various modifications see [120], [161], [31], [66], [132], [137].

It is important for us to emphasize that the finding of the upper price $\widetilde{\mathbb{C}}^*(F)$ of American options on arbitrage-free complete markets reduces to solving the *optimal stopping* problem: to find the optimal time τ^* (if it exists), such that

$$\mathsf{E}_{\widetilde{\mathsf{P}}} F_{\tau^*} = \sup_{\tau \in \mathfrak{M}_N} \mathsf{E}_{\widetilde{\mathsf{P}}} F_\tau. \tag{11.82}$$

4. The theory of optimal stopping is the key theme of a number of monographs (see, *e.g.*, [159], [48], [139]). In many modern textbooks, special chapters are devoted to this theory, [137], [160]. In mathematical statistics, the methods of the theory of optimal stopping rules are widely used for "statistical sequential analysis", which, in essence, was the "ground floor" for this theory (see [172], [33]).

5. With minor changes, the assertions of Theorem 11.7 translate to the continuous-time case, although the techniques of proofs become more complicated (for details, see, *e.g.*, [161; Chap. VIII, § 2], where the corresponding literature is cited and examples are given on pricing of American options in arbitrage-free complete markets).

11.3 Duality and Symmetry of the Semimartingale Models

1. In Sec. 11.1, for the CRR-model of the financial markets we considered the relationship between the rational prices $\mathbb{P}_N$ and $\mathbb{C}_N$ of the standard put and call options of European type with the pay-off functions $F_N = (K - S_N)^+$ and $F_N = (S_N - K)^+$.

This relationship is described by the so-called *call-put parity* formula:

$$\mathbb{P}_N = \mathbb{C}_N - S_0 + K(1+r)^{-N} \tag{11.83}$$

which is the simple corollary of the identity $(K - S_N)^+ = (S_N - K)^+ - S_N + K$ integrated with respect to the martingale measure $\widetilde{\mathsf{P}}_N$ (see Theorem 11.1).

The importance of the formula (11.83) is that it permits to reduce the work of the calculation of the rational prices, namely, if you know $\mathbb{C}_N$ then you know $\mathbb{P}_N$, and vice versa.

For American options the similar formula does not exist. However, we shall see in the next section that there exists an analogue in the form of the so-called *call-put duality formula*, which is based on the ideas of the change of measures. In the several words: *duality formulae* are based on the *duality principle* which states that

> the calculation of the rational price for a *call option* in the model $S = e^H$, where H is a semimartingale (with respect to the initial probability measure P) is equivalent to the calculation of the rational price for a *put option* in a suitable *dual model* $S' = e^{H'}$ and under the *dual* probability measure P'.

Below we give (following [61]) definitions of these dual models (S, P) and (S', P') of markets and, first of all, we study the relationship between their probability characteristics. Applications to the call-put duality formulae (again following [61]) will be given in Sec. 11.4.

2. In all our subsequent considerations a stochastic basis $(\Omega, \mathcal{F}, (\mathcal{F}_t)_{t\le T}, \mathsf{P})$ is assumed given. The process $S = (S_t)_{t\le T}$ of prices will have the form

$$S = e^H, \tag{11.84}$$

where $H = (H_t)_{t\ge 0}$ is a semimartingale with $H_0 = 0$, whose triplet of predictable characteristics is

$$\mathbb{T}(H \mid \mathsf{P}) = (B, C, \nu)$$

with respect to the some truncation function $h = h(x)$ (Sec. 3.2).

We have already emphasized several times that, together with the representation (11.84), it is very useful to operate with the representation

$$S_t = \mathcal{E}(\widetilde{H})_t, \tag{11.85}$$

where $\mathcal{E}(\widetilde{H})_t$ is the *stochastic exponential* and

$$\widetilde{H}_t = \mathcal{L}(e^H)_t \tag{11.86}$$

is the *stochastic logarithm* of the process e^H (see Sec. 4.1).

Denote by $\widetilde{\mathbb{T}}(\widetilde{H} \mid \mathsf{P}) = (\widetilde{B}, \widetilde{C}, \widetilde{\nu})$ the triplet of predictable characteristics of $\widetilde{H}$. Recall that connection between $\widetilde{\mathbb{T}}(H \mid \mathsf{P})$ and $\widetilde{\mathbb{T}}(\widetilde{H} \mid \mathsf{P})$ is given by the formulae (4.13) and (4.14) which play a crucial role in the sequel. Recall also that for any triplet (B, C, ν) the ("internal") characteristics C and ν do not depend on h. However, the first characteristic $B = B(h)$ depends on h and for two different truncation functions h and h'

$$B(h) - B(h') = (h - h') * \nu$$

(see (3.31)).

3. Denote by $\mathcal{M}_{\rm loc}(\mathsf{P})$ the class of local martingales given on the stochastic basis $(\Omega, \mathcal{F}, (\mathcal{F}_t)_{t\le T}, \mathsf{P})$.

From Chap. 3 (Sec. 3.2, Subsec. 3) we know that

$$H \in \mathcal{M}_{\rm loc}(\mathsf{P}) \iff \begin{pmatrix} B + (x - h(x)) * \nu = 0 \\ \text{(up to an evanescent set)} \end{pmatrix}. \tag{11.87}$$

Similarly,

$$\widetilde{H} \in \mathcal{M}_{\rm loc}(\mathsf{P}) \iff \begin{pmatrix} \widetilde{B} + (x - h(x)) * \widetilde{\nu} = 0 \\ \text{(up to an evanescent set)} \end{pmatrix}. \tag{11.88}$$

Our subsequent considerations will be based on the following assumption.

Assumption $\mathbb{ES}$. *The process $I_{\{x>1\}}e^x * \nu$ has a bounded variation.*

This assumption is equivalent to the property that the semimartingale H is $\mathbb{E}$xponentially $\mathbb{S}$pecial, *i.e.*, the process $S = e^H$ is a special semimartingale (see Sec. 3.2 and [111]).

Taking into account the formulae (4.13), we can rewrite (11.88) in the form

$$\widetilde{H} \in \mathcal{M}_{\mathrm{loc}}(\mathsf{P}) \iff B + \frac{c}{2} + (e^x - 1 - h(x)) * \nu = 0 \ (\mathsf{P}\text{-a.s.}). \tag{11.89}$$

Since

$$\widetilde{H} \in \mathcal{M}_{\mathrm{loc}}(\mathsf{P}) \iff \mathcal{E}(\widetilde{H}) \in \mathcal{M}_{\mathrm{loc}}(\mathsf{P}),$$

we find from (11.89) that

$$S = e^H = \mathcal{E}(\widetilde{H}) \in \mathcal{M}_{\mathrm{loc}}(\mathsf{P}) \iff B + \frac{c}{2} + (e^x - 1 - h(x)) * \nu = 0 \ (\mathsf{P}\text{-a.s.}). \tag{11.90}$$

4. From now on we assume that with respect to the initial ("physical") measure P the process S is not only a local martingale ($S \in \mathcal{M}_{\mathrm{loc}}(\mathsf{P})$) but also a *martingale* ($S \in \mathcal{M}(\mathsf{P})$) on $[0, T]$.

Hence $\mathsf{E}S_T = 1$, and nonnegativity of S allows one to introduce a new, *dual*, measure P' such that

$$d\mathsf{P}' = S_T \, d\mathsf{P}. \tag{11.91}$$

It is clear that

$$d(\mathsf{P}'|\mathcal{F}_t) = S_t \, d(\mathsf{P}|\mathcal{F}_t)$$

and, because $S > 0$ (P-a.s.) we get from (11.91) that

$$d\mathsf{P} = \frac{1}{S_T} \, d\mathsf{P}'. \tag{11.92}$$

Put $S' = 1/S$, $H' = -H$, then

$$S' = e^{H'}.$$

The process S' is called a *dual process* (with respect to S). The pair (S', P') is called a *dual model.*

The following simple, but in many respect useful, formula plays a key role for the duality principle.

Lemma 11.1. *Suppose that $S = e^H \in \mathcal{M}(\mathsf{P})$,* i.e., *$S$ is a P-martingale. Then the process $S' = e^{H'} \in \mathcal{M}(\mathsf{P}')$,* i.e., *$S'$ is a P'-martingale.*

Proof. We will use the following general result [100; Chap. III, Proposition 3.8]: *If* $Z = \dfrac{d\mathsf{P}'}{d\mathsf{P}}$, *then*

$$S' \in \mathcal{M}(\mathsf{P}') \Leftrightarrow S'Z \in \mathcal{M}(\mathsf{P}).$$

In our case $Z = S$ and $S'S = 1$. Hence $S' \in \mathcal{M}(\mathsf{P}')$. □

5. The following proposition is critical for the application of the duality principle to calculation of option prices.

Theorem 11.8. *Suppose that* $S = e^H \in \mathcal{M}(\mathsf{P})$, *where* H *is a semimartingale with the triplet* $\mathbb{T}(H \mid \mathsf{P}) = (B, C, \nu)$, *and let the truncation function* $h = h(x)$ *satisfy the antisymmetry property* $h(-x) = -h(x)$. *Then the triplet* $\mathbb{T}(H' \mid \mathsf{P}')$ *for the dual model* (S', P'), *where* $S' = e^{H'}$, $H' = -H$, $d\mathsf{P}' = S_T\, d\mathsf{P}$, *can be expressed via the triplet* $\mathbb{T}(H \mid \mathsf{P})$ *of the* (S, P)*-model by the following formulae*:

$$\begin{aligned} B' &= -B - C - h(x)(e^x - 1) * \nu, \\ C' &= C, \\ I_A(x) * \nu' &= I_A(-x)e^x * \nu, \qquad A \in \mathcal{B}(\mathbb{R} \setminus \{0\}). \end{aligned} \tag{11.93}$$

Proof. We prefer to give two proofs, since these different proofs contain some additional useful relationships between different triplets obtained under different transformations of the processes and measures.

The following scheme gives an idea about the structure of the proofs:

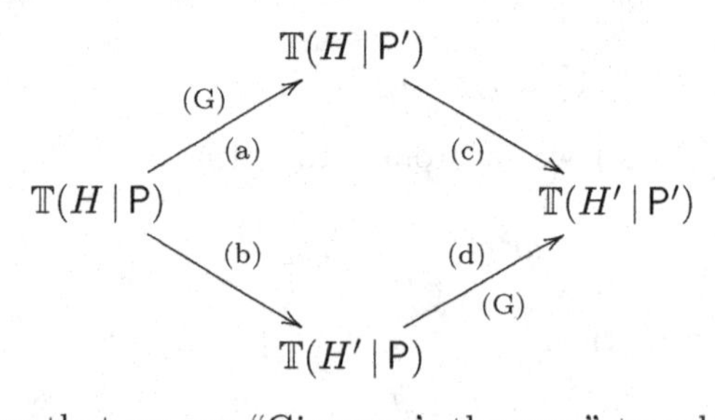

where (G) means that we use "Girsanov's theorem" to calculate the right-hand triplet from the left-hand one.

(a) $\mathbb{T}(H \mid \mathsf{P}) \overset{(G)}{\longrightarrow} \mathbb{T}(H \mid \mathsf{P}')$. Let $\mathbb{T}(H \mid \mathsf{P}') = (B^+, C^+, \nu^+)$. Girsanov's theorem for semimartingales (Theorem 6.1) states that

$$\begin{aligned} B^+ &= B + \beta^+ \cdot C + h(x), \\ C^+ &= C, \\ \nu^+ &= Y^+ \cdot \nu. \end{aligned} \tag{11.94}$$

Here $\beta^+ = \beta_t^+(\omega)$ and $Y^+ = Y_t^+(\omega; t, x)$ are defined by (6.24) and (6.25) (see also [100; Chap. III, Theorem 3.24]) and can for our case be written in the following way:

$$\langle S^c, H^c \rangle = \langle S_- \beta^+ \rangle \cdot C, \tag{11.95}$$

$$Y^+ = M^{\mathsf{P}}_{\mu^H}\left(\frac{S}{S_-} \,\Big|\, \widetilde{\mathcal{P}}\right). \tag{11.96}$$

Here $\widetilde{\mathcal{P}} = \mathcal{P} \otimes \mathcal{B}(\mathbb{R})$ is the σ-field of predictable sets in $\widetilde{\Omega} = \Omega \times [0, T] \times \mathbb{R}$ and $M^{\mathsf{P}}_{\mu^H} = \mu^H(\omega; dt, dx)\, \mathsf{P}(d\omega)$ is the positive measure on $(\widetilde{\Omega}, \mathcal{F} \otimes \mathcal{B}([0, T]) \otimes \mathcal{B}(\mathbb{R}))$ defined by

$$M^{\mathsf{P}}_{\mu^H}(W) = \mathsf{E}(W * \mu^H)_T$$

for measurable nonnegative functions $W = W(\omega; t, x)$ given on $\Omega \times [0, T] \times \mathbb{R}$. The conditional expectation $M^{\mathsf{P}}_{\mu^H}(S/S_- \mid \widetilde{\mathcal{P}})$ in (11.96) is by definition the $M^{\mathsf{P}}_{\mu^H}$-a.s. unique $\widetilde{\mathsf{P}}$-measurable function Y^+ with the property

$$M^{\mathsf{P}}_{\mu^H}\left(\frac{S}{S_-} U\right) = M^{\mathsf{P}}_{\mu^H}(Y^+ U) \tag{11.97}$$

for all nonnegative $\widetilde{\mathsf{P}}$-measurable functions $U = U(\omega; t, x)$.

It turns out that for our special case $S = e^H$, where evidently $S/S_- = e^{\Delta H}$, one can take the following version of β^+ and Y^+:

$$\beta^+ \equiv 1, \qquad Y^+ = e^x. \tag{11.98}$$

Indeed, in (11.95), for $S = e^H$ applying Itô's formula to e^H we get

$$(e^H)^c = \int_0^{\cdot} e^{H_{s-}}\, dH_s^c. \tag{11.99}$$

Hence, for the predictable quadratic covariance $\langle S^c, H^c \rangle$ (see Sec. 3.1) we find the representation

$$\begin{aligned} \langle S^c, H^c \rangle &= \langle (e^H)^c, H^c \rangle = \left\langle \int_0^{\cdot} e^{H_-}\, dH^c, H^c \right\rangle \\ &= \int_0^{\cdot} e^{H_-}\, d\langle H^c \rangle = \int_0^{\cdot} e^{H_-}\, dC = S_- \cdot C. \end{aligned} \tag{11.100}$$

In other words,

$$\langle S_- \beta^+ \rangle \cdot C = \int_0^{\cdot} e^{H_-} \beta^+\, dC. \tag{11.101}$$

From (11.100), (11.101), and (11.95) we see that one can take $\beta^+ \equiv 1$.

To prove that one can choose $Y^+ = e^x$ we need to verify (11.97) with this version of Y^+.

Taking into account that μ^H is the random measure of jumps of H, we have:

$$\begin{aligned}
M^{\mathsf{P}}_{\mu^H}(e^x U) &= \mathsf{E}\left[\int_0^T \int_{\mathbb{R}} e^x U(\omega; t, x)\, \mu^H(\omega; dt, dx)\right] \\
&= \mathsf{E}\left[\sum_{0<t\le T} e^{\Delta H_t(\omega)} U(\omega; t, \Delta H_t(\omega)) I(\Delta H_t(\omega) \neq 0)\right] \\
&= \mathsf{E}\left[\int_0^T \int_{\mathbb{R}\setminus\{0\}} \frac{S_t(\omega)}{S_{t-}(\omega)} U(\omega; t, x)\, \mu^H(\omega; dt, dx)\right] \\
&= M^{\mathsf{P}}_{\mu^H}\left(\frac{S}{S_-} U\right).
\end{aligned}$$

Thus, in (11.97) one can take $Y = e^x$.

Consequently, choosing $\beta^+ \equiv 1$ and $Y = e^x$, we obtain from (11.94) that

$$\begin{aligned}
B^+ &= B + C + h(x)(e^x - 1) * \nu, \\
C^+ &= C, \\
\nu^+ &= e^x \cdot \nu.
\end{aligned} \tag{11.102}$$

Remark 11.2. It is reasonable to note that for the *discrete time* case the relation $d\nu^+ = e^x\, d\nu$ can be proved (with the obvious notation) in the following simple way.

Let $h_n = \Delta H_n$, and let $\mu_n = \mu_n(\omega; \cdot)$ be the random measure of jumps at time n, *i.e.*,

$$\mu_n(\omega; A) = I(h_n(\omega) \in A), \qquad A \in \mathcal{B}(\mathbb{R} \setminus \{0\}).$$

For this discrete time case the compensator $\nu_n = \nu_n(\omega; \cdot)$ of $\mu_n = \mu_n(\omega; \cdot)$ has the simple form

$$\nu_n(\omega; A) = \mathsf{P}(h_n \in A \,|\, \mathcal{F}_{n-1})(\omega)$$

(see Sec. 9.2). Put $\nu_n^+(\omega; A) = \mathsf{P}'(h_n \in A \,|\, \mathcal{F}_{n-1})(\omega)$. Then from the generalized Bayed formula or the conversion formula for conditional expectations (Lemma 6.2)

$$\begin{aligned}
\nu_n^+(\omega; A) &= \mathsf{E}'\big[I_A(h_n) \,|\, \mathcal{F}_{n-1}\big](\omega) \\
&= \mathsf{E}\big[I_A(h_n) e^{h_n} \,|\, \mathcal{F}_{n-1}\big](\omega) = \int_A e^x\, \nu(\omega; dx).
\end{aligned}$$

Therefore

$$\nu_n^+ \ll \mu_n \qquad \text{and} \qquad \frac{d\nu_n^+}{d\nu_n}(\omega; x) = e^x \quad (\nu_n\text{-a.s.}).$$

(b) $\mathbb{T}(H \mid \mathsf{P}) \longrightarrow \mathbb{T}(H' \mid \mathsf{P})$. Since $H' = -H$, we get that $\mathbb{T}(H' \mid \mathsf{P}) = \mathbb{T}(-H \mid \mathsf{P})$. Let $\mathbb{T}(-H \mid \mathsf{P}) = (B^-, C^-, \nu^-)$. It is clear that

$$-H_t = \int_0^t \varphi_s \, dH_s, \quad \text{where } \varphi_s \equiv -1.$$

So, for calculation of the triplet $\mathbb{T}(-H \mid \mathsf{P})$ we can use formulae (4.41) which give

$$\begin{aligned} B^- &= -B, \\ C^- &= C, \\ I_A(x) * \nu^- &= I_A(-x) * \nu, \qquad A \in \mathcal{B}(\mathbb{R} \setminus \{0\}). \end{aligned} \tag{11.103}$$

(c) $\mathbb{T}(H \mid \mathsf{P}') \longrightarrow \mathbb{T}(H' \mid \mathsf{P}')$. The triplet $\mathbb{T}(H \mid \mathsf{P}') = (B^+, C^+, \nu^+)$ is given by (11.102). Then from (11.103) we get

$$\begin{aligned} B' &= -B^+ = -B - C - h(x)(e^x - 1) * \nu, \\ C' &= C^+ = C, \\ I_A(x) * \nu' &= I_A(-x) * \nu^+ = I_A(-x) e^x * \nu. \end{aligned} \tag{11.104}$$

The formulae (11.93) follow from (a) and (c).

(d) $\mathbb{T}(H' \mid \mathsf{P}) \overset{(G)}{\longrightarrow} \mathbb{T}(H' \mid \mathsf{P}')$. Here $\mathbb{T}(H' \mid \mathsf{P}) = \mathbb{T}(-H \mid \mathsf{P}) = (B^-, C^-, \nu^-)$ and $\mathbb{T}(H' \mid \mathsf{P}') = \mathbb{T}(-H \mid \mathsf{P}') = (B', C', \nu')$.

Similarly to the case (a),

$$B' = B^- + \beta^- \cdot C^- + h(x)(Y^- - 1) * \nu^- \tag{11.105}$$

$$C' = C^- \tag{11.106}$$

$$\nu' = Y^- \cdot \nu^-, \tag{11.107}$$

where $\beta^- = \beta_t^-(\omega)$ and $Y^- = Y_t^-(\omega; t, x)$ are given by

$$\langle S^c, (-H)^c \rangle = (S_- \beta^-) \cdot C^- \tag{11.108}$$

and

$$Y^- = M^{\mathsf{P}}_{\mu^{-H}}\Big(\frac{S}{S_-} \,\Big|\, \widetilde{\mathcal{P}}\Big). \tag{11.109}$$

(cf. (11.95) and (11.96)). We have

$$\begin{aligned} \langle S^c, (-H)^c \rangle &= \langle (e^H)^c, -H^c \rangle = \Big\langle \int_0^{\cdot} e^{H_-} \, dH^c, -H^c \Big\rangle = -\int_0^{\cdot} e^{H_-} \, d\langle H^c \rangle \\ &= -\int_0^{\cdot} e^{H_-} \, d\langle (-H)^c \rangle = -\int_0^{\cdot} e^{H_-} \, dC^- = (-S_-) \cdot C^-. \end{aligned} \tag{11.110}$$

Comparing (11.108) and (11.110), we see that one can take $\beta^- \equiv -1$.

Similarly to the calculations in (a) we derive

$$\begin{aligned}
M^{\mathsf{P}}_{\mu^{-H}}(e^{-x}U) &= \mathsf{E}\left[\int_0^T \int_{\mathbb{R}} e^{-x}U(\omega;t,x)\,\mu^{-H}(\omega;dt,dx)\right] \\
&= \mathsf{E}\left[\sum_{0<t\le T} e^{-\Delta(-H_t(\omega))}U(\omega;t,\Delta(-H_t(\omega)))I\{\Delta(-H_t(\omega))\neq 0\})\right] \\
&= \mathsf{E}\left[\int_0^T \int_{\mathbb{R}} \frac{S_t(\omega)}{S_{t-}(\omega)}\,U(\omega;t,x)\,\mu^{-H}(\omega;dt,dx)\right] \\
&= M^{\mathsf{P}}_{\mu^{-H}}\left(\frac{S}{S_-}U\right).
\end{aligned}$$

Therefore one can take $Y^- = e^{-x}$ in (11.109), and from (11.105)–(11.107) and (11.103) we find that

$$\begin{aligned}
B' &= -B - C + h(x)(e^{-x}-1)*\nu^-, \\
C' &= C, \qquad\qquad (11.111)\\
\nu' &= e^{-x}\cdot\nu^-,
\end{aligned}$$

where ν^- is such that $I_A(x)*\nu^- = I_A(-x)*\nu$ for all $A \in \mathcal{B}(\mathbb{R}\setminus\{0\})$. Hence

$$I_A(x)*\nu' = I_A(x)*(e^{-x}\cdot\nu') = I_A(-x)e^x*\nu. \qquad (11.112)$$

As was mentioned before the formulation of the theorem, we assume that the truncation function $h = h(x)$ is antisymmetric ($h(-x) = -h(x)$). Thus,

$$h(x)(e^{-x}-1)*\nu^- = h(-x)(e^x-1)*\nu = -h(x)(e^x-1)*\nu. \qquad (11.113)$$

From (11.111)–(11.113) we find that the triplet $\mathbb{T}(H'\,|\,\mathsf{P}') = (B', C', \nu')$ is given by (11.93). The theorem is proved. □

Corollary 11.3. *Let H be a Lévy process with the triplet $\mathbb{T}_{\mathrm{loc}}(H\,|\,\mathsf{P}) = (b, c, F)$ of local characteristics* (Sec. 5.2). *Let $S = e^H \in \mathcal{M}_{\mathrm{loc}}(\mathsf{P})$. Then by* (11.90) *or* (7.14) *(under assumption $\int_{|x|>1} e^x\,F(dx) < \infty$)*

$$b + \frac{c}{2} + \int_{\mathbb{R}} (e^x - 1 - h(x))\,F(dx) = 0$$

and $S \in \mathcal{M}(\mathsf{P})$ (see (7.14)).

The process $S' = e^{H'} \in \mathcal{M}_{\mathrm{loc}}(\mathsf{P}')$ and its triplet $\mathbb{T}_{\mathrm{loc}}(H'\,|\,\mathsf{P}') = (b', c', F')$ has the following form (with the assumption $h(x) = -h(-x)$):

$$\begin{aligned}
b' &= -b - c - \int h(x)(e^x-1)\,F(dx), \\
c' &= c, \\
F'(A) &= \int_{\mathbb{R}} I_A(-x)e^x\,F(dx), \qquad A \in \mathcal{B}(\mathbb{R}\setminus\{0\}).
\end{aligned}$$

6. Now we consider a series of simple examples that show how to calculate the triplet $\mathbb{T}(H' \mid \mathsf{P}')$ from the triplet $\mathbb{T}(H \mid \mathsf{P})$. For more complicated (Lévy) models (for which Assumption $\mathbb{ES}$ does hold and therefore Theorem 11.8 can be used to calculate the characteristics of the dual process S') see [61; § 3]).

Example 11.1. (Brownian case.) Suppose that $S = e^H$ and $\nu = 0$. Then $S \in \mathcal{M}_{\rm loc}(\mathsf{P})$ if and only if $B + C/2 = 0$ (see (11.90)). By Theorem 11.8 the triplet $\mathbb{T}(H' \mid \mathsf{P}') = (B', C', 0)$ has the form

$$B' = -(B + C),$$
$$C' = C.$$

So, $B' + C'/2 = -(B + C/2) = 0$ that implies $S' \in \mathcal{M}_{\rm loc}(\mathsf{P}')$. In particular, if

$$S_t = e^{\sigma W_t - \sigma^2 t/2},$$

i.e., $H_t = \sigma W_t - \sigma^2 t/2$, where $W = (W_t)_{0 \le t \le T}$ is a standard Brownian motion (Wiener process), then

$$B_t = -\frac{\sigma^2}{2}\,t, \quad C_t = t. \tag{11.114}$$

Evidently, here $B + C/2 = 0$ which implies that $S \in \mathcal{M}_{\rm loc}(\mathsf{P})$. In fact, here $S \in \mathcal{M}(\mathsf{P})$ and $dS_t = \sigma S_t\, dW_t$. The process $S' = e^{H'} = e^{-H}$ has the stochastic differential

$$dS'_t = -\sigma S'_t (dW_t - \sigma\, dt). \tag{11.115}$$

Since $S' \in \mathcal{M}_{\rm loc}(\mathsf{P}')$, from (11.115) and (4.7) we conclude that the process $W'_t = W_t - \sigma t$, $0 \le t \le T$, is a P'-local martingale. Of course, this is a particular case of the Girsanov theorem (Sec. 6.3). In fact, W' is a P'-martingale (as follows from Lévy's characterization of the Brownian motion: if a continuous process $X = (X_t)_{t \ge 0} \in \mathcal{M}_{\rm loc}(\mathsf{P})$ with $X_0 = 0$ is such that $X_t^2 - t \in \mathcal{M}_{\rm loc}(\mathsf{P})$, then X is a standard Brownian motion). So,

$$dS'_t = -\sigma S'_t\, dW'_t, \tag{11.116}$$

$$S'_t = e^{-\sigma W'_t - \sigma^2 t/2}, \tag{11.117}$$

and

$$B'_t = -\frac{\sigma^2}{2}\,t, \quad C'_t = t. \tag{11.118}$$

From (11.114) and (11.118) we see that for models $S = e^H$ and $S' = e^{H'}$ the triplets $\mathbb{T}(H \mid \mathsf{P})$ and $\mathbb{T}(H' \mid \mathsf{P}')$ coincide:

$$B = B' \quad \text{and} \quad C = C'. \tag{11.119}$$

This property of symmetry make it reasonable to say that a model (S, P) and its dual model (S', P') have the *symmetry property* if their triplets $\mathbb{T}(H \mid \mathsf{P})$ and $\mathbb{T}(H' \mid \mathsf{P}')$ coincide.

Remark 11.3. For many cases the triplets of semimartingales define their distributions uniquely. For example, such is the case of processes with independent increments in the canonical setting; see [100] for this and other examples. So, the given definition of the symmetry is an analog of the property that the probability laws of the models (S, P) and (S', P') coincide.

Example 11.2. (Poissonian case.) Suppose that S has the form $S = e^H$ with

$$H_t = \alpha\pi_t - \lambda(e^\alpha - 1)t, \qquad \alpha \neq 0 \tag{11.120}$$

where $\pi = (\pi_t)_{0 \le t \le T}$ is a Poisson process with parameter $\lambda > 0$, $\mathsf{E}\pi_t = \lambda t$. (Compare with the process $(Z_t)_{t \ge 0}$ in Example 6.2 on page 130.) Take $h(x) = 0$ (this is possible, since for the Poissonian case the Lévy measure is "sitting" in one point and thus in the Kolmogorov–Lévy–Khinchin formula, see Sec. 4.2, the truncation—which guarantees the convergence of the integral—is not needed).

From (11.120) we find that the triplet $\mathbb{T}(H \mid \mathsf{P}) = (B, C, \nu)$ has the form

$$\begin{aligned} B_t &= -\lambda(e^\alpha - 1)t, \\ C_t &= 0, \\ \nu(dt, dx) &= \lambda I_{\{\alpha\}}(dx)\,dt. \end{aligned} \tag{11.121}$$

It follows from (11.90) that

$$S = e^H \in \mathcal{M}_{\mathrm{loc}}(\mathsf{P}) \iff B + (e^x - 1) * \nu = 0. \tag{11.122}$$

By (11.121)

$$B_t + (e^x - 1) * \nu_t = -\lambda(e^\alpha - 1)t + \lambda(e^\alpha - 1)t = 0.$$

Hence, by (11.122) $S \in \mathcal{M}_{\mathrm{loc}}(\mathsf{P})$ and, in fact, $S \in \mathcal{M}(\mathsf{P})$, since H is a process with independent increments (see the corollary to Theorem 11.7 on page 234).

So, one can define the measure P' with $d\mathsf{P}' = S_T\, d\mathsf{P}$.

By Theorem 11.8

$$B'_t = \lambda(e^\alpha - 1)t \tag{11.123}$$

and

$$\nu'(dx, dt) = I_{\{\alpha\}}(dx)\lambda e^\alpha\, dt. \tag{11.124}$$

From these formulae it follows, by the way, that

$$B'_t + (e^x - 1) * \nu'_t = \lambda(e^\alpha - 1)t + (e^{-\alpha} - 1)\lambda e^\alpha t = 0.$$

So, by criterion (11.90) (with $C = 0$, $h = 0$) we find that $S' = e^{H'} \in \mathcal{M}_{\mathrm{loc}}(\mathsf{P}')$.

Example 11.3. (Discrete time, CRR-model.) In the Cox–Ross–Rubinstein model (CRR-model) described in (10.16) the asset prices are modeled by $S_n = e^{H_n}$, where $H_n = h_1 + \cdots + h_n$, $n \geq 1$, $H_0 = 0$, where $(h_n)_{n\geq 1}$ is a sequence of P-i.i.d. random variables h_n with values $\log \lambda$ and $\log(1/\lambda)$, $\lambda > 1$.

If the probability measure P is such that

$$\mathsf{P}\Big(h_n = \log\frac{1}{\lambda}\Big) = \frac{\lambda}{1+\lambda} \quad \text{and} \quad \mathsf{P}(h_n = \log\lambda) = \frac{1}{1+\lambda},$$

then we find

$$\mathsf{E}e^{h_n} = 1.$$

This property implies that P is a martingale measure for the sequence $S = (S_n)_{n\geq 0}$. One can claim more, that the measure P is a unique martingale measure for the CRR-model (see, for example, [161; Example 2 on pages 477–480]).

Given a truncation function $h(x) = x$ and a martingale measure P, we find the triplet $\mathbb{T}(H \mid \mathsf{P}) = (B, 0, \nu)$:

$$\Delta B_n = \mathsf{E}h_n = \frac{1-\lambda}{1+\lambda}\log\lambda \tag{11.125}$$

and

$$\begin{aligned} \nu_n(\{\log\lambda\}) &= \mathsf{P}(h_n = \log\lambda) = \frac{1}{1+\lambda}, \\ \nu_n\Big(\Big\{\log\frac{1}{\lambda}\Big\}\Big) &= \mathsf{P}\Big(h_n = \log\frac{1}{\lambda}\Big) = \frac{\lambda}{1+\lambda}. \end{aligned} \tag{11.126}$$

Note that (11.125) and (11.126) imply

$$\Delta B_n + (e^x - 1 - x) * \nu_n = 0. \tag{11.127}$$

Comparing this property with (11.90), we obtain by a different way the property $S \in \mathcal{M}_{\mathrm{loc}}(\mathsf{P})$ (which in the considered case is equivalent to the property $S \in \mathcal{M}(\mathsf{P})$).

It is easy to derive from (11.93) that

$$\begin{aligned} \Delta B'_n &= \Delta B_n, \\ \nu'_n &= \nu_n \end{aligned} \tag{11.128}$$

for all $n \geq 1$. These formulae clearly show that for CRR-model the symmetry property does hold.

For the dual market (S', P'), we have the martingale property $S' \in \mathcal{M}(\mathsf{P}')$. (See again (11.90) and (11.128) for the direct proof of the inclusion $S' \in \mathcal{M}_{\mathrm{loc}}(\mathsf{P}')$ which is equivalent here to the property $S' \in \mathcal{M}(\mathsf{P})$.)

11.4 Call-Put Duality in Option Pricing. II. Lévy Models

1. In this section we give several examples which illustrate how the duality principle together with the results of Theorem 11.8 can be used for calculating the rational prices of different options.

Let $S = (S_t)_{0 \le t \le T}$ be the price process and $F_T = F_T(S)$ the pay-off of the option. Here $F_T(S) = f_T(S_t, 0 \le t \le T)$ is a $\mathcal{F}_T^S$-measurable functional, where $\mathcal{F}_T^S = \sigma(S_t, 0 \le t \le T)$. We assume that on the given market there exists a bank account with zero interest rate. (This restriction is imposed for simplicity only. The case of a positive interest rate and dividends can be investigated in a similar way.)

Let us make some comments about using martingale measures in option pricing.

From Chap. 10, Sec. 10.1, we know that in a complete arbitrage-free market—where the martingale measure, say P, is unique—the rational (or fair, or arbitrage-free) price of the options is given by the formula (10.26) in the discrete time case and by the formula

$$\mathbb{C}_T = \mathsf{E} F_T$$

in the continuous time case (see Theorem 10.8(a)).

In incomplete arbitrage-free markets—where the martingale measure is not unique—there exists a difficult problem of finding a suitable, reasonable such measure. There are many different approaches to choosing a martingale measure, *e.g.*, in the sense of minimization of a distance from the given measure P (L^2-distance, Hellinger distance, f-divergence, ...; see Sec. 7.4) or in the sense of constructing the simplest possible measure (*e.g.*, Esscher transforms; see Sec. 6.4). The practitioner's point of view is that the "good", "reasonable" choice of this measure should be the result of a calibration to market prices of plain vanilla options.

In the sequel we assume that the initial measure P is already a martingale measure. The value $\mathsf{E}F_T$, where expectation is taken with respect to this measure P, is called a *quasi-rational* price (on the incomplete market).

2. European call and put options. In the case of a standard *call option* the pay-off function is given by

$$F_T = (S_T - K)^+, \qquad K > 0, \tag{11.129}$$

whereas for a *put option*

$$F_T = (K - S_T)^+, \qquad K > 0. \tag{11.130}$$

The corresponding (quasi-rational, or P-quasi-rational) option prices are defined by the formulae

$$\mathbb{C}_T(S;K) = \mathsf{E}(S_T - K)^+ \tag{11.131}$$

and

$$\mathbb{P}_T(K;S) = \mathsf{E}(K - S_T)^+, \tag{11.132}$$

where E is the expectation with respect to the initial martingale measure P.

From (11.131) for $S = e^H$ we get

$$\begin{aligned} \mathbb{C}_T(S;K) &= \mathsf{E}\left[S_T \frac{F_T}{S_T}\right] = \mathsf{E}'\left[\frac{F_T}{S_T}\right] = \mathsf{E}'(1 - KS'_T)^+ \\ &= K\mathsf{E}'\Big(\frac{1}{K} - S'_T\Big)^+ = K\mathsf{E}'\Big(K' - S'_T\Big)^+, \end{aligned} \tag{11.133}$$

where $K' = 1/K$.

Comparing (11.133) and (11.132) leads to the following result.

Theorem 11.9. *For the standard call option and the standard put option the rational prices (on a complete market) or quasi-rational prices (on an incomplete market) satisfy the duality relations:*

$$\frac{1}{K}\,\mathbb{C}_T(S;K) = \mathbb{P}'_T(K';S'), \tag{11.134}$$

$$\frac{1}{K}\,\mathbb{P}_T(K;S) = \mathbb{C}'_T(S';K'); \tag{11.135}$$

here $K' = 1/K$ and $\mathbb{P}'_T(K';S')$, $\mathbb{C}_T('S';K')$ are the corresponding prices for put and call options on (S',P')-market.

Corollary 11.4. *Call and put prices in the dual (S,P)- and (S',P')-markets satisfy the "call-call parity"*

$$\mathbb{C}_T(S;K) = K\mathbb{C}'_T(S';K') + 1 - K \tag{11.136}$$

and the "put-put parity"

$$\mathbb{P}_T(K;S) = K\mathbb{P}'_T(K';S') + K - 1. \tag{11.137}$$

Proof of this corollary follows directly from the call-put parity property

$$\mathbb{C}_T(S;K) = \mathbb{P}_T(K;S) + 1 - K \tag{11.138}$$

(cf. (11.83)) and the formulae (11.134), (11.135).

Remark 11.4. The formulae (11.134) and (11.135) are especially useful if the dual market (S',P') has "simpler" structure than the underlying market (S,P). Having the triplet $\mathbb{T}(H\,|\,\mathsf{P})$ for the (S,P)-market, we calculate the triplet $\mathbb{T}(H'\,|\,\mathsf{P}')$ for the dual (S',P')-market, after that we calculate the prices $\mathbb{P}'(K';S')$ and $\mathbb{C}'(S';K')$ and, finally, we find prices the $\mathbb{C}(S;K)$ and $\mathbb{P}(K;S)$ using the formulae (11.134), (11.135).

3. Floating strike lookback call and put options. For a floating strike lookback call option on the (S, P)-market with $S \in \mathcal{M}(\mathsf{P})$ we get (assuming $\alpha \geq 1$) that

$$\begin{aligned}
\mathbb{C}_T(S; \alpha \inf S) &= \mathsf{E}\Big(S_T - \alpha \inf_{t \le T} S_t\Big)^+ = \mathsf{E}\left[S_T\left(1 - \frac{\alpha \inf_{t\le T} S_t}{S_t}\right)^+\right] \\
&= \mathsf{E}'\Big(1 - \alpha \exp\Big\{\inf_{t\le T} H_t - H_T\Big\}\Big)^+ \\
&= \mathsf{E}'\Big(1 - \alpha \exp\Big\{H'_T - \sup_{t\le T} H'_t\Big\}\Big)^+ \\
&= \alpha \mathsf{E}'\Big(\alpha' - \exp\Big\{H'_T - \sup_{t\le T} H'_t\Big\}\Big)^+, \qquad (11.139)
\end{aligned}$$

where $\alpha' = 1/\alpha$.

To simplify further the last expression, assume that the process $H' = (H'_t)_{0\le t\le T}$ satisfied the following *reflection principle*:

$$\mathrm{Law}\Big(\sup_{t\le T} H'_t - H'_T \,|\, \mathsf{P}'\Big) = \mathrm{Law}\Big(-\inf_{t\le T} H'_t \,|\, \mathsf{P}'\Big). \qquad (11.140)$$

It is known that for Lévy processes this property does hold (see, *e.g.*, [121; Lemma 3.5]).

From (11.139), (11.140) we get (with $\alpha' = 1/\alpha$)

$$\begin{aligned}
\mathbb{C}_T(S; \alpha \inf S) &= \mathsf{E}'\Big(\alpha' - \exp\Big\{\inf_{t\le T} H'_t\Big\}\Big)^+ \\
&= \mathsf{E}'\Big(\alpha' - \inf_{t\le T} S'_t\Big)^+ = \mathbb{P}'_T(\alpha'; \inf S'). \qquad (11.141)
\end{aligned}$$

Similarly, assuming fulfilled the reflection principle

$$\mathrm{Law}\Big(H'_T - \inf_{t\le T} H'_t \,|\, \mathsf{P}'\Big) = \mathrm{Law}\Big(\sup_{t\le T} H'_t \,|\, \mathsf{P}'\Big), \qquad (11.142)$$

which also holds for Lévy processes (see the same lemma in [121]), we find that

$$\beta' \mathbb{P}_T(\beta \sup S; S) = \mathbb{C}'_T(\sup X'; \beta'), \qquad (11.143)$$

where $0 < \beta \le 1$ and $\beta' = 1/\beta$.

Hence, at least for Lévy processes we have the following result.

Theorem 11.10. *Consider a (S, P)-market with $S = e^H$, where H is a Lévy process. Then the calculation of the prices $\mathbb{C}_T(S; \alpha \inf S)$ and $\mathbb{P}_T(\beta \sup S; S)$ ($\alpha \ge 1$ and $0 < \beta \le 1$) with floating strikes $\alpha \inf S$ and $\beta \sup S$ can be reduced via formulae* (11.142), (11.143) *to the calculation of the prices $\mathbb{P}'_T(\alpha'; \inf S')$ and $\mathbb{C}'_T(\sup S'; \beta')$ (with $\alpha' = 1/\alpha$ and $\beta' = 1/\beta$) which are prices of lookback put and call options with fixed strikes.*

4. Floating strike Asian options. Suppose again that $S \in \mathcal{M}(\mathsf{P})$ and consider the prices

$$\mathbb{C}_T\Big(S; \frac{1}{T}\int S\Big) = \mathsf{E}\Big(S_T - \frac{1}{T}\int_0^T S_t\,dt\Big)^+ \tag{11.144}$$

and

$$\mathbb{P}_T\Big(\frac{1}{T}\int S; S\Big) = \mathsf{E}\Big(\frac{1}{T}\int_0^T S_t\,dt - S_T\Big)^+. \tag{11.145}$$

We have from (11.144):

$$\begin{aligned}\mathbb{C}_T\Big(S; \frac{1}{T}\int S\Big) &= \mathsf{E}\Big[S_T\Big(1 - \frac{1}{T}\int_0^T \frac{S_t}{S_T}\,dt\Big)^+\Big] = \mathsf{E}'\Big(1 - \frac{1}{T}\int_0^T \frac{S'_T}{S'_t}\,dt\Big)^+ \\ &= \mathsf{E}'\Big(1 - \frac{1}{T}\int_0^T \exp\{H'_T - H'_t\}\,dt\Big)^+ \\ &= \mathsf{E}'\Big(1 - \frac{1}{T}\int_0^T \exp\{H'_T - H'_{T-u}\}\,du\Big)^+. \end{aligned} \tag{11.146}$$

If the process H' is a Lévy process, then it is known [121; Lemma 3.4] that

$$\operatorname{Law}\big(H'_T - H'_{(T-t)-}; 0 \le t < T \,|\, \mathsf{P}'\big) = \operatorname{Law}\big(H'_t; 0 \le t < T \,|\, \mathsf{P}'\big). \tag{11.147}$$

From (11.146) and (11.147) we conclude that

$$\begin{aligned}\mathbb{C}_T\Big(S; \frac{1}{T}\int S\Big) &= \mathsf{E}'\Big(1 - \frac{1}{T}\int_0^T \exp\{H'_u\}\,du\Big)^+ \\ &= \mathsf{E}'\Big(1 - \frac{1}{T}\int_0^T S'_u\,du\Big)^+ = \mathbb{P}'_T\Big(1; \frac{1}{T}\int S'\Big). \end{aligned} \tag{11.148}$$

Similarly,

$$\mathbb{P}_T\Big(\frac{1}{T}\int S; S\Big) = \mathbb{C}'_T\Big(\frac{1}{T}\int S'; 1\Big). \tag{11.149}$$

Thus, we have the following result.

Theorem 11.11. *Let H be a Lévy process. Then the calculation of prices $\mathbb{C}'_T(S; \frac{1}{T}\int S')$ and $\mathbb{P}_T(\frac{1}{T}\int S; S)$ of Asian call and put options with floating strikes $\frac{1}{T}\int S$ can be reduced via formulae (11.148) and (11.149) to the calculation of the prices $\mathbb{P}'_T(1; \frac{1}{T}\int S')$ and $\mathbb{C}'_T(\frac{1}{T}\int S'; 1)$ of fixed strike Asian put and call options.*

5. Standard call and put options of American type. The general theory of pricing of American options on the complete (S, P)-market states (see [161; Chaps. VI, VIII]) that for (nonnegative) pay-off function $(F_t)_{0\le t\le T}$ with $T < \infty$ the price $\widehat{\mathbb{V}}_T(S)$ of the American option is given by

$$\widehat{\mathbb{V}}_T(S) = \sup_{\tau\in\mathfrak{M}_T} \mathsf{E}F_\tau, \tag{11.150}$$

where $\mathfrak{M}_T$ is the class of stopping times τ such that $0 \le \tau \le T$.

For a standard call option, $F_t = e^{-\lambda t}(S_t - K)^+$ with some $\lambda \ge 0$. For a standard put option we have $F_t = e^{-\lambda t}(K - S_t)^+$, where again $\lambda \ge 0$. Let

$$\widehat{\mathbb{C}}_T(S;K) = \sup_{\tau\in\mathfrak{M}_T} \mathsf{E}e^{-\lambda\tau}(S_\tau - K)^+,$$

$$\widehat{\mathbb{P}}_T(K;S) = \sup_{\tau\in\mathfrak{M}_T} \mathsf{E}e^{-\lambda\tau}(K - S_\tau)^+.$$

As in the case of European options, we find for $f_t = (S_t - K)^+$ that

$$\begin{aligned}
\widehat{\mathbb{C}}_T(S;K) &= \sup_{\tau\in\mathfrak{M}_T} \mathsf{E}e^{-\lambda\tau} f_\tau = \sup_{\tau\in\mathfrak{M}_T} \mathsf{E}e^{-\lambda\tau}\Big(f_\tau \frac{S_T}{S_T}\Big) \\
&= \sup_{\tau\in\mathfrak{M}_T} \mathsf{E}'e^{-\lambda\tau}(f_\tau S'_T) = \sup_{\tau\in\mathfrak{M}_T} \mathsf{E}'e^{-\lambda\tau}\Big(f_\tau \mathsf{E}'(S'_T \,|\, \mathcal{F}_\tau)\Big) \\
&= \sup_{\tau\in\mathfrak{M}_T} \mathsf{E}'e^{-\lambda\tau}(f_\tau S'_\tau) = \sup_{\tau\in\mathfrak{M}_T} \mathsf{E}'e^{-\lambda\tau}(S_\tau - K)^+ S'_\tau \\
&= \sup_{\tau\in\mathfrak{M}_T} \mathsf{E}'e^{-\lambda\tau}(1 - KS'_\tau)^+ = K \sup_{\tau\in\mathfrak{M}_T} \mathsf{E}'e^{-\lambda\tau}(K' - S'_\tau)^+ \\
&= K\widehat{\mathbb{P}}'_T(K';S').
\end{aligned}$$

Thus, for American standard call options

$$\frac{1}{K}\widehat{\mathbb{C}}_T(S;K) = \widehat{\mathbb{P}}'_T(K';S'),$$

and similarly for American standard put options

$$\frac{1}{K}\widehat{\mathbb{P}}_T(K;S) = \widehat{\mathbb{C}}'_T(S';K').$$

Chapter 12

Conditionally Brownian and Lévy Processes. Stochastic Volatility Models

12.1 From Black–Scholes Theory of Pricing of Derivatives to the Implied Volatility, Smile Effect and Stochastic Volatility Models

1. The two classical—Bachelier and Black–Scholes (or Samuelson–Black–Merton–Scholes)—models of the financial markets in which prices $S = (S_t)_{t \geq 0}$ are described by the formulae

$$S_t = S_0 + \mu t + \sigma B_t \tag{12.1}$$

and

$$S_t = S_0 \exp\Big\{\Big(\mu - \frac{\sigma^2}{2}\Big)t + \sigma B_t\Big\}, \tag{12.2}$$

respectively, have played remarkable roles both in formation and development of mathematical finance and in elaboration of stochastic methodology of trading on real financial markets.

At the heart of these models (linear and exponential, respectively) lies the idea that the stochasticity displayed by financial markets is generated by a Brownian motion $B = (B_t)_{t \geq 0}$. Independence of increments and Gaussianity of the latter permitted the mathematical analysis of these models; Bachelier's and Black–Scholes' formulae for the rational prices of standard European call and put options, on the one hand, and formulae for hedging (replicating) strategies, on the other, are prime examples of such an analysis.

Models (12.1) and (12.2) have the following discrete-time analogues:

$$S_n = S_0 + \mu n + \sigma \Sigma_n$$

and

$$S_n = S_0 e^{\mu n + \sigma \Sigma_n},$$

where $\Sigma_n = \varepsilon_1 + \cdots + \varepsilon_n$ (a sum of independent Gaussian, $\mathcal{N}(0,1)$, random variables); these analogues together with their generalizations were considered in Chap. 9.

2. For the model (12.2), used by F. Black & M. Scholes [32] and R. Merton [130] in their theory of pricing of derivatives, the formulae (11.60) and (11.62) for rational prices $\mathbb{C}_T$ and $\mathbb{P}_T$ of call and put options and formulae (see (11.71) and (11.72)) for the hedging portfolio $(\beta_t^*, \gamma_t^*)_{t\le T}$ (for call options) are widely applied on financial markets.

Recall that in the case, where the bank account interest rate $r = 0$ and $B_0 = 1$, the values $\mathbb{C}_T$ and $\mathbb{P}_T$ are determined by

$$\mathbb{C}_T = \mathsf{E}_{\widetilde{\mathsf{P}}_T} F_T, \quad \text{where} \quad F_T = (S_T - K)^+$$

and

$$\mathbb{P}_T = \mathsf{E}_{\widetilde{\mathsf{P}}_T} F_T, \quad \text{where} \quad F_T = (K - S_T)^+;$$

here $\widetilde{\mathsf{P}}_T$ is the martingale measure (with respect to which the price $S = (S_t)_{t\le T}$ is a martingale); T is the maturity and K is the exercise price, both fixed when writing the option contract.

From Sec. 11.1 it is clear that to construct the portfolio $(\beta_t^*, \gamma_t^*)_{t\le T}$ one has to operate, in addition to prices $\mathbb{C}_T$ and $\mathbb{P}_T$, with their dynamical analogs $\mathbb{C}_{[t,T]}$ and $\mathbb{P}_{[t,T]}$, which are rational prices of call and put options with payoff functions $F_T = (S_T - K)^+$ and $F_T = (K - S_T)^+$ under assumption that the option starts at t and at that time the "information" $\mathcal{F}_t$ on prices S_u, $u \le t$, is available.

Since the process $S = (S_t)_{t\le T}$ is Markovian and the function F_T depends only on S_T, the values $\mathbb{C}_{[t,T]}$ and $\mathbb{P}_{[t,T]}$ have the form $C(t, S_t)$ and $P(t, S_t)$, where the function $C(t, S_t)$ is given by (11.68) and $P(t, S_t)$ is determined from the call-put parity

$$P(t,x) = C(t,x) - S_t + K$$

(equivalently, $P(t, S_t)$ can be determined by (11.62), where one should replace S_0 by S_t and T by $T - t$; recall that for simplicity we assume now that $r = 0$).

From the formulae for $C(t,x)$ and $P(t,x)$ we see that these functions depend, in addition to t and x, on T, K, and σ. These parameters being now important for us, we will write

$$C_{\mathrm{BS}}(t,x;T,K;\sigma) \quad \text{and} \quad P_{\mathrm{BS}}(t,x;T,K;\sigma)$$

instead of $C(t,x)$ and $P(t,x)$, where "BS" stands for "Black–Scholes". The parameters T and K are fixed by the conditions of the option contract, whereas the parameter σ characterizes "activity", "fluctuability", "changeability", in other words—*volatility* of prices $S = (S_t)_{t \le T}$.

In the next subsection we will discuss the ideas which led to the notion of *implied volatility*, which in turn initiated the construction, for $S = (S_t)_{t \le T}$, of models with "stochastic volatility". Stochastic volatility is now one of the basic concepts for constructing models which fit adequately the empirical data related to both the *underlying* financial instruments (bank accounts, bonds, stock, *etc.*) and the *derivative* financial instruments (options, future contracts, warrants, swaps, spreads, *etc.*).

3. In addition to the rational prices $C_{\mathrm{BS}}(t, x; T, K; \sigma)$ of call options, consider the prices $C_{\mathrm{obs}}(t, x; T, K)$ which are *really observable* on the call option market. If we proceed from the assumption that the Black–Scholes model is a good pattern for the dynamics of prices S, then in this model σ should be such that, at least approximately,

$$C_{\mathrm{obs}}(t, x; T, K) \approx C_{\mathrm{BS}}(t, x; T, K; \sigma). \tag{12.3}$$

However, it was noticed that the value

$$\widetilde{\sigma} = \widetilde{\sigma}(t, x; T, K) \tag{12.4}$$

determined from (12.3) is far from being constant (which conflicts with the assumption of "fidelity" of the Black–Scholes model).

Remark 12.1. The Black–Scholes formula implies that $\partial \mathbb{C}_{\mathrm{BS}}/\partial \sigma > 0$. Therefore if $C_{\mathrm{obs}}(t, x; T, K) > C_{\mathrm{BS}}(t, x; T, K; 0)$, then the equation

$$C_{\mathrm{BS}}(t, x; T, K; \sigma) = C_{\mathrm{obs}}(t, x; T, K)$$

can be solved with respect to σ.

The function $\widetilde{\sigma} = \widetilde{\sigma}(t, x; T, K)$ is called *implied volatility*, and there are numerous works where one can find graphs illustrating the behavior of this function $\widetilde{\sigma}(T, K)$ with different values of strike price K and expiration time T (under fixed t and x).

It turned out that for many financial markets and options on them the surface $\widetilde{\sigma}(T, K)$ demonstrates the *smile effect*. The heart of this phenomenon consists in the following. Let us fix T, in addition to t and x. Consider the form of the corresponding function $\widetilde{\sigma}(K)$ for different values of the strike price K. Recall that, in compliance with terminology adopted in financial literature, for a call option with payoff function $(x - K)^+$ one distinguishes three groups of values of K (in comparison with x):

$K \gg x$: so-called *out-of-the-money* case (option brings losses for buyer);

$K \sim x$: *at-the-money* case (option with zero gain);

$K \ll x$: *in-the-money* case (option brings a gain).

(*Vice versa*, for put options the case $K \gg x$ is in-the-money, the case $K \ll x$ is out-of-the-money.)

The *smile effect* describes the behavior of the function $\widetilde{\sigma}(K)$: this function attains its minimal value at the point $K = K_{\min}$ close to x ($K_{\min} \sim x$), it increases when K decreases from $K_{\min}$ ("left branch"). With K increasing on the right of $K_{\min}$, the "right branch" often increases as well. As observed in [75], until the famous crisis of 1987 this was so for a great number of options. After that in many cases the tendency of decrease of the "right branch" is revealed (this is typically described as a downward sloping skew; see, *e.g.*, Fig. 2.1 in [75]).

The analogous smile effect occurs for $\widetilde{\sigma}(T)$ with t, x, K fixed.

All said above on the smile effect suggests that the Black–Scholes model with a *constant volatility* is not adequate for the probability-statistical structure of observable prices $S = (S_t)_{t \geq 0}$. This is corroborated also by numerous statistical studies of one-dimensional distributions of returns

$$h_t^{(\Delta)} = \log\bigl(S_t / S_{t-\Delta}\bigr)$$

and of covariances $\operatorname{cov}(h_t^{(\Delta)}, h_s^{(\Delta)})$.

To summarize the above considerations, we list the so-called *stylized features of prices* (for more detail see, *e.g.*, [161; Chap. IV, §§ 2a–2d] and [26; Chap. 1]).

(A) The empirical one-dimensional distribution of the logarithms of relative price changes $h_{t_1}^{(\Delta)}, h_{t_2}^{(\Delta)}, \dots$ ($t_k = k\Delta$)—which, for statistical analysis, are assumed identically distributed—is such that its density $\widehat{p}^{(\Delta)}(x)$ differs, by its form, from the normal (Gaussian) density: it often has a positive skewness coefficient (which means that $\widehat{p}^{(\Delta)}(x)$ is asymmetric) and a large, increasing as $\Delta \downarrow 0$, kurtosis coefficient (which implies the "heavy tails", *i.e.*, the density $\widehat{p}^{(\Delta)}(x)$ decreases more slowly than the normal density as $|x| \to \infty$).

(B) Introduce the *autocorrelation function* of the sequence $(h_{t_k}^{(\Delta)})_{k \geq 0}$:

$$\rho^{(\Delta)}(n) = \frac{\mathsf{E} h_{t_k}^{(\Delta)} h_{t_{k+n}}^{(\Delta)} - \mathsf{E} h_{t_k}^{(\Delta)} \mathsf{E} h_{t_{k+n}}^{(\Delta)}}{\sqrt{\mathsf{D} h_{t_k}^{(\Delta)} \mathsf{D} h_{t_{k+n}}^{(\Delta)}}},$$

where $t_k = k\Delta$, $k \geq 0$. Denote by $\hat{\rho}^{(\Delta)}(n)$ its empirical estimate.

The statistical analysis shows that the empirical autocorrelation function $\hat{\rho}^{(\Delta)}(n)$ takes *negative* values for small $n\Delta$ and most of its values are close to zero. (A simple model for which the autocorrelation function has such structure can be found in [161; p. 356].) At the same time the empirical autocorrelation function $\widehat{R}^{(\Delta)}(n)$ of the sequence $(|h_{t_k}^{(\Delta)}|)_{k\geq 0}$ is large for small n and decreases rather slowly when n increases [161; p. 363].

(C) The properties of $\widehat{R}^{(\Delta)}(n)$ formulated above show that if $|h_{t_k}^{(\Delta)}|$ is large, then the subsequent value $|h_{t_{k+1}}^{(\Delta)}|$ is likely to be also large, and if $|h_{t_k}^{(\Delta)}|$ is small, then $|h_{t_{k+1}}^{(\Delta)}|$ tends to be small as well. This "grouping" property of the observation data bears the name *clustering effect* (see, *e.g.*, [161; Chap. IV, § 3e]).

These and several other features of the prices (such as, *e.g.*, the leverage effect (see Subsec. 9 below and, for more details, [74], [26], [161]) cannot be explained within the Samuelson–Black–Merton–Scholes model $S_t = e^{H_t}$, where $H_t = \mu t + \sigma B_t$ and the increments $H_{t_k} - H_{t_{k-1}}$ are independent and Gaussian. Hence there has been numerous attempts in the literature to alter that model in order to catch the properties like (A)–(C), as well as the smile effect.

4. Diagrams I and II—in which, for simplicity, the drift terms are assumed zero—show schematically ways of construction of stochastic models aimed to describe adequately the statistical data of financial indices.

In the previous chapters the transformations $\sigma \cdot B$ and $B \circ T$ were discussed in detail, which is justified, in particular, by the need to model adequately the dynamics of prices $S = (S_t)_{t\geq 0}$.

In Diagram II the Brownian motion B used before is replaced by a wider class of processes, namely, by Lévy processes L, that provides broader means for construction of models reflecting properties of real prices which are observable on the underlying asset markets but cannot be explained within purely Brownian models.

All processes $\sigma{\cdot}B$, $\sigma{\cdot}L$, $B{\circ}T$, $L{\circ}T$ are semimartingales. Therefore, after having discussed models where the driving processes are Brownian motion and Lévy processes, it is natural to consider exponential semimartingale models. Exponential semimartingales form a very wide class of positive stochastic processes (see Chap. 3) and play a crucial role in martingale characterization of no arbitrage (Chap. 10).

Certainly, exponential semimartingales do not exhaust all interesting classes of processes, for example, they do not cover models based on frac-

Diagram I

Exponential Brownian Model:

$S = S_0 e^{\sigma B}$

B is a Brownian motion, $\sigma > 0$ is a *constant*

Exponential Integral Representation Model:

$S = S_0 e^{\sigma \cdot B}$

B is a Brownian motion, $\sigma = (\sigma(t))_{t \geq 0}$ is a *stochastic volatility*

Exponential Change of Time Representation:

$S = S_0 e^{B \circ T}$

B is a Brownian motion, $T = (T(t))_{t \geq 0}$ is a *change of time*

Diagram II

Exponential Lévy Model:

$S = S_0 e^{\sigma L}$

L is a Lévy process, $\sigma > 0$ is a *constant*

 $\Downarrow$

Exponential Integral Representation Model:

$S = S_0 e^{\sigma \cdot L}$

L is a Lévy process, $\sigma = (\sigma(t))_{t \geq 0}$ is a *stochastic volatility*

Exponential Change of Time Representation:

$S = S_0 e^{L \circ T}$

L is a Lévy process, $T = (T(t))_{t \geq 0}$ is a *change of time*

tal Brownian motion which have many properties resembling observable properties of real statistical data.

5. The *fractional* (or *fractal*) *Brownian motion* $B^{\mathbf{H}} = (B_t^{\mathbf{H}})_{t\geq 0}$ is a Gaussian process with continuous trajectories, $B_0^{\mathbf{H}} = 0$, $\mathsf{E}B_t^{\mathbf{H}} = 0$, $t \geq 0$, and its autocorrelation function is given by

$$\mathsf{E}B_s^{\mathbf{H}}B_t^{\mathbf{H}} = \frac{1}{2}\Big(|s|^{2\mathbf{H}} + |t|^{2\mathbf{H}} - |t-s|^{2\mathbf{H}}\Big), \tag{12.5}$$

where $\mathbf{H}$ stands for the *Hurst exponent*, $0 < \mathbf{H} < 1$. If $\mathbf{H} = 1/2$, then $B^{\mathbf{H}}$ is a standard Brownian motion. The fractional Brownian motion has the property of self-similarity:

$$\mathrm{Law}\big(B_{at}^{\mathbf{H}}, t \geq 0\big) = \mathrm{Law}\big(a^{\mathbf{H}}B_t^{\mathbf{H}}, t \geq 0\big) \quad \text{for all} \quad a > 0,$$

that follows from the Gaussianity of $B^{\mathbf{H}}$ and the property

$$\mathrm{Law}\big(B_{as}^{\mathbf{H}}, B_{at}^{\mathbf{H}}\big) = \mathrm{Law}\big(a^{\mathbf{H}}B_s^{\mathbf{H}}, a^{\mathbf{H}}B_t^{\mathbf{H}}\big)$$

which in turn follows, by (12.5), from the identity

$$\mathsf{E}B_{as}^{\mathbf{H}}B_{at}^{\mathbf{H}} = \mathsf{E}\big(a^{\mathbf{H}}B_s^{\mathbf{H}}\big)\big(a^{\mathbf{H}}B_t^{\mathbf{H}}\big).$$

Let us fix some $0 < \mathbf{H} < 1$ and let

$$\beta_n = B_n^{\mathbf{H}} - B_{n-1}^{\mathbf{H}}, \qquad n \geq 1.$$

It is evident that the sequence $(\beta_n)_{n\geq 1}$ is a Gaussian sequence with $\mathsf{E}\beta_n = 0$, $\mathsf{E}\beta_n^2 = 1$, and its covariance is given by

$$\rho^{\mathbf{H}}(n) = \frac{1}{2}\Big(|n+1|^{2\mathbf{H}} - 2|n|^{2\mathbf{H}} - |n-1|^{2\mathbf{H}}\Big).$$

Consequently,

$$\rho^{\mathbf{H}}(n) \sim \mathbf{H}(2\mathbf{H}-1)|n|^{2\mathbf{H}-2}, \qquad n \to \infty. \tag{12.6}$$

If $\mathbf{H} = 1/2$, then $\rho^{\mathbf{H}}(n) = 0$ for $n \neq 0$, and the sequence $(\beta_n)_{n\geq 1}$ is the classical "white noise", *i.e.*, the sequence of independent Gaussian random variables such that $\mathsf{E}\beta_n = 0$ and $\mathsf{E}\beta_n^2 = 1$.

If $\mathbf{H} \neq 1/2$, then we see from (12.6) that the correlation function decreases as $n \to \infty$ rather slowly, namely, as $|n|^{-2(1-\mathbf{H})}$. This phenomenon is usually interpreted as a "long memory" or a "strong aftereffect". (In [11] a wide class of processes with "long range dependency" was constructed via a superposition of Ornstein–Uhlenbeck type processes.)

The cases $0 < \mathbf{H} < 1/2$ and $1/2 < \mathbf{H} < 1$ are substantially different.

If $0 < \mathbf{H} < 1/2$, then for $n \neq 0$ the covariance is negative, $\rho^{\mathbf{H}}(n) < 0$, and $\sum_{n=0}^{\infty} |\rho^{\mathbf{H}}(n)| < \infty$.

If $1/2 < \mathbf{H} < 1$, then for $n \neq 0$ the covariance is positive, $\rho^{\mathbf{H}}(n) > 0$, and $\sum_{n=0}^{\infty} \rho^{\mathbf{H}}(n) = \infty$.

The positive covariance means that positive (negative) values of β_n are expected to be followed also by positive (negative) values β_{n+1}. Thus, the model based on the fractal Brownian motion with $\frac{1}{2} < \mathbf{H} < 1$ displays the *clustering* effect (as was mentioned above, this effect appears also in the behavior of returns $h_n^{(\Delta)} = \log(S_{t_n}/S_{t_{n-1}})$ $(= \log S_{t_n} - \log S_{t_{n-1}})$, where $t_n - t_{n-1} = \Delta$).

On the other hand, if $0 < \mathbf{H} < \frac{1}{2}$ (*i.e.*, the covariance is negative), then the positive (negative) values β_n are expected to be followed by negative (positive) values β_{n+1}. Such a behavior ("up and down"), called *strong intermittency* or *mean reversion*, is inherent to many processes which describe stochastic volatility.

All these properties justify the interest to modeling prices $S = (S_t)_{t \geq 0}$ in the form

$$S_t = S_0 e^{B_T^{\mathbf{H}}}, \qquad t \geq 0,$$

where $B^{\mathbf{H}} = (B_t^{\mathbf{H}})_{t \geq 0}$ is a fractal Brownian motion.

Next we will consider several modifications of the exponential Brownian model from Diagram I on page 264.

6. One of the first models aimed to rectify the exponential Brownian model in such a way as to imitate the smile effect was proposed by R. Merton who replaced the volatility σ in (12.2) by a (deterministic) function $\sigma = \sigma(t)$ depending on time t.

The geometric Brownian motion $S = (S_t)_{t \geq 0}$ defined in (12.2) can also be determined as solution to the stochastic differential equation

$$dS_t = S_t(\mu \, dt + \sigma \, dB_t). \tag{12.7}$$

In the Merton model, this equation is substituted by

$$dS_t = S_t(\mu \, dt + \sigma(t) \, dB_t). \tag{12.8}$$

A solution to the latter equation is given by

$$S_t = S_0 \exp\left\{ \int_0^t \left(\mu - \frac{1}{2}\sigma^2(s)\right) ds + \int_0^t \sigma(s) \, dB_s \right\}. \tag{12.9}$$

Taking into account the notation introduced above, we get

$$S_t = S_0 \exp\left\{ \mu t - \frac{1}{2} \langle \sigma \cdot B \rangle_t + (\sigma \cdot B)_t \right\}, \tag{12.10}$$

where $\langle \sigma \cdot B \rangle = (\langle \sigma \cdot B \rangle_t)_{t \geq 0}$ stands for the quadratic characteristic of the square-integrable martingale $(\sigma \cdot B) = ((\sigma \cdot B)_t)$ with $(\sigma \cdot B)_t = \int_0^t \sigma(s)\, dB_s$. (We assume that $\int_0^t \sigma^2(s)\, ds < \infty$, $t \geq 0$.)

In the model (12.10) with deterministic volatility there is no difficulty to find rational prices and hedging strategies for standard call and put options: it suffices to substitute $\sqrt{\overline{\sigma}_T^2}$, where (see [130])

$$\overline{\sigma}_T^2 = \frac{1}{T} \int_0^T \sigma^2(s)\, ds, \tag{12.11}$$

for σ in the Black–Scholes formulae (11.60) and (11.62), and to substitute $\sqrt{\overline{\sigma}_{T-t}^2}$, where

$$\overline{\sigma}_{T-t}^2 = \frac{1}{T-t} \int_t^T \sigma^2(s)\, ds, \tag{12.12}$$

for σ in the formulae (11.71) and (11.72).

In such models with *deterministic volatility* σ there is no smile effect across strikes (for fixed T and fixed $\overline{\sigma}_T^2$), since there was no such effect in the Black–Scholes model. However, the smile effect appears for *different maturities* which is rather natural because the volatility depends on time.

In the framework of models with nonstochastic volatility, one can obtain the smile effect across strikes, assuming that $\sigma = \sigma(t, x)$, *i.e.*, that the volatility depends not only on time but also on a phase variable x. In other words, instead of the model (12.8) one should consider the model of the form

$$dS_t = S_t\big(\mu\, dt + \sigma(t, S_t)\, dB_t\big), \tag{12.13}$$

for which the advanced theory of existence and uniqueness of both strong and weak solutions is developed.

However, there is an annoying circumstance, namely, the prices and volatility turn out to be perfectly correlated and this contradicts the statistical observations showing that the correlation should be negative but not equal to -1.

7. The above refinement of the exponential Brownian model was a result of conditions imposed on volatility:

$$\sigma \longrightarrow \sigma(t) \longrightarrow \sigma(t, x), \tag{12.14}$$

where $x \in \mathbb{R}_+$ (see (12.8) and (12.13)). One can go further, assuming, for example, that the volatility depends not only on t and S_t but also on all preceding values S_u, $u \leq t$, *i.e.*, one can consider models of the form

$$dS_t = S_t\big(\mu\, dt + \sigma(t; S_u, u \leq t)\, dB_t\big). \tag{12.15}$$

Such models, where the volatility $\sigma(t; x_u, u \leq t)$ depends on all "past" observed prices, are both very natural (volatility depends on and reflects the whole history of price evolution) and valuable, since the corresponding markets are complete. However, the study of such models is not advanced because of analytical difficulties arising when operating with the equation (12.15), where the coefficient of dB_t depends on the "past". (Some results concerning existence of solutions to equations of such type with coefficient at dt and dB_t being functionals of the "past" are given in [125; Chap. 4, § 4].)

8. Most of modern models of "stochastic volatility" are built in another way. Namely, one assumes that the volatility (being itself "volatile") is generated by a source of randomness which is different from $B = (B_t)_{t\geq 0}$, say, by $Z = (Z_t)_{t\geq 0}$.

In the framework of "Brownian models", the simplest way is to think of Z as another Brownian motion $\widetilde{B} = (\widetilde{B}_t)_{t\geq 0}$ which may be correlated with $B = (B_t)_{t\geq 0}$:

$$d\langle B, \widetilde{B}\rangle_t = \rho\, dt,$$

where $\langle B, \widetilde{B}\rangle$ is the predictable quadratic covariance of B and $\widetilde{B}$ (see Chap. 3, Sec. 3.1).

As to the volatility $\sigma(t)$, it is convenient to assume [51] that

$$\sigma(t) = f(Y_t), \tag{12.16}$$

where $f(y)$ is a positive function (for example, e^y or $\sqrt{|y|}$) and $Y = (Y_t)_{t\geq 0}$ is a diffusion process which satisfies the equation

$$dY_t = b(t, Y_t)\, dt + a(t, Y_t)\, d\widetilde{B}_t. \tag{12.17}$$

Instead of (12.17) one could assume that the process $Y = (Y_t)_{t\geq 0}$ belongs to a larger class, namely it is an Itô process with the differential

$$dY_t = b(t, \omega)\, dt + a(t, \omega)\, d\widetilde{B}_t, \tag{12.18}$$

where $b(t, \omega)$ and $a(t, \omega)$ are $\mathcal{F}_t$-measurable functions (for every $t > 0$) such that (P-a.s.)

$$\int_0^t |b(s, \omega)|\, ds < \infty, \qquad \int_0^t a^2(s, \omega)\, ds < \infty, \qquad t > 0.$$

(As usual, all considerations presume given a filtered probability space $(\Omega, \mathcal{F}, (\mathcal{F})_{t\geq 0}, \mathsf{P})$.)

Certainly, the coefficients $b(t, \omega)$ and $a(t, \omega)$ of the equation (12.17) must be "adjusted" to the behavior of volatility. Here arises some difficulty, since the volatility is not observed directly: one has to estimate it through indirect information, *e.g.*, by the use of multipower variations [25] or by implied volatility. Nevertheless, indirect observations can allow one to make important conclusions about properties of volatility itself. One of the most important properties of volatility is *mean reversion*, that is, return of the process towards the mean (of invariant distribution arising after a long period of observation).

The simplest (but not least important) example of a mean reverting process $Y = (Y_t)_{t \geq 0}$ is the Ornstein–Uhlenbeck process satisfying the equation

$$dY_t = (\alpha - \beta Y_t)\,dt + \gamma\,d\widetilde{B}_t, \qquad Y_0 = y \in \mathbb{R}, \qquad \alpha > 0, \quad \beta > 0, \quad \gamma > 0 \tag{12.19}$$

(compare with the more general case (1.28)). This process takes values in $\mathbb{R}$, and one can take $\sigma(t)$ equal to $f(Y_t)$, where, *e.g.*, $f(y) = e^y$.

An example of models, where Y has mean reversion property and is nonnegative, is given by the CRC-model (Cox–Ingersoll–Ross):

$$dY_t = (\alpha - \beta Y_t)\,dt + \gamma\sqrt{Y_t}\,d\widetilde{B}_t, \qquad Y_0 > 0, \qquad \alpha > 0, \quad \beta > 0, \quad \gamma > 0. \tag{12.20}$$

Note that all such processes $Y = (Y_t)_{t \geq 0}$ have *continuous* trajectories. In Sec. 12.2 and Sec. 12.3 we shall consider the models (for volatility and prices) with *discontinuous* trajectories.

9. Statistical investigations often reveal the already mentioned *leverage-effect*, which consists in *negative* correlation between stock prices and the volatility of stocks.

Assume that the prices $S = (S_t)_{t \geq 0}$ obey the equation

$$dS_t = S_t(\mu\,dt + \sigma(t)\,dB_t), \tag{12.21}$$

where $\sigma(t) = e^{Y_t}$ and $Y = (Y_t)_{t \geq 0}$ satisfies (12.19).

Then, by Itô's formula,

$$\begin{aligned} d\sigma(t) &= \sigma(t)\,dY_t + \tfrac{1}{2}\sigma(t)\gamma^2\,dt \\ &= \left(\alpha - \beta Y_t + \tfrac{1}{2}\sigma(t)\gamma^2\right)dt + \sigma(t)\gamma\,d\widetilde{B}_t, \end{aligned} \tag{12.22}$$

which together with (12.21) imply that

$$d\langle S, \sigma\rangle_t = \sigma(t)^2 S_t \gamma\, d\langle B, \widetilde{B}\rangle_t. \tag{12.23}$$

From this we see that to have negative correlation between prices and volatility we must assume that the driving Brownian motions B and $\widetilde{B}$ are also negatively correlated, *i.e.*, $d\langle B, \widetilde{B}\rangle_t = \rho\,dt$, where $\rho < 0$.

10. The characteristic feature of stochastic volatility models considered above is that they were constructed on basis of integral representation of the type

$$\sigma \cdot B. \tag{12.24}$$

The general theory of such integral representations driven by either Brownian motion or other processes (and random measures) was expounded in Chap. 2.

In the next section we will consider stochastic volatility models of the form

$$B \circ T, \tag{12.25}$$

i.e., models of change of time in a Brownian motion.

12.2 Generalized Inverse Gaussian Subordinator and Generalized Hyperbolic Lévy Motion: Two Methods of Construction, Sample Path Properties

1. In the previous section, when considering the representation $S_t = S_0 e^{H_t}$ of prices $S = (S_t)_{t \geq 0}$ we assumed that the return process $H = (H_t)_{t \geq 0}$ is a diffusion. Trajectories of such a process are continuous functions.

However, statistical analysis of financial data, as we have already mentioned, shows it reasonable also to consider models where trajectories of the return process $H = (H_t)_{t \geq 0}$ are "purely discontinuous" functions. Among processes with such a property are Lévy processes without Gaussian component studied in detail in Chap. 5.

In the present section, we intend to describe certain models based on Lévy processes, which were proposed and successfully applied by O. E. Barndorff-Nielsen, E. Eberlein and their co-authors for "capturing well" statistical features of observable data from financial source (see their works in the Bibliography at the end of the book).

It must be stressed, however, that Lévy processes by themselves do not capture the highly significant and important time-wise dependencies in financial time series.

2. Recall that in Chap. 9 we used (to model the increments $H_{t+1} - H_t$) the variables

$$h = \mu + \beta\sigma^2 + \sigma\varepsilon, \tag{12.26}$$

where μ and β are parameters, the random variables σ^2 and ε are independent, ε has standard normal distribution, $\mathcal{N}(0,1)$, and σ^2 has an

$$\mathrm{IG}(a,b) \quad (\textit{Inverse Gaussian distribution})$$

or, more generally, a

$$\mathbb{G}\mathrm{IG}(a,b,\nu) \quad (\mathbb{G}\textit{eneralized Inverse Gaussian distribution}).$$

The densities $p(s;a,b)$ and $p(s;a,b,\nu)$ of these distributions are given (see (9.40), (9.42)) by

$$p(s;a,b) = c_1(a,b)s^{-3/2}e^{-(as+b/s)/2}, \qquad s \geq 0, \tag{12.27}$$

where

$$c_1(a,b) = \sqrt{\frac{b}{2\pi}}\, e^{\sqrt{ab}},$$

and

$$p(s;a,b,\nu) = c_2(a,b,\nu)s^{\nu-1}e^{-(as+b/s)/2}, \qquad s \geq 0, \tag{12.28}$$

where

$$c_2(a,b,\nu) = \frac{(a/b)^{\nu/2}}{2K_\nu(\sqrt{ab})}$$

and $K_\nu(y)$ is a modified Bessel function of the third kind and index ν.

It is clear that

$$\mathbb{G}\mathrm{IG}(a,b,-\tfrac{1}{2}) = \mathrm{IG}(a,b), \qquad a \geq 0, \quad b > 0. \tag{12.29}$$

Other important examples of $\mathbb{G}$IG-distributions are

- Positive Hyperbolic distribution H^+:

$$\mathbb{G}\mathrm{IG}(a,b,1) = \mathrm{H}^+(a,b), \quad a > 0, \quad b > 0; \tag{12.30}$$

- Gamma distribution:

$$\mathbb{G}\mathrm{IG}(a,0,\nu) = \mathrm{Gamma}(a,\nu), \quad a > 0, \quad \nu > 1. \tag{12.31}$$

(See Table 9.1 on page 185.)

3. Remarkably, distributions of the class $\mathbb{G}\mathrm{IG}(a,b,\nu)$ are *infinitely divisible* [14], [87]; consequently, one can construct a Lévy process $T = T(t)$, $t \geq 0$, such that $T(0) = 0$ and

$$\mathrm{Law}(T(1)) = \mathrm{Law}(\sigma^2), \tag{12.32}$$

$$\mathsf{E}e^{i\lambda T(t)} = \left(\mathsf{E}e^{i\lambda\sigma^2}\right)^t. \tag{12.33}$$

Trajectories of the process $T = T(t)$, $t \geq 0$, associated with σ^2 are nondecreasing functions. Such processes are commonly named *subordinators.*

In the case $\nu = -1/2$ (when σ^2 has the Inverse Gaussian distribution $\mathrm{IG}(a,b)$) one can construct the process $T = T(t)$ *explicitly*—by using the method (proposed in Chap. 9) of constructing an $\mathrm{IG}(a,b)$-distribution from a Wiener process $W = (W_s)_{s\geq 0}$. According to Sec. 9.4, $p(s;a,b)$ is the density of the random variable

$$T^A(B) = \inf\{s \geq 0 : As + W_s = B\}, \tag{12.34}$$

where $A^2 = a > 0$ and $B^2 = b > 0$.

Consider, for $t \geq 0$, the random variables

$$T^*(t) \equiv T^A(Bt) = \inf\{s \geq 0 : As + W_s = Bt\}, \tag{12.35}$$

or, equivalently,

$$T^*(t) \equiv T^{\sqrt{a}}(\sqrt{b}t) = \inf\{s \geq 0 : \sqrt{a}s + W_s = \sqrt{b}t\}. \tag{12.36}$$

According to (9.38)

$$\mathsf{E}e^{-\lambda T^A(B)} = \exp\left\{AB\left(1 - \sqrt{1 + \frac{2\lambda}{A^2}}\right)\right\}. \tag{12.37}$$

Herefrom we see that for any $t \geq 0$

$$\begin{aligned}\mathsf{E}e^{-\lambda T^*(t)} = \mathsf{E}e^{-\lambda T^A(Bt)} &= \exp\left\{A(Bt)\left(1 - \sqrt{1 + \frac{2\lambda}{A^2}}\right)\right\} \\ &= \left(\mathsf{E}\exp\{-\lambda T^A(B)\}\right)^t = \left(\mathsf{E}\exp\{-\lambda\sigma^2\}\right)^t. \end{aligned} \tag{12.38}$$

For the Lévy process $T = T(t)$, $t \geq 0$, introduced above we have (see (12.33))

$$\mathsf{E}e^{-\lambda T(t)} = \left(\mathsf{E}e^{-\lambda\sigma^2}\right)^t. \tag{12.39}$$

From (12.38) and (12.39) we can conclude that the Lévy processes $T^* = T^*(t)$, $t \geq 0$, and $T = T(t)$, $t \geq 0$, are in fact stochastically indistinguishable.

Thus, starting from a random variable σ^2 with $\mathrm{IG}(a,b)$-distribution, one can construct, by two different ways, a Lévy subordinator, which will be denoted by $T = T(t)$, $t \geq 0$.

In the case of processes $\mathbb{G}\mathrm{IG}(a,b,\nu)$ (except for $\nu = -1/2$), this method of constructing subordinators explicitly already does not work.

The Lévy processes (subordinators) $T = T(t)$, $t \geq 0$, constructed by means of the distributions $\mathrm{IG}(a,b)$ and $\mathbb{G}\mathrm{IG}(a,b)$ will be denoted by

$$\mathrm{L(IG)} \quad \text{and} \quad \mathrm{L}(\mathbb{G}\mathrm{IG})$$

and referred to as

Inverse Gaussian Lévy subordinator

and

$\mathbb{G}$*eneralized Inverse Gaussian Lévy subordinator.*

4. Now we will construct the return process $H = (H_t)_{t \geq 0}$ such that its increments $H_{t+1} - H_t$ have the same distributions as the random variable $h = \mu + \beta\sigma^2 + \sigma\varepsilon$ (see (12.26)).

If σ^2 has $\mathbb{G}\mathrm{IG}(a, b, \nu)$ ($\mathbb{G}$eneralized Inverse Gaussian) distribution and ε is a standard normal, $\mathcal{N}(0,1)$, random variable which *does not depend* on σ^2, then the distribution of h, by its construction, is a normal variance-mean mixture $\mathsf{E}_{\sigma^2}\mathcal{N}(\mu + \beta\sigma^2, \sigma^2)$ (see (9.29), (9.30)). For this reason it was denoted by $\mathbb{N} \circ \mathbb{G}\mathrm{IG} = \mathbb{N} \circ \mathbb{G}\mathrm{IG}(a, b, \mu, \beta, \nu)$ to evoke $\mathbb{N}$ormal/$\mathbb{G}$eneralized Inverse Gaussian distribution.

This "long" name was superseded by a "shorter" one, $\mathbb{G}$H ($\mathbb{G}$eneralized Hyperbolic distribution, $\mathbb{G}\mathrm{H}(a, b, \mu, \beta, \nu)$). According to Sec. 9.4, this name comes from the fact that the $\mathbb{G}\mathrm{IG}(a, b, 1)$-distribution (for $a > 0$, $b > 0$, $\nu = 1$) is the Positive Hyperbolic distribution $\mathrm{H}^+ = \mathrm{H}^+(a, b)$ (see Table 9.1 on page 185) and the distribution $\mathbb{N} \circ \mathbb{G}\mathrm{IG}(a, b, 1) = \mathbb{N} \circ \mathrm{H}^+(a, b)$ is commonly called the *hyperbolic* distribution. This fact clarifies why, in the general case of arbitrary ν, the distribution $\mathbb{N} \circ \mathbb{G}\mathrm{IG}(a, b, \nu)$ got the name *generalized hyperbolic distribution* ($\mathbb{G}$eneralized Hyperbolic). Thus, it is natural to write

$$\mathbb{G}\mathrm{H} = \mathbb{N} \circ \mathbb{G}\mathrm{IG}. \tag{12.40}$$

The distribution $\mathbb{N} \circ \mathbb{G}\mathrm{IG}$ is *infinitely divisible* (property A* on page 188), hence there exists a Lévy process $H^* = (H_t^*)_{t \geq 0}$ (defined on a "sufficiently rich" probability space) such that

$$\mathsf{E} \exp\{i\lambda H_t^*\} = (\mathsf{E} \exp\{i\lambda h\})^t, \tag{12.41}$$

where $\mathrm{Law}(h)$ is the $\mathbb{G}$H-distribution. Thus, $\mathrm{Law}(H_{t+1}^* - H_t^*) = \mathrm{Law}(h)$.

Remark 12.2. Notice that the distribution $\mathrm{Law}(H_{t+\Delta}^* - H_t^*)$ for $\Delta \neq 1$ is not generally of the same type as $\mathrm{Law}(h)$. However, as was mentioned in Chap. 9 (Subsec. 8 of Sec. 9.4), this property holds if $a \geq 0$, $b > 0$, $\nu = -1/2$ (*i.e.*, when $\mathbb{G}\mathrm{IG} = \mathrm{IG}$) or if $a > 0$, $b = 0$, $\nu > 0$ (*i.e.*, when $\mathbb{G}\mathrm{IG} = \mathrm{Gamma}$).

One of the remarkable properties of $\mathbb{G}$H-processes $H^* = (H^*_t)_{t\geq 0}$ is that they admit lucid, constructive versions.

Namely, consider processes $H = (H_t)_{t\geq 0}$, given by

$$H_t = \mu t + \beta T(t) + B_{T(t)}, \tag{12.42}$$

where $T = T(t)$, $t \geq 0$, is a L($\mathbb{G}$IG)-process constructed above which has the property (12.39) and where $B = (B_t)_{t\geq 0}$ is a Brownian motion which does not depend on $T = T(t)$, $t \geq 0$.

For such processes it follows from (12.39) that

$$\begin{aligned}
\mathsf{E}e^{i\lambda H_t} &= e^{i\lambda\mu t}\,\mathsf{E}e^{i\lambda(\beta T(t)+B_{T(t)})} = e^{i\lambda\mu t}\,\mathsf{E}\mathsf{E}\big[e^{i\lambda(\beta T(t)+B_{T(t)})}\,\big|\,T(t)\big] \\
&= e^{i\lambda\mu t}\,\mathsf{E}\Big(e^{i\lambda\beta T(t)}\mathsf{E}\big[e^{i\lambda B_{T(t)}}\mid T(t)\big]\Big) = e^{i\lambda\mu t}\,\mathsf{E}\big(e^{i\lambda\beta T(t)}\,e^{-\lambda^2 T(t)/2}\big) \\
&= e^{i\lambda\mu t}\,\big(\mathsf{E}e^{-(\lambda^2/2-i\lambda\beta)\sigma^2}\big)^t = \big(\mathsf{E}e^{i\lambda(\mu+\beta\sigma^2+\sigma\varepsilon)}\big)^t \\
&= \big(\mathsf{E}e^{i\lambda h}\big)^t.
\end{aligned} \tag{12.43}$$

The process $H^* = (H^*_t)_{t\geq 0}$ is a Lévy process by construction. The process $H = (H_t)_{t\geq 0}$ defined in (12.42) is also a Lévy process (it follows from the proof of Theorem 8.5). Then it follows from (12.43) that $\mathrm{Law}(H^*_t) = \mathrm{Law}(H_t)$ for any $t > 0$. The independence and homogeneity of increments of the processes H^* and H imply that their infinite dimensional distributions coincide. Thus, $\mathrm{Law}(H^*) = \mathrm{Law}(h)$; in other words, each of these processes is a modification (version) of the other.

Using notation adopted above (see, for example, Introduction), one can rewrite (12.42) in the form

$$h = \mu\,\mathrm{Leb} + \beta T + B \circ T, \tag{12.44}$$

where Leb stands for "Lebesgue".

The following schema summarizes the ways of constructing the Lévy processes H^* and H (this schema uses notation L($\mathbb{N}\circ\mathbb{G}$IG) and $\mathbb{B}\circ$L($\mathbb{G}$IG) reflecting the character of these constructions):

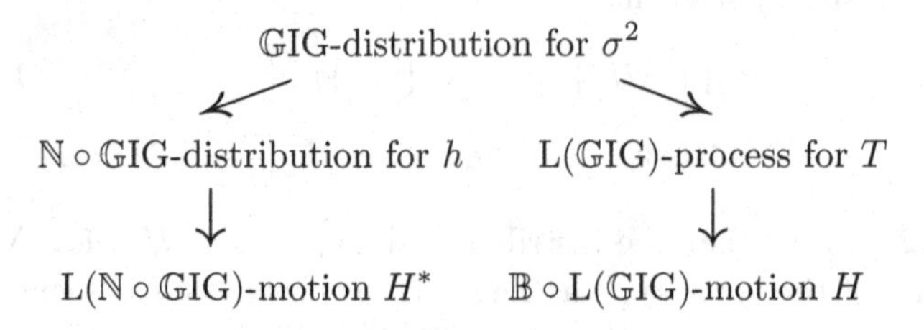

Identity (12.40) explains both the notation

$$\mathrm{L}(\mathbb{G}\mathrm{H}) = \mathrm{L}(\mathbb{N}\circ\mathbb{G}\mathrm{IG})$$

and naming processes H^* and their modifications H as

Generalized Hyperbolic Lévy motions (or processes).

Thus,

$$\mathrm{L}(\mathbb{G}\mathrm{H}) = \mathrm{L}(\mathbb{N} \circ \mathbb{G}\mathrm{IG}) = \mathbb{B} \circ \mathrm{L}(\mathbb{G}\mathrm{IG}).$$

Particular cases of these processes are

$$\begin{aligned}
\mathrm{L}(\mathbb{N}\mathrm{IG}) &= \mathrm{L}(\mathbb{N} \circ \mathrm{IG}) &&= \mathbb{B} \circ \mathrm{L}(\mathrm{IG}),\\
\mathrm{L}(\mathbb{H}) &= \mathrm{L}(\mathbb{N} \circ \mathrm{H}^+) &&= \mathbb{B} \circ \mathrm{L}(\mathrm{H}^+),\\
\mathrm{L}(\mathbb{V}\mathrm{G}) &= \mathrm{L}(\mathbb{N} \circ \mathrm{Gamma}) &&= \mathbb{B} \circ \mathrm{L}(\mathrm{Gamma}).
\end{aligned}$$

Processes of the classes L(ℍ) and L(𝕍G) are commonly named *Hyperbolic* and 𝕍*ariance Gamma* Lévy processes, respectively.

12.3 Distributional and Sample-path Properties of the Lévy Processes L(𝔾IG) and L(𝔾H)

1. We start with a question about the structure of one-dimensional distributions of the processes $T = T(t)$, $t \geq 0$, constructed by means of the infinitely divisible distribution $\mathbb{G}\mathrm{IG} = \mathbb{G}\mathrm{IG}(a, b, \nu)$ of the volatility σ^2. We confine ourselves to consider the three most important cases (see Table 9.1 on page 185):

(a) $\nu = -1/2$, $a \geq 0$, $b > 0$;

(b) $\nu = 1$, $a > 0$, $b > 0$;

(c) $\nu > 0$, $a > 0$, $b = 0$.

In the case $\nu = -1/2$, $a \geq 0$, $b > 0$ (IG-distribution) the density $p_{\mathrm{IG}}(s; a, b)$ of the one-dimensional probabilities $\mathsf{P}(T(1) \leq s)$ is of the form

$$p(s; a, b, -1/2) = \sqrt{\frac{b}{2\pi}}\, e^{\sqrt{ab}}\, s^{-3/2}\, e^{-(as+b/s)/2}, \qquad s > 0 \tag{12.45}$$

(see (9.40)). This implies that as $s \to \infty$ ("tail behavior")

$$p_{\mathrm{IG}}(s; a, b) \sim \sqrt{\frac{b}{2\pi}}\, e^{\sqrt{ab}}\, s^{-3/2}\, e^{-as/2}; \tag{12.46}$$

consequently, the (density of) IG-distribution has "heavier (upper) tail" than the normal distribution. At the same time (12.45) implies that $\mathsf{E}(T(1))^n < \infty$ for all $n \geq 1$. Thus, the "tails" of the IG-distribution are not so "heavy" yet as, for example, the "tails" of Cauchy's distribution (see page 120). Symbolically, we write it as

$$\mathrm{Tails(Cauchy)} \gg \mathrm{Tail(IG)} \gg \mathrm{Tails}(\mathcal{N}). \tag{12.47}$$

In the case $\nu = 1$, $a > 0$, $b > 0$ (H^+-distribution) the density $p_{\mathrm{H}^+}(s; a, b)$ of the distribution $\mathsf{P}(T(1) \le s)$ is of the form

$$p(s; a, b, 1) = \frac{\sqrt{a/b}}{2K_1(\sqrt{ab})}\, e^{-(as+b/s)/2}. \tag{12.48}$$

Comparing (12.46) with (12.48) shows that

$$\mathrm{Tails(Cauchy)} \gg \mathrm{Tail(H^+)} \gg \mathrm{Tails(IG)}. \tag{12.49}$$

In the case $\nu > 1$, $a > 0$, $b = 0$ (Gamma-distribution) the density $p_{\mathrm{Gamma}}(s; a, \nu)$ is of the form

$$p(s; a, 0, \nu) = \frac{(a/b)^\nu}{\Gamma(\nu)}\, s^{\nu-1}\, e^{-as/2}. \tag{12.50}$$

Here all moments $\mathsf{E}(T(1))^n$ are finite: $\mathsf{E}(T(1))^n < \infty$ for all $n \ge 1$, and are given by (9.64).

As was mentioned in Chap. 9 (Subsec. 8 of Sec. 9.4), in the class of $\mathbb{G}$IG-distributions the IG- and Gamma-distributions are characterized by the property that for $t \ne 1$ the variables $T(t)$ belongs to the same class of distributions (IG- and Gamma), see (12.63) and (12.67).

2. Now we address the structure of one-dimensional distributions of the processes of class $\mathbb{G}\mathrm{H} = \mathbb{G}\mathrm{H}(a, b, \nu, \beta, \mu)$.

In the case $\nu = -1/2$, $a \ge 0$, $b > 0$, $\beta \in \mathbb{R}$, $\mu \in \mathbb{R}$ the corresponding $\mathbb{G}$H-distribution is called Normal Inverse Gaussian ($\mathbb{N} \circ \mathrm{IG}$). According to Table 9.2 on page 193, the density of this distribution is given by (9.89):

$$p^*(x; a, b, \mu, \beta, -1/2) = \frac{\alpha b}{\pi}\, e^{\sqrt{ab}}\, \frac{K_1(\alpha\sqrt{b + (x-\mu)^2})}{\sqrt{b + (x-\mu)^2}}\, e^{\beta(x-\mu)} \tag{12.51}$$

for $x \in \mathbb{R}$, where $\alpha = \sqrt{a + \beta^2}$.

Since

$$K_1(x) \sim \sqrt{\frac{\pi}{2}}\, x^{-1/2}\, e^{-x}, \qquad x \to \infty \tag{12.52}$$

(see (9.100)), it follows that as $x \to \pm\infty$

$$p^*(x; a, b, \mu, \beta, -1/2) \sim \frac{\sqrt{\alpha}\, b}{\sqrt{2\pi}}\, \frac{e^{\beta(x-\mu) - \alpha\sqrt{b + (x-\mu)^2}}}{(b + (x-\mu)^2)^{3/4}}. \tag{12.53}$$

In the case $\nu = 1$, $a > 0$, $b > 0$, $\beta \in \mathbb{R}$, $\mu \in \mathbb{R}$ the $\mathbb{G}$H-distribution is called Hyperbolic ($\mathbb{H}$) or Normal Positive Hyperbolic ($\mathbb{N} \circ \mathrm{H}^+$ or $\mathbb{N} \circ \mathrm{PH}$). The density of this distribution is

$$p^*(x; a, b, \mu, \beta, 1) = \frac{a}{2b\alpha K_1(\sqrt{ab})}\, e^{\beta(x-\mu) - \alpha\sqrt{b + (x-\mu)^2}} \tag{12.54}$$

(see (9.88)).

In the case $\nu > 0$, $a > 0$, $b = 0$, $\beta \in \mathbb{R}$, $\mu \in \mathbb{R}$ the corresponding 𝔾H-distribution is the Normal Gamma (ℕ ∘ Gamma), called also 𝕍ariance Gamma (𝕍Gamma, 𝕍G). Its density is given by

$$p^*(x; a, 0, \mu, \beta, \nu) = \frac{a^\nu}{\sqrt{\pi}\,\Gamma(\nu)\,(2\alpha)^{\nu-1/2}} |x-\mu|^{\nu-1/2} \times K_{\nu-1/2}(\alpha|x-\mu|)\, e^{\beta(x-\mu)} \tag{12.55}$$

(see (9.92)).

Taking into account (9.100), we find that as $x \to \pm\infty$

$$p^*(x; a, 0, \mu, \beta, \nu) \sim \left(\frac{a}{2\alpha}\right)^\nu \frac{1}{\Gamma(\nu)} |x-\mu|^{\nu-1} e^{-\alpha|x-\mu|+\beta(x-\mu)}. \tag{12.56}$$

It follows from (12.53), (12.54), and (12.56) that

$$\text{Tails}(\mathbb{V}\text{G}) \gg \text{Tails}(\mathbb{H}) \gg \text{Tails}(\mathbb{N} \circ \text{IG}). \tag{12.57}$$

3. The structure of general Lévy processes was considered in Sec. 5.3 and Sec. 5.4.

The Lévy processes $T = T(t)$, $t \geq 0$, of the class L(𝔾IG) are subordinators and $\text{Law}(T(1)) = \text{Law}(\sigma^2)$, where $\text{Law}(\sigma^2)$ belongs to the class $\mathbb{G}\text{IG} = \mathbb{G}\text{IG}(a, b, \nu)$. The sample functions $T(t)$, $t \geq 0$, with $T(0) = 0$ are right continuous, nondecreasing (thus having limits from the left).

Fluctuational properties of such Lévy processes (which have no continuous component) are completely determined by their "space-time" Lévy measure $\nu(dt, dy) = dt\, F(dy)$, where $F = F(dy)$ is the "space" Lévy measure of 𝔾IG-distributions of the variables σ^2.

The measure F has the density $f = f(y)$ $(F(dy) = f(y)\,dy,\ y > 0)$ given by (9.50).

- In the case $\nu = -1/2$, $a \geq 0$, $b > 0$ (IG-distribution) the density is given by

$$f_{\text{IG}}(y) = \sqrt{\frac{b}{2\pi}}\, e^{-ay/2}\, y^{-3/2}, \qquad y > 0 \tag{12.58}$$

(see (9.54)).

From here we see that

$$F_{\text{IG}}(\mathbb{R}_+) = \int_0^\infty f_{\text{IG}}(y)\,dy = \infty, \tag{12.59}$$

i.e., this is a case of "infinite activity" (see Sec. 5.4), characterized by the property that with probability one the process $T = T(t)$, $t \geq 0$, has *infinitely many jumps* on any time interval $(0,t]$, $t>0$, and consequently—since $\sum_{0\le s\le t}\Delta T(s)<\infty$ (P-a.s.) for all $t>0$—the trend (upwards) of this process is mainly composed by a "great number of small jumps".

• In the case $\nu=1$, $a>0$, $b>0$ (H^+-distribution) the density $f_{\mathrm{H}^+}(y)$ of the Lévy measure is given by

$$f(y)=\frac{e^{-ay/2}}{y}\left[\frac{1}{2}\int_0^\infty e^{-uy/(2b)}\,g_\nu(u)\,du+\max(0,\nu)\right] \tag{12.60}$$

(see (9.50)).

• In the case $\nu>1$, $a>0$, $b=0$ (Gamma-distribution) the density is given by a simple formula (see (9.50) and (9.55)):

$$f_{\mathrm{Gamma}}(y)=\frac{\nu\, e^{-ay/2}}{y}, \tag{12.61}$$

which shows that this is again an "infinite activity" case.

4. The very construction of L(𝔾H)-processes $H=(H_t)_{t\ge0}$, implies that the distribution of the variable H_1 coincides with the distribution of $h=\mu+\beta\sigma^2+\sigma\varepsilon$ (see (9.28)) whose density $p^*(x;a,b,\mu,\beta,\nu)$ is given by (9.77):

$$p^*(x;a,b,\mu,\beta,\nu)=c_3(a,b,\beta,\nu)\frac{K_{\nu-1/2}\big(\alpha\sqrt{b+[x-\mu]^2}\big)}{\big(\sqrt{b+[x-\mu]^2}\big)^{1/2-\nu}}\,e^{\beta(x-\mu)}. \tag{12.62}$$

• In the case $\nu=-1/2$, $a\ge0$, $b>0$, $\beta\in\mathbb{R}$, $\mu\in\mathbb{R}$ the corresponding 𝔾H-distribution is the Normal Inverse Gaussian (ℕ ∘ IG) distribution with density given by (12.51). An important property of the processes $H=(H_t)_{t\ge0}$, of the class L(ℕ ∘ IG) is that for any $t\neq1$ the distribution of H_t is of the same type as the distribution of H_1 (see Subsec. 8 in Sec. 9.4). If $p_t^*(x;a,b,\mu,\beta,-1/2)$ denotes the density of H_t, then

$$p_t^*(x;a,b,\mu,\beta,-1/2)=p^*(x;a,bt^2,\mu t,\beta,-1/2). \tag{12.63}$$

(It follows directly from the property $\mathsf{E}e^{i\lambda H_t}=(\mathsf{E}e^{i\lambda h})^t$ and the formula (9.91) with $\theta=i\lambda$.) This property—of H_t (and, more generally, of $H_{t+\Delta}-H_\Delta$ with $\Delta\ge0$) to belong to the same type for all t—distinguishes the processes L(ℕ ∘ IG) from the general class L(𝔾H) and makes them preferable when one works with empirical data.

From (12.62) and the asymptotic expansion (9.100) for $K_1(y)$ as $y\to\infty$, one can deduce that $p_t^*(x;a,b,0,\beta,-1/2)$ as $x\to\pm\infty$ is of the order $|x|^{-3/2}e^{-\alpha|x|+\beta x}$, where $\alpha=\sqrt{a+\beta^2}\ge|\beta|$, since $a\ge0$.

• In the case $\nu = 1$, $a > 0$, $b > 0$, $\beta \in \mathbb{R}$, $\mu \in \mathbb{R}$ the corresponding 𝔾H-distribution is the Normal Positive Hyperbolic ($\mathbb{N} \circ \mathrm{H}^+$), or briefly ℍyperbolic (ℍ) distribution.

By construction, the density $p^*(x; a, b, \mu, \beta, 1)$ of the probability distribution $\mathsf{P}(H_1 \leq x)$ is given by (9.88):

$$p^*(x; a, b, \mu, \beta, 1) = \frac{a}{2b\alpha_1 K_1(\sqrt{ab})} \times \exp\{-\alpha\sqrt{b + (x-\mu)^2} + \beta(x-\mu)\}. \qquad (12.64)$$

It is important to emphasize that—in contrast to formula (9.77) which contains *two* Bessel's functions ($K_{\nu-1/2}(\cdot)$ and $K_\nu(\cdot)$)—in the considered case of the hyperbolic distribution the formula (12.63) deals with only one Bessel's function $K_1(\sqrt{ab})$, the second, $K_{1/2}(\alpha\sqrt{b + |x-\mu|^2})$, having been reduced—thanks to (9.101)—to elementary functions:

$$K_{1/2}(y) = \sqrt{\frac{\pi}{2}}\, y^{-1/2}\, e^{-y}, \qquad y > 0. \qquad (12.65)$$

This fact simplifies considerably the work with the density (12.63). For example, when calculating the values of this density, say in n points $x_1, \ldots, x_n$, in the general case (9.77) one has to calculate the values of the Bessel function $K_{\nu-1/2}(\cdot)$ at these points. However, in the case $\nu = 1$ these calculations reduce to finding values of elementary functions.

Notice, once more, that the graph of the function $\log p^*(x; a, b, \mu, \beta, 1)$ is a hyperbola, whence comes the term 'hyperbolic distribution'.

Observe that (contrary to the preceding case $\nu = -1/2$) the distribution $\mathsf{P}(H_t \leq x)$ is not anymore of the same type as the distribution $\mathsf{P}(H_1 \leq x)$. One should keep this fact in mind when dealing with Hyperbolic Lévy processes.

• In the case $\nu > 0$, $a > 0$, $b = 0$, $\beta \in \mathbb{R}$, $\mu \in \mathbb{R}$ the corresponding 𝔾H-distribution is the Normal Gamma ($\mathbb{N} \circ$ Gamma) distribution (called also a 𝕍ariance Gamma (𝕍G) distribution).

By construction of the L($\mathbb{N} \circ$ Gamma)-process $H = (H_t)_{t \geq 0}$ the density $p^*(x; a, 0, \mu, \beta, \nu)$ of the probability distribution $\mathsf{P}(H_1 \leq x)$ is given by (9.92):

$$p^*(x; a, 0, \mu, \beta, \nu) = \frac{a^\nu}{\sqrt{\pi}\,\Gamma(\nu)\,(2\alpha)^{\nu-1/2}}\, |x-\mu|^{\nu-1/2} \times K_{\nu-1/2}(\alpha|x-\mu|)\, e^{\beta(x-\mu)}. \qquad (12.66)$$

It is interesting that in the considered case the distribution $\mathsf{P}(H_t \leq x)$ is (as in the case of L($\mathbb{N} \circ$ IG)-processes) of the same type as the distribution

$\mathsf{P}(H_1 \le x)$: if we denote by $p_t^*(x; a, 0, \mu, \beta, \nu)$ the density of the distribution $\mathsf{P}(H_t \le x)$, then

$$p_t^*(x; a, 0, \mu, \beta, \nu) = p^*(x; a, 0, \mu t, \beta, \nu t). \tag{12.67}$$

(It follows directly from the property $\mathsf{E}e^{i\lambda H_T} = (\mathsf{E}e^{i\lambda h})^t$ and the formula (9.93) with $\theta = i\lambda$.)

Asymptotic

$$K_{\nu-1/2}(y) \sim \sqrt{\pi/(2y)}\, e^{-y}, \qquad y \to \infty,$$

where $\nu - 1/2 > 0$ (see (9.100)), and the formula (12.66) allow us to conclude that when $x \to \pm\infty$ and $\mu = 0$ the density $p^*(x; a, 0, 0, \beta, \nu)$ is of the order $|x|^{\nu-1} e^{-\alpha|x|+\beta x}$, where $\alpha = \sqrt{a + \beta^2} > |\beta|$ (since $a > 0$).

Comparing this asymptotic with the asymptotic $|x|^{-3/2} e^{-\alpha|x|+\beta x}$ in the case $\nu = -1/2$ and asymptotic $e^{-\alpha|x|+\beta x}$ in the case $\nu = 1$ (see (12.64)) shows that

$$\text{Tails}(\mathbb{N} \circ \text{Gamma}) \gg \text{Tails}(\mathbb{H}) \gg \text{Tails}(\mathbb{N} \circ \text{IG}) \tag{12.68}$$

(cf. (12.49) and (12.57)).

5. Finally, consider the structure of trajectories of the Lévy processes $H = (H_t)_{t \ge 0}$, which belong to the class of Generalized Hyperbolic Lévy ($\mathrm{L}(\mathbb{GH})$) processes.

Recall that in Chap. 5 we considered the basic properties and methods of analysis of general Lévy processes $X = (X_t)_{t \ge 0}$.

The *analytical* method of study of such processes is based on studying the characteristic function $\mathsf{E}e^{i\theta X_t}$, $\theta \in \mathbb{R}$, $t \ge 0$. According to the Kolmogorov–Lévy–Khinchin formula (Sec. 4.2),

$$\mathsf{E}e^{i\theta X_t} = e^{t\varkappa(\theta)}, \tag{12.69}$$

where the local Fourier cumulant function $\varkappa(\theta)$ is given by

$$\varkappa(\theta) = i\theta b - \frac{\theta^2}{2} c + \int \left(e^{i\theta y} - 1 - i\theta h(y)\right) F(dy) \tag{12.70}$$

(see (5.11), (5.12)).

In (12.70), b and c are constants ($b \in \mathbb{R}$, $c \in \mathbb{R}_+$) and $F = F(dy)$ is the Lévy measure, *i.e.*, a positive measure on $\mathbb{R}$ which satisfies $F(\{0\}) = 0$ and integrates $y^2 \wedge 1$:

$$\int_{\mathbb{R}} (y^2 \wedge 1)\, F(dy) < \infty. \tag{12.71}$$

The *stochastic* method of study of the Lévy processes $X = (X_t)_{t\geq 0}$ is based on the canonical representation

$$X_t = bt + X_t^c + \int_0^t \int h(y)\, d(\mu - \nu) + \int_0^t \int (y - h(y))\, d\mu \qquad (12.72)$$

(see (5.41)), where $X_t^c = \sqrt{c}\, B_t$, $t \geq 0$, $B = (B_t)_{t\geq 0}$ is a standard Brownian motion, $h = h(y)$ is a truncation function (Sec. 3.2), and $\mu = \mu(\omega; dt, dy)$ is the measure of jumps:

$$\mu(\omega; dt, dy) = \sum_{s>0} I(\Delta X_s(\omega) \neq 0)\, \delta(s, \Delta X_s(\omega))(dt, dy), \qquad (12.73)$$

where $\Delta X_s(\omega) = X_s(\omega) - X_{s-}(\omega)$ and $\delta = \delta(a)$ is the Dirac measure "sitting" at a point a $(= (s, \Delta X_s(\omega)))$.

As usual, $\nu = \nu(dt, dy)$ is a "time-space" Lévy measure (*i.e.*, a compensator of the jump measure μ). For Lévy processes we have

$$\nu(dt, dy) = dt\, F(dy),$$

where $F = F(dy)$ is the "space" Lévy measure which have appeared in the formula (12.70) for the cumulant function $\varkappa(\theta)$.

The process X^c and measure μ which appear in the canonical representation (12.72) are independent, and μ is a Poisson random measure (see (5.8)) whose properties are completely determined by the properties of the Lévy measure $F = F(dy)$.

For the Lévy processes $X = (X_t)_{t\geq 0}$ we have $e^{i\theta X_t} = e^{t\varkappa(\theta)} = (\mathsf{E} e^{i\theta X_1})^t$. Consequently, if $X_t = H_t$, where $H = (H_t)_{t\geq 0}$ is a process of the class L(𝔾H), then its Lévy measure $F^* = F^*(dy)$ (which determines the properties of the jump component of the process H) can be calculated from (12.69) and (12.70), where, by definition of characteristic function,

$$\mathsf{E} e^{i\theta H_1} = \int_{-\infty}^{\infty} e^{i\theta x}\, p^*(x; a, b, \mu, \beta, \nu)\, dx, \qquad (12.74)$$

with the density $p^*(x; a, b, \mu, \beta, \nu)$ of the random variable H_1 (which has the 𝔾H-distribution) given by (9.77).

Thus, to find the triplet (b, c, F), one should represent the integral on the right-hand side of (12.74) in the form $e^{\varkappa(\theta)}$, where $\varkappa(\theta)$ is given by (12.70).

Such a representation was obtained by Halgreen in [90]. It builds on the following formula for characteristic function $\mathsf{E} e^{i\theta H_1}$ (with $\alpha = \sqrt{a + \beta^2}$):

$$\mathsf{E} e^{i\theta H_1} = e^{i\mu\theta} \left[\frac{a}{\alpha^2 - (\beta + i\theta)^2}\right]^{\nu/2} \frac{K_\nu\big(\sqrt{b[\alpha^2 - (\beta + i\theta)^2]}\big)}{K_\nu(\sqrt{ab})}. \qquad (12.75)$$

Putting the right-hand side of (12.75) equal to $e^{\varkappa(\theta)}$, one can get, after simple but rather long calculations (see, *e.g.*, [141]), that in the representation (12.70) of $\varkappa(\theta)$ the constant b equals μ, the constant c is zero and the Lévy measure $F^* = F^*(dy)$ has the density $f^*(y)$ given by (9.80), which (in view of (9.52)) can be rewritten in the form

$$f^*(y) = \frac{e^{by}}{|y|}\left\{\int_0^\infty \frac{\exp\{-\sqrt{\alpha^2+2u}\,|y|\}}{\pi^2 u\big[J^2_{|\nu|}(\sqrt{2bu}) + N^2_{|\nu|}(\sqrt{2bu})\big]}\,du + \nu e^{-\alpha|y|}\max(\nu,0)\right\}, \qquad u \in \mathbb{R}, \tag{12.76}$$

where J_ν and N_ν are Bessel functions.

The fact that for $\mathbb{G}$H-distributions $c = 0$ displays that Lévy processes L($\mathbb{G}$H) have no (Gaussian) component, *i.e.*, each L($\mathbb{G}$H)-process is a *purely discontinuous* process.

To model financial indexes $S_t = S_0 e^{H_t}$, $t \geq 0$, one cannot restrict oneself to continuous processes $H = (H_t)_{t\geq 0}$ (*i.e.*, to Black–Scholes type models). The class L($\mathbb{G}$H) provides an alternative approach based on purely discontinuous processes $H = (H_t)_{t\geq 0}$. A great number of recent investigations (see, *e.g.*, [23], [26], [59], [62]) demonstrate effectiveness of such processes for constructing models of the form $S = S_0 e^H$ adequately fitting many features of statistical data.

6. Consider now properties of trajectories of generalized hyperbolic Lévy processes L($\mathbb{G}$H), again for the three cases: $\nu = -1/2$, $\nu = 1$, and $\nu > 0$ (with $b = 0$).

Case $\nu = -1/2$, $a \geq 0$, $b > 0$, $\beta \in \mathbb{R}$, $\mu \in \mathbb{R}$ (H_1 has $\mathbb{N} \circ$ IG, Normal Inverse Gaussian, distribution). The corresponding Lévy process $H = (H_t)_{t\geq 0}$, is called a Normal Inverse Gaussian Lévy process. Since this process (being a L($\mathbb{G}$H)-process) is purely discontinuous, the properties of its trajectories are completely determined by the density $f^* = f^*(y)$ of its Lévy measure:

$$f^*(y) = e^{\beta y}\,\frac{\alpha\sqrt{b}}{\pi|y|}\,K_1(\alpha|y|), \tag{12.77}$$

where $\alpha = \sqrt{a+\beta^2}$. This simple formula can be deduced from (12.76), if one takes into account that, according to (9.98),

$$J^2_{1/2}(u) + N^2_{1/2}(u) = (\pi u)^{-1}.$$

Formulae (12.77) and (9.99) imply that

$$f^*(y) \sim \frac{\sqrt{b}}{\pi}\,\frac{1}{y^2} \quad \text{as } y \to 0. \tag{12.78}$$

Therefore

$$F^*(\mathbb{R}) = \int_{-\infty}^{\infty} f^*(y)\,dy = \infty.$$

This means that (cf. (12.59)) we deal with the "infinite activity" case, *i.e.*, on any time interval $(0,t]$, $t > 0$, the process $H = (H_t)_{t\geq 0}$, has, with probability one, *infinitely many jumps.*

In the considered case, $\int_{-\infty}^{\infty}(|y| \wedge 1)f^*(y)\,dy = \infty$, and consequently (see (5.27), (5.28)), the trajectories of $\mathrm{L}(\mathbb{N} \circ \mathrm{IG})$-processes have *unbounded variation.*

Case $\nu = 1$, $a > 0$, $b > 0$, $\beta \in \mathbb{R}$, $\mu \in \mathbb{R}$ ($\mathbb{H} = \mathbb{N} \circ \mathrm{H}^+$, Hyperbolic or Normal Positive Hyperbolic distribution for H_1). To find the density $f^*(y)$ of Lévy's measure for the corresponding Hyperbolic Lévy process, one can rely on the following general result [141]: if we denote $\rho^*(y) = y^2 f^*_{\mathbb{GH}}(y)$, where $f^*_{\mathbb{GH}}(y)$ is the density of the Lévy measure of the $\mathbb{GH}$-distribution, then

$$\rho^*(y) = \frac{\sqrt{b}}{\pi} + \frac{\nu + 1/2}{2}\,|y| + \frac{\sqrt{b}\beta}{\pi}\,y + o(|y|), \qquad y \to 0. \tag{12.79}$$

Thus, in the considered case, where $\nu = 1$, the asymptotics of the density $f^*(y)$ is given by the same formula (12.79), as in the case $\nu = -1/2$, so that the structural properties (infinite activity and unboundedness) are the same as in the case $\nu = -1/2$.

Case $\nu > 0$, $a > 0$, $b = 0$, $\beta \in \mathbb{R}$, $\mu \in \mathbb{R}$ ($\mathbb{N} \circ$ Gamma or $\mathbb{V}$ariance Gamma distribution). Formula (12.79) implies that the density $f^*(y)$ of the Lévy measure has the following asymptotics:

$$f^*(y) \sim \frac{\nu + 1/2}{2}\,\frac{1}{|y|} \quad \text{as } y \to 0. \tag{12.80}$$

Hence $F^*(\mathbb{R}) = \infty$; nevertheless, in contrast to the preceding cases, here we have $\int_{-1}^{1} |y| f^*(y)\,dy < \infty$. Thus, $\mathbb{V}$ariance Gamma Lévy processes possess the infinite-activity property but have *bounded* variation.

12.4 On Some Others Models of the Dynamics of Prices. Comparison of the Properties of Different Models

1. From all what precedes we see that the main models which describe dynamics of the prices, $S_t = S_0 e^{H_t}$, $t \geq 0$, of underlying financial instruments, various indexes and so on, rely on assumption that the process $H = (H_t)_{t\geq 0}$ has the following structure:

$$H_t = \mu t + \beta \int_0^t \sigma^2(s)\,ds + \int_0^t \sigma(s)\,dB_s, \tag{12.81}$$

or

$$H_t = \mu t + \beta T(t) + B_{T(t)}. \tag{12.82}$$

In the case $\mu = \beta = 0$ (see Diagram I on page 264), (12.81) and (12.82) were written in the form

$$H = \sigma \cdot B \tag{12.83}$$

and

$$H = B \circ T, \tag{12.84}$$

where $\sigma = (\sigma(t))_{t\geq 0}$ is the process of stochastic volatility and $T = (T(t))_{t\geq 0}$ is a random change of time.

Here the driving process is a Brownian motion—which has so rich structure that it allows one to construct models embodying various stylized features of prices (many of which were described in Sec. 12.1, Subsec. 1).

In mathematical finance, there are numerous approaches to description of volatility processes $\sigma = (\sigma(t))_{t\geq 0}$. Several of them are illustrated in (12.14). A popular model of stochastic volatility is $\sigma(t) = \sqrt{Y_t}$, where $Y = (Y_t)_{t\geq 0}$ is a process solving (12.20) (CIR model):

$$dY_t = (\alpha - \beta Y_t)\,dt + \gamma\sqrt{Y_t}\,d\widetilde{B}_t, \tag{12.85}$$

here $Y_0 > 0$, $\alpha > 0$, $\beta > 0$, $\gamma > 0$, and $\widetilde{B} = (\widetilde{B}_t)_{t\geq 0}$ is a Brownian motion. For other *continuous* models see, *e.g.*, [51; Table 15.1].

The authors of [23], [26] proposed to consider $\sigma(t) = \sqrt{Y_t}$ with $Y = (Y_t)_{t\geq 0}$ a non-Gaussian Ornstein–Uhlenbeck type process which solves the linear stochastic differential equation

$$dY_t = -\lambda Y_t\,dt + dZ(\lambda t), \qquad Y_0 = 0$$

(cf. (12.19)), where $Z = (Z(t))_{t\geq 0}$ is a Lévy process with nonnegative increments, *i.e.*, a subordinator. The corresponding change of time $T = (T(t))_{t\geq 0}$ is given by

$$T(t) = \int_0^t Y_s\,ds = \frac{1}{\lambda}\big(Z(\lambda t) - Y_t + Y_0\big). \tag{12.86}$$

2. In the models considered in Sec. 12.2 and Sec. 12.3 the role of subordinator $T = T(t)$, $t \geq 0$, was played by a Lévy process L($\mathbb{G}$IG), and the corresponding process $H = (H_t)_{t\geq 0}$ given by (12.82) was a generalized hyperbolic Lévy process L(𝔾H).

Notice that in (12.82) the processes B and T were independent, so that one can say that the process H is generated by *two* independent *Brownian motions*.

In this respect, it is interesting to compare models (12.81), (12.82) with the following BNS model proposed in [24]:

$$\begin{aligned} H_t &= \mu t + \beta \int_0^t \beta\sigma^2(s)\, ds + \int_0^t \sigma(s)\, dB_s + \rho\, dZ_{\lambda t}, \\ d\sigma^2(t) &= -\lambda\sigma^2(t)\, dt + dZ_{\lambda t}, \qquad \sigma^2(0) = 0, \end{aligned} \tag{12.87}$$

where $B = (B_t)_{t\geq 0}$ is a Brownian motion and $Z = (Z_t)_{t\geq 0}$ is a subordinator which does not depend on B. The parameter ρ is called the *leverage parameter*, $\rho \in \mathbb{R}$. If $\rho = 0$, then the process H has continuous trajectories. If $\rho \neq 0$, then the trajectories of H exhibit jumps. (For detailed discussion of model (12.87) see [24], [26].)

3. Let us discuss one more model of the process $H = (H_t)_{t\geq 0}$ which has properties similar to the L(𝔾H) process.

Consider a 𝕍ariance Gamma Lévy process $H = (H_t)_{t\geq 0}$ (which belongs to the class L(𝔾H)). By (1.43) (see also [127], [128]), the Lévy density $f^*(y)$ of this process is given by

$$f^*(y) = \frac{C}{|y|} \exp\{Ay - B|y|\} \tag{12.88}$$

with some constants $A, B, C > 0$. The authors of [38] proposed to consider densities of the form

$$f^*(y) = \frac{\mathrm{C}}{|y|^{1+\mathrm{Y}}} \exp\left\{\frac{\mathrm{G}-\mathrm{M}}{2}\, y - \frac{\mathrm{M}+\mathrm{G}}{2}\, |y|\right\}, \tag{12.89}$$

where $\mathrm{C} > 0$, $\mathrm{G} \geq 0$, $\mathrm{M} \geq 0$, and $\mathrm{Y} < 2$. (If $\mathrm{Y} \leq 0$, then it is assumed that $\min(\mathrm{G}, \mathrm{M}) > 0$.)

Let $(b, 0, F^*(dy))$ be a triplet, where

$$b = \int \left(h(y) f^*(y) - yI(|y| \leq 1)\, \frac{\mathrm{C}}{|y|^{1+\mathrm{Y}}}\, e^{-|y|} \right) dy \tag{12.90}$$

and

$$F^*(dy) = f^*(y)\, dy \tag{12.91}$$

($f^*(y)$ being given by (12.89)).

The function $e^{\varkappa(\theta)}$, where

$$\varkappa(\theta) = ib\theta + \int \big(e^{i\theta y} - 1 - i\theta h(y)\big) f^*(y)\, dy$$

and b and $f^*(y)$ are defined in (12.90) and (12.89), respectively, is the characteristic function of an infinite divisible distribution called a *CGMY* (Carr–Geman–Madan–Yor) distribution. The Lévy process L(CGMY) constructed upon this distribution is purely discontinuous (contains no Brownian component). The fluctuation properties of this process are completely determined by the parameter Y.

If $\mathrm{Y} \in [0,2)$, then—similarly to the case of stable processes (see Sec. 5.4, formula (5.39) with $0 < \alpha < 2$) and to the case of L($\mathbb{V}$G)-process ($\mathrm{Y} = 0$ in (12.89); see (12.88))—we are in the "infinite activity" situation (*i.e.*, with probability one the process has an infinite number of jumps on any finite interval $(0,t]$, $t > 0$).

If $\mathrm{Y} \in (0,1)$ (respectively, $\mathrm{Y} \in [1,2)$), then the trajectories have bounded (respectively, unbounded) variation.

Notice that the density of the Lévy measure of the CGMY-distribution is described by *four* parameters. In fact, the latter density is obtained from the density of an α-stable distribution (with $\mathrm{Y} = \alpha$) by multiplying it by decreasing (as $x \to \pm\infty$) exponential factors.

The class L(CGMY) can be extended to the so-called *Generalized tempered stable processes*, whose Lévy density $f^*(y)$ is determined now by *six* parameters:

$$f^*(y) = \frac{\mathrm{C}_-}{|y|^{1+\alpha_-}}\, e^{-\lambda_-|y|} I(y<0) + \frac{\mathrm{C}_+}{|y|^{1+\alpha_+}}\, e^{-\lambda_+|y|} I(y \geq 0), \qquad (12.92)$$

where $\alpha_+ < 2$ and $\alpha_- < 2$ (see [51; formula (4.27)]). The formula (12.92) shows that the density $f^*(y)$ has different asymptotics as $y \uparrow 0$ and as $y \downarrow 0$.

4. In the following table we give a comparative analysis of the financial models $S = S_0 e^H$ for different processes H. Recall that, by the notation in Sec. 9.4,

$\mathbb{GH} = \mathbb{N} \circ \mathbb{GIG}$ is the Generalized Hyperbolic distribution (see (9.73))

with its particular cases:

$\mathbb{N} \circ \mathrm{IG}$	Normal Inverse Gaussian distribution;
$\mathbb{H} = \mathbb{N} \circ \mathbb{GH}^+$	Hyperbolic distribution or Normal Positive Hyperbolic distribution;
$\mathbb{N} \circ$ Gamma ($\mathbb{V}$G)	Normal Gamma distribution or Variance Gamma distribution.

Tracing [26], [45], and [51], we consider the following features observable on the markets:

I. The marginal distributions of increments of the processes H are skewed.
II. The marginal distributions of increments of the processes H have heavy tails.
III. The increments of the processes H are stationary in time.
IV. The increments of H over disjoint intervals are not correlated.
V. The absolute values of increments of H over disjoint intervals are positively correlated (the effect of "clustering", "volatility persistence").

Properties of the models $S = S_0 e^H$

Model	I	II	III	IV	V	VI
$H = \mu\,\mathrm{Leb} + \sigma B$ (Black–Scholes)	−	−	+	+	−	+
H is $\mathrm{L}(\mathbb{GH})$ (*e.g.*, $\mathrm{L}(\mathbb{N}\circ\mathrm{IG})$, $\mathrm{L}(\mathbb{H})$, $\mathrm{L}(\mathbb{VG})$)	+	+	+	+	−	+
H is L(CGMY)	+	+	+	+	−	+
$H = \sigma\cdot\widetilde{H}$ (*i.e.*, $H_t = \int_0^t \sigma(s)\,d\widetilde{H}_s,\ t\geq 0$) where $\sigma^2(s) = Y_s$, $Y = (Y_s)_{s\geq 0}$ is a CIR (12.85), BNS (12.87) $\widetilde{H}$ is $\mathrm{L}(\mathbb{GH})$ (*e.g.*, $\mathrm{L}(\mathbb{N}\circ\mathrm{IG})$, $\mathrm{L}(\mathbb{H})$, $\mathrm{L}(\mathbb{VG})$)	+	+	+	+	+	+
$H = \widetilde{H}\circ T$ (*i.e.*, $H_t = \widetilde{H}_{T(t)},\ t\geq 0$) where $T(t) = \int_0^t Y_s\,ds$, $Y = (Y_s)_{s\geq 0}$ is a CIR (12.85), BNS (12.87) $\widetilde{H}$ is L(CGMY)	+	+	+	+	+	+

All the model in the above table are arbitrage free (for an appropriate choice of the model parameters).

Among other important stylized features we mention

VI. The leverage-effect which means a negative correlation between the stock prices and their volatility (see Subsec. 9 of Sec. 12.1). This property can be modeled by using either the CIR process (12.85) or the Ornstein–Uhlenbeck type process as in the BNS model (12.87), for the processes $Y = (Y_s)_{s\geq 0}$ in the above table.

5. We conclude our exposition of various financial models and of the properties of their distributions with some remarks on parallels between certain notions in mathematical theory of finance and in mathematical theory of turbulence.

At first glance, there seems little in common between these disciplines; first of all, because their objects of investigation are different. Namely, in Mathematical Finance we observe the values of prices $S = S_0 e^H$ (equivalently, the values $H = \log(S/S_0)$), and the main characteristic of their changeability is *volatility*, which, in a wide sense, is understood as their quadratic variation ($[S]$ or $[H]$; see Chap. 3). As for Turbulence, there the observable variables are velocities $u(t,x)$ or their increments $\Delta u(t,x) = u(t+\Delta,x) - u(t,x)$ (measured, say, in the direction of the mean flow) for a fixed state x. In this set-up the main characteristic of changeability is *intermittency*, which is defined as the (surrogate) energy dissipation rate per unit mass around position x,

$$e_r(x) = r^{-1}\int_{x-r/2}^{x+r/2}\left(\frac{\partial u}{\partial x}\right)^2 dx,$$

measured over a symmetric interval of length r and centered at x (since u is considered as being stationary in time, t is suppressed in the notation for e_r).

In spite of the difference of the notions of prices (in Finance) and velocities (in Turbulence), their *probability-statistical* properties have a lot of commonality. Thus, we noticed many a time that the empirical densities of distributions of the variables $H_{t+\Delta} - H_t$ $(= \log(S_{t+\Delta}/S_t))$, at least for small Δ, in many cases are well approximated by $\mathrm{N}\circ\mathrm{IG}$ (Normal Inverse Gaussian) distributions; see Remark 12.2 on page 273). The very remarkable feature is that the aforementioned variables $u(t+\Delta,x) - u(t,x)$, at least for small Δ, are also well approximated by $\mathrm{N}\circ\mathrm{IG}$-distributions (thus their densities have "semi-heavy tails" in the sense that they admit the representation $\mathrm{const}\cdot|x|^{\rho_\pm}\exp\{-\sigma_\pm|x|\}$ as $x\to\pm\infty$, where $\rho_\pm \geq 0$, $\sigma_\pm \geq 0$).

For more detailed discussion of these and other properties as well as of analogies between Finance and Turbulence see the classical papers [115], [116], [117], [134], [135], [118]) [13], and also [20], [21], [22], [164] and references therein.

Afterword

To conclude the present text on "Change of Time and Change of Measure" we would like to make some final remarks.

In Finance, the *change of time* is related directly with the notion of *volatility* which is the fundamental notion characterizing the "temperature" or "variability" of financial markets. The change of time is often referred to as an *operational time* or *business time*, and it expresses the intensity of the fluctuations of the prices. Note that in Turbulence volatility—referred to in that context as intermittency—is equally of key importance and gives rise to a Fundamental change of time, see, *e. g.*, [13], [22].

Change of measure is important, since this notion gives a possibility to transfer the purely economic notion of arbitrage to the mathematical notion of martingale measures. Recall also formulae (10.22) and (10.23) which give representations for the lower and upper prices. It is important that these formulae involve infimum and supremum over all martingale measures. In the general theory of the financial risks (coherent, convex, *etc.*) the corresponding infima and suprema are taken over certain subsets of measures which are determined by the character of the chosen risks.

In the Financial Industry, there are three approaches for analyzing the financial markets and determining the optimal strategies of investing and trading.

Within the first, classical, so-called *Fundamental Approach* the decisions are based on fundamental economic principles. "Fundamentalists" try to explain "why the stock price moves", and their decisions depend on the state of the economy at large and are based on the calculation of the proper stock price in view of its estimated future value. "Fundamentalists" build their analysis under assumption that the market operators are acting "rationally."

The advocates of the second approach, *Technical Analysis*, concentrate on the local peculiarities of the market and emphasize "mass behavior", market moods as a crucial factor. They start their analysis from an idea that the movement of the stock prices is a product of "Supply and Demand"; they predict the future movements taking into account the history of the prices.

Finally, the third, *Quantitative Analysis*, approach—which we support in this monograph—is the "most mathematical" part of the Finance. It is to this analysis that the term 'mathematical finance' is commonly attributed, and its recommendations on investing and trading are based on constructing adequate mathematical models for prices fitting well the prices which are really observable on financial markets. Typical examples of such probability-statistical models are Bachelier's model, the Samuelson–Black–Merton–Scholes model, the Cox–Ross–Rubinstein model, linear models AR, MA, ARMA, nonlinear discrete-time models ARCH, GARCH with stochastic volatility, diffusion models with stochastic volatility, BNS type models, *etc.* This is why we devote much place to the martingale and semimartingale theory, stochastic integration, stochastic exponents, and stochastic differential equations.

A special subclass of semimartingales is formed by the Lévy processes, which are widely adopted to model prices of financial instruments. In the concluding chapter 12 we describe the models of price dynamics based on the hyperbolic Lévy processes, which—as the statistical analysis shows—fit aspects of the really observed financial data well. Together with [153], [51], [17], [26], the handbooks [143], [2], [50] are good sources on distributions and models in Finance.

Bibliography

[1] Abramowitz, M. and Stegun, I. A. (1970). *Handbook of Mathematical Functions* (Dover Publ., New York).

[2] Andersen, T. G., Davis, R. A., Kreiß, J.-P., and Mikosch, T. (eds.) (2009). *Handbook of Financial Time Series* (Birkhäuser, Boston).

[3] Ansel, J.-P. and Stricker, C. (1994). Couverture des actifs contingents et prix maximum, *Ann. Inst. H. Poincaré Probab. Statist.* **30**, 2, pp. 303–315.

[4] Bachelier, L. (1900). Théorie de la spéculation, *Ann. École Norm. Sup.* **17**, pp. 21–86.

[5] Back, K. (1991). Asset pricing for general processes, *J. Math. Econom.* **20**, 4, pp. 371–395.

[6] Barlow, M. T. (1982). One-dimensional stochastic differential equations with no strong solution, *J. London Math. Soc.* **26**, pp. 335–347.

[7] Barndorff-Nielsen, O. E. (1994). A note on electrical networks and the inverse Gaussian distribution, *Adv. in Appl. Probab.* **26**, 1, pp. 63–67.

[8] Barndorff-Nielsen, O. E. (1997). Normal inverse Gaussian distributions and stochastic volatility modeling, *Scand. J. Statist.* **24**, 1, pp. 1–13.

[9] Barndorff-Nielsen, O. E. (1998). Probability and statistics: self-decomposability, finance and turbulence, in L. Accardi and C. C. Heyde (eds.), *Probability Towards* 2000: *Proceedings of a Symposium held* 2–5 *October* 1995 *at Columbia University*, Lecture Notes in Statist., Vol. 128 (Springer-Verlag, New York), pp. 47–57.

[10] Barndorff-Nielsen, O. E. (1998). Processes of normal inverse Gaussian type, *Finance Stoch.* **2**, 1, pp. 41–68.

[11] Barndorff-Nielsen, O. E. (2001). Superposition of Ornstein–Uhlenbeck type processes, *Theory Probab. Appl.* **45**, 2, pp. 175–194.

[12] Barndorff-Nielsen, O. E., Blæsild, P., and Halgreen, C. (1978). First hitting time models for the generalized inverse Gaussian distribution, *Stochastic Process. Appl.* **7**, 1, pp. 49–54.

[13] Barndorff-Nielsen, O. E., Blæsild, P. and Schmiegel, J. (2004). A parsimonious and universal description of turbulent velocity increments. *Eur. Phys. J. B* **41**, pp. 345–363.

[14] Barndorff-Nielsen, O. E. and Halgreen, C. (1977). Infinite divisibility of the

hyperbolic and generalized inverse Gaussian distributions, *Z. Wahrscheinlichkeitstheor. verw. Geb.* **38**, 4, pp. 309–311.

[15] Barndorff-Nielsen, O. E., Kent, J., and Sørensen, M. (1982). Normal variance-mean mixtures and z-distributions, *Internat. Statist. Rev.* **50**, 2, pp. 145–159.

[16] Barndorff-Nielsen, O. E. and Koudou, A. E. (1998). Trees with random conductivities and the (reciprocal) inverse Gaussian distribution, *Adv. in Appl. Probab.* **30**, 2, pp. 409–424.

[17] Barndorff-Nielsen, O. E., Mikosch, T., and Resnick, S. (eds.) (2001). *Lévy Processes. Theory and Applications* (Birkhäuser, Boston).

[18] Barndorff-Nielsen, O. E., Pedersen, J., and Sato, K. (2001). Multivariate subordination, self-decomposability and stability, *Adv. in Appl. Probab.* **33**, 1, pp. 160–187.

[19] Barndorff-Nielsen, O. E. and Prause, K. (2001). Apparent scaling, *Finance Stoch.* **5**, 1, pp. 103–113.

[20] Barndorff-Nielsen, O. E. and Schmiegel, J. (2007). Ambit processes; with applications to turbulence and tumour growth, in F. E. Benth, G. D. Nunno, T. Linstrøm, B. Øksendal, and T. Zhang (eds.), *Stochastic Analysis and Applications: The Abel Symposium* 2005 (Springer, Heidelberg), pp. 93–124.

[21] Barndorff-Nielsen, O. E. and Schmiegel, J. (2007). A stochastic differential equation framework for the timewise dynamics of turbulent velocities, *Theory Probab. Appl.* **52**, 3, pp. 372–388.

[22] Barndorff-Nielsen, O. E. and Schmiegel, J. (2008). Time change, volatility and turbulence, in A. Sarychev, A. Shiryaev, M. Guerra, and M. da Rosario Grossinho (eds.), Proceedings of the *Workshop on Mathematical Control Theory and Finance*, Lisbon, 2007 (Springer, Berlin), pp. 29–53.

[23] Barndorff-Nielsen, O. E. and Shephard, N. (2001). Modelling by Lévy processes for financial econometrics, in O. E. Barndorff-Nielsen, T. Mikosch, and S. Resnick (eds.), *Lévy Processes. Theory and Applications* (Birkhäuser, Boston), pp. 283–318.

[24] Barndorff-Nielsen, O. E. and Shephard, N. (2001). Non-Gaussian Ornstein–Uhlenbeck-based models and some of their uses in financial economics, *J. Roy. Statist. Soc. Ser. B* **63**, 2, pp. 167–241.

[25] Barndorff-Nielsen, O. E. and Shephard, N. (2003). Realized power variation and stochastic volatility models, *Bernoulli* **9**, 2, pp. 243–265.

[26] Barndorff-Nielsen, O. E. and Shephard, N. (2007). *Financial Volatility in Continuous Time* (Cambridge Univ. Press, Cambridge).

[27] Bellini, F. and Frittelli, M. (2002). On the existence of minimax martingale measures, *Math. Finance* **12**, 1, pp. 1–21.

[28] Bertoin, J. (1996). *Lévy Processes* (Cambridge Univ. Press., Cambridge).

[29] Bertoin, J. (1999). Subordinators: examples and applications, in P. Bernard (ed.), *Lectures on Probability Theory and Statistics*: *École d'Eté de Probabilités de Saint-Flour* XXVII, 1997, Lecture Notes in Math., Vol. 1717 (Springer, Berlin), pp. 1–91.

[30] Bibby, M. and Sørensen, M. (2003). Hyperbolic processes in Finance, in S. T. Rachev (ed.), *Handbook of Heavy Tailed Distributions in Finance*

(Elsevier), pp. 212–248.

[31] Bingham, N. H. and Kiesel, R. (2004). *Risk-neutral Valuation: Pricing and Hedging of Financial Derivatives*, 2nd edn. (Springer, London).

[32] Black, F. and Scholes, M. (1973). The pricing of options and corporate liabilities, *J. Polit. Econ.* **81**, 3, pp. 637–659.

[33] Blackwell, D. and Girshik, M. A. (1954). *Theory of Games and Statistical Decisions* (Wiley, New York; Chapman & Hall, London).

[34] Bochner, S. (1955). *Harmonic Analysis and the Theory of Probability* (Univ. of California Press).

[35] Bollerslev, T. (1986). Generalized autoregressive conditional heteroscedasticity, *Econometrics* **31**, pp. 307–327.

[36] Bondesson, L. (1992). *Generalized Gamma Convolutions and Related Classes of Distributions and Densities*, Lecture Notes in Statist., Vol. 76 (Springer-Verlag, New York).

[37] Calvet, L. and Fisher, A. (2002). Multifractality in asset returns: Theory and evidence, *Rev. Econom. Statist.* **3**.

[38] Carr, P., Geman, H., Madan, D. B., and Yor, M. (2003). Stochastic volatility for Lévy processes, *Math. Finance* **13**, 3, pp. 345–382.

[39] Carr, P. and Wu, L. (2004). Time-changed Lévy processes and option pricing, *J. Financial Econom.* **71**, 1, pp. 113–141.

[40] Cherny, A. S. (2001). No-arbitrage and completeness for the linear and exponential models based on Lévy processes, Research Report 33, Centre for Mathematical Physics and Stochastics, Aarhus University, Aarhus, 24 pp.

[41] Cherny, A. S. (2002). Families of consistent probability measures, *Theory Probab. Appl.* **46**, 1, pp. 118–121.

[42] Cherny, A. S. and Engelbert, H.-J. (2004). *Singular Stochastic Differential Equations*, Lecture Notes in Math., Vol. 1858 (Springer, Berlin).

[43] Cherny, A. S. and Shiryaev, A. N. (2000). On criteria for the uniform integrability of Brownian stochastic exponentials, Research Report 38, Centre for Mathematical Physics and Stochastics, Aarhus University, Aarhus, 13 pp.

[44] Cherny, A. S. and Shiryaev, A. N. (2002). A vector stochastic integral and the fundamental theorem of asset pricing, *Proc. Steklov Inst. Math.* **237**, pp. 6–49.

[45] Cherny, A. S. and Shiryaev, A. N. (2002). Change of time and measure for Lévy processes, Lecture Notes 13, Centre for Mathematical Physics and Stochastics, Aarhus University, Aarhus, 46 pp.

[46] Cherny, A. S. and Shiryaev, A. N. (2005). On stochastic integrals up to infinity and predictable criteria for integrability, *Lecture Notes in Math.* **1857**, pp. 165–185.

[47] Cherny, A. S., Shiryaev, A. N., and Yor, M. (2003). Limit behavior of the "horizontal-vertical" random walk and some extensions of the Donsker–Prokhorov invariance principle, *Theory Probab. Appl.* **47**, 3, pp. 377–394.

[48] Chow, Y. S., Robbins, H., and Siegmund D. (1991). *The Theory of Optimal Stopping* (Dover, New York).

[49] Clark, P. K. (1973). A subordinated stochastic process model with finite

variance for speculative prices, *Econometrica* **41**, 1, pp. 135–155.
[50] Cont, R. (ed.) (2010). *Encyclopedia of Quntitative Finance* (Wiley, Chichester).
[51] Cont, R. and Tankov, P. (2004). *Financial Modelling with Jump Processes* (Chapman & Hall / CRC).
[52] Cox, J. C., Ross, S. A., and Rubinstein, M. (1979). Option pricing: a simplified approach, *J. Fin. Econ.* **7**, 3, pp. 229–263.
[53] Dalang, R. C., Morton, A., and Willinger, W. (1990). Equivalent martingale measures and no-arbitrage in stochastic securities market models, *Stochastics Stochastics Rep.* **29**, 2, pp. 185–201.
[54] Davis, M. H. A. (1997). Option pricing in incomplete markets, in M. Dempster and S. Pliska (eds.),*Mathematics of Derivative Securities*, Publ. Newton Inst., Vol. 15 (Cambridge Univ. Press, Cambridge), pp. 216–226.
[55] De Souza, C. and Smirnov, M. (2004). Dynamic leverage, *J. Portfolio Management* **31**, 1.
[56] Delbaen, F. and Schachermayer, W. (2006). *The Mathematics of Arbitrage* (Springer-Verlag).
[57] Dupire, B. (1994). Pricing with a smile, *RISK* **7**, 1, pp. 18–20.
[58] Eberlein, E. (2001). Application of generalized hyperbolic Lévy motions to finance, in O. E. Barndorff-Nielsen, T. Mikosch, and S. Resnick (eds.), *Lévy Processes—Theory and Applications* (Birkhäuser, Boston), pp. 319–336.
[59] Eberlein, E. and Keller, U. (1995). Hyperbolic distributions in finance, *Bernoulli* **1**, 3, pp. 281–299.
[60] Eberlein, E., Keller, U., and Prause, K. (1998). New insights into smile, mispricing and value at risk: the hyperbolic model, *J. Bus.* **71**, 3, pp. 371–406.
[61] Eberlein, E., Papapantoleon, A., and Shiryaev, A. N. (2008). On the duality principle in option pricing: semimartingale setting, *Finance Stoch.* **12**, 2, pp. 265–292.
[62] Eberlein, E. and Prause, K. (2002). The generalized hyperbolic model: Financial derivatives and risk measures, in H. Geman *et al.* (eds.), *Mathematical Finance—Bachelier Congress 2000* (Springer, Heidelberg), pp. 245–267.
[63] Eberlein, E. and Raible, S. (1999). Term structure models driven by general Lévy processes, *Math. Finance* **9**, 1, pp. 31–53.
[64] Eberlein, E. and Raible, S. (2001). Some analytic facts on the generalized hyperbolic model, in C. Casacuberta *et al.* (eds.), *3rd European Congress of Mathematics*, Vol. II, Progr. Math., Vol. 202 (Birkhäuser, Basel), pp. 367–378.
[65] Einstein A. (1905). Über die von der molekularkinetischen Theorie der Wärme geforderte Bewegung von in ruhenden Flüssigkeiten suspendierten Teilchen, *Ann. Phys. (4)* **17**, pp. 549–560.
[66] Elliott, R. J. and Kopp, P. E. (1999). *Mathematics of Financial Markets* (Springer, New York).
[67] Encyclopaedia of mathematics. An updated and annotated transl. of the Soviet "Mathematical Encyclopaedia". Vol. 1–6. (Kluwer, Dordrecht).
[68] Engelbert, H.-J. and Schmidt, W. (1989). Strong Markov continuous local

martingales and solutions of one-dimensional stochastic differential equations. I, *Math. Nachr.* **143**, pp. 167–184.

[69] Engelbert, H.-J. and Schmidt, W. (1989). Strong Markov continuous local martingales and solutions of one-dimensional stochastic differential equations. II, *Math. Nachr.* **144**, pp. 241–281.

[70] Engelbert, H.-J. and Schmidt, W. (1991). Strong Markov continuous local martingales and solutions of one-dimensional stochastic differential equations. III, *Math. Nachr.* **151**, pp. 149–197.

[71] Engle, R. F. (1982). Autoregressive conditional heteroscedasticity with estimates of the variance of United Kingdom inflation, *Econometrica* **50**, 4, pp. 987–1008.

[72] Esscher, F. (1932). On the probability function in the collective theory of risk, *Scand. Actuarietidskrift* **15**, pp. 175–179.

[73] Feller, W. (1971). *An Introduction to Probability Theory and Its Applications*, Vol. II (Wiley, New York).

[74] Föllmer, H. and Schied, A. (1995). *Stochastic Finance. An Introduction in Discrete Time*, de Gruyter Studies in Mathematics, Vol. 27 (de Gruyter); 2nd edn., 2004.

[75] Fouque, J.-P., Papanicolaou, G., and Sincar, K. R. (2001). *Derivatives in Financial Markets with Stochastic Volatility* (Cambridge Univ. Press).

[76] Frisch, U. (1995). *Turbulence. The Legacy of A. N. Kolmogorov* (Cambridge Univ. Press, Cambridge).

[77] Fujiwara, T. and Miyahara, Y. (2003). The minimal entropy martingale measures for geometric Lévy processes, *Finance Stoch.* **7**, 4, pp. 509–531.

[78] Galtchouk, L. (2003). On the reduction of a multidimensional continuous martingale to a Brownian motion, *Sém. de Probabilités XXXVII*, Lecture Notes in Math., Vol. 1832, pp. 90–93.

[79] Geman, H. and Madan, D. (2004). Pricing in incomplete markets: From absence of good deals to acceptable risk, in G. Szegö (ed.), *Risk Measures for the 21st Century* (Wiley), pp. 451–473.

[80] Geman, H., Madan D., and Yor, M. (2001). Time changes for Lévy processes, *Math. Finance* **11**, 1, pp. 79–96.

[81] Geman, H., Madan D., and Yor, M. (2002). Stochastic volatility, jumps, and hidden time changes, *Finance Stoch.* **6**, 1, pp. 63–90.

[82] Gikhman, I. I. and Skorokhod, A. V. (1996). *Random Processes* (Dover, Mineola). (Translation from the 1965 Russian edition.)

[83] Girsanov, I. V. (1962). On transforming a certain class of stochastic processes by absolutely continuous substitution of measures, *Theory Probab. Appl.* **5(1960)**, 3, pp. 285–301.

[84] Gnedenko, B. V. and Kolmogorov, A. N. (1954). *Limit Distributions for Sums of Independent Random Variables* (Addison-Wesley, Cambridge, Mass.).

[85] Goll, T. (2001). Derivative pricing and logarithmic portfolio optimization in incomplete markets, Dissertation, Albert-Ludwigs-Universität, Freiburg i. Br.

[86] Goll, T. and Kallsen, J. (2003). A complete explicit solution to the log-

optimal portfolio problem, *Ann. Appl. Probab.* **13**, 2, pp. 774–799.

[87] Gradshteyn, I. S. and Ryzhik, I. M. (2007). *Tables of Integrals, Series, and Products*, 7th edn. (Elsevier/Academic Press, Amsterdam).

[88] Graversen, S. E. and Shiryaev, A. N. (2000). An extension of P. Lévy's distributional properties to the case of a Brownian motion with drift, *Bernoulli* **6**:4, pp. 615–620.

[89] Grigoriev, P. G. (2005). On low dimensional case in the fundamental asset pricing theorem with transaction costs, *Statist. Decisions* **23**, pp. 33–48.

[90] Halgreen, C. (1979). Self-decomposability of the generalized inverse Gaussian and hyperbolic distributions, *Z. Wahrscheinlichkeitstheor. verw. Geb.* **47**, 1, pp. 13–17.

[91] Harrison, J. M. and Pliska, S. R. (1981). Martingales and stochastic integrals in the theory of continuous trading, *Stochastic Process. Appl.* **11**, 3, pp. 215–260.

[92] Hubalek, F. and Sgarra, C. (2005). Esscher transforms and the minimal entropy martingale measure for exponential Lévy models, Research Report 13, Thiele Centre for Applied Mathematics in Natural Science, Aarhus.

[93] Huff, B. W. (1969). The strict subordination of differential processes, *Sankhyā Ser. A* **31**, pp. 403–412.

[94] Itô, K. (1951). On stochastic differential equations, *Mem. Amer. Math. Soc.* **4**, pp. 1–51.

[95] Itô K. (2004). *Stochastic Processes*, O. E. Barndorff-Nielsen and K. Sato (eds.) (Springer-Verlag, Berlin).

[96] Jacob, N. and Schilling, R. L. (1996). Subordination in the sense of S. Bochner—an approach through pseudo-differential operators, *Math. Nachr.* **178**, pp. 199–231.

[97] Jacobsen, M. and Yor, M. (2002). Multi-selfsimilar Markov processes on $\mathbb{R}^n_+$ and their Lamperti representaitons, Research Report 28, Centre for Mathematical Physics and Stochastics, Aarhus University, Aarhus.

[98] Jacod, J. (1979). *Calcul Stochastique et Problèmes de Martingales*, Lecture Notes in Math., Vol. 714 (Springer).

[99] Jacod, J. and Shiryaev, A. N. (1998). Local martingales and the fundamental asset pricing theorems in the discrete-time case, *Finance Stoch.* **2**, 3, pp. 259–273.

[100] Jacod, J. and Shiryaev, A. N. (2003). *Limit Theorems for Stochastic Processes*, 2nd edn., rev. and expanded (Springer-Verlag, Berlin).

[101] Janicki, A. and Weron, A. (1994). *Simulation and Chaotic Behavior of alpha-Stable Stochastic Processes* (Dekker, New York).

[102] Jeanblanc, M., Pitman J., and Yor, M. (2002). Self-similar processes with independents increments associated with Lévy and Bessel processes, *Stochastic Process. Appl.* **100**, pp. 223–231.

[103] Jeanblanc, M., Yor, M., and Chesney, M. (2009). *Mathematical Methods for Financial Markets* (Springer-Verlag, Berlin–New York).

[104] Jørgensen, B. (1982). *Statistical Properties of the Generalized Inverse Gaussian Distributions*, Lecture Notes in Statist., Vol. 9 (Springer-Verlag, New York).

[105] Kabanov, Yu. M. and Kramkov, D. O. (1994). No-arbitrage and equivalent martingale measures: An elementary proof o fthe Harrison–Pliska theorem, *Theory Probab. Appl.* **39**, 3, pp. 523–527.

[106] Kabanov, Yu. M. and Kramkov, D. O. (1998). Asymptotic arbitrage in large financial markets, *Finance Stoch.* **2**, 2, pp. 143–172.

[107] Kabanov, Yu. M., Liptser, R. Sh., and Shiryaev, A. N. (1980). On the representation of integer-valued random measures and local martingales by means of random measures with deterministic compensators, *Math. USSR, Sb.* **39**, pp. 267–280.

[108] Kallenberg, O. (2002). *Foundation of Modern Probability,* 2nd edn. (Springer).

[109] Kallsen, J. (2006). A didactic note on affine stochastic volatility models, in Yu. Kabanov, R. Liptser, and J. Stoyanov, *From Stochastic Calculus to Mathematical Finance* (Springer, Berlin), pp. 343–368.

[110] Kallsen, J. and Shiryaev, A. N. (2001). Time change representation of stochastic integrals, *Theory Probab. Appl.* **46**, 3, pp. 522–528.

[111] Kallsen, J. and Shiryaev, A. N. (2002). The cumulant process and Esscher's change of measure, *Finance Stoch.* **6**, 4, pp. 397–428.

[112] Keller, U. (1997). Realistic modeling of financial derivatives. Dissertazion, Albert-Ludwigs-Universität, Freiburg i. Br.

[113] Kendall, M. and Stuart, A. (1969). *The Advanced Theory of Statistics. Vol.* 1: *Distribution Theory,* 3rd edn. (Hafner, New York).

[114] Kolmogorov, A. N. (1931). Über die analytischen Methoden in der Wahrscheinlichkeitsrechnung, *Math. Ann.* **104**, pp. 415–458.; English translation: On analytical methods in probability theory, *Selected Works of A. N. Kolmogorov. Vol.* II: *Probability Theory and Mathematical Statistics,* Math. Appl., **26**, ed. by A. N. Shiryaev (Kluwer, Dordrecht), 1992, 62–108.

[115] Kolmogorov, A. N. (1941). Lokal'naya structura turbulentnosti v neszhimaemoj vyazkoj zhidkosti pri ochen' bol'shikh chislakh Reynoldsa (Russian) [Local structure of turbulence in an incompressible viscous fluid at very high Reynolds numbers], *Dokl. Akad. Nauk SSSR* **30**:4, pp. 299–303.; German translation in: Statistische Theorie der Turbulenz (Akademie-Verlag, Berlin), 1958, pp. 71–76..

[116] Kolmogorov, A. N. (1941). K vyrozhdeniyu izotropnoj turbulentnosti v neszhimaemoj vyazkoj zhidkosti (Russian) [On degeneration of isotropic turbulence in an incompressible viscous fluid], *Dokl. Akad. Nauk SSSR* **31**:6, pp. 538–541.; German translation in: Statistische Theorie der Turbulenz (Akademie-Verlag, Berlin), 1958, pp. 147–150..

[117] Kolmogorov, A. N. (1941). Rasseyanie energii pri lokalno izotropnoj turbulentnosti (Russian) [Dissipation of energy for the locally isotropic turbulence], *Dokl. Akad. Nauk SSSR* **32**:1, pp. 19–21.; German translation in: Statistische Theorie der Turbulenz (Akademie-Verlag, Berlin), 1958, pp. 77–81..

[118] Kolmogorov, A. N. (1962). Precisions sur la structure locale de la turbulence dans un fluid visqueux aux nombres de Reynolds élev'es (Russian, French), *Mécanique de la Turbulence: Actes du Colloque International du CNRS*

(*Marseille,* 1961)(Éditions du CNRS, Paris), pp. 447–458.; English translation: A refinement of previous hypotheses concerning the local structure of turbulence in a viscous incompressible fluid at high Reynolds number, *J. Fluid Mech.* **13**:1 (1962), pp. 82–85..

[119] Kolmogoroff, A. (1933). *Grundbegriffe der Wahrscheinlichkeitsrechnung*, Ergeb. Math. Grenzgeb., Vol. 2, no. 3 (Springer, Berlin). English transl.: Kolmogorov, A. N. (1956). *Foundations of the Theory of Probability* (Chelsea Publ. Co., New York).

[120] Kramkov, D. O. (1996). Optional decomposition of supermartingales and hedging contingent claims in incomplete security markets, *Probab. Theory Related Fields* **105**, 4, pp. 459–479.

[121] Kyprianou, A. E. (2006). *Introductory Lectures in Fluctuations of Lévy Processes with Applications* (Springer, New York).

[122] Lamperti, J. (1972). Semi-stable Markov processes, *Z. Wahrscheinlichkeitstheor. verw. Geb.* **22**, 2, pp. 205–225.

[123] Liese, F. and Vajda, I. (1987). *Convex Statistical Distances* (Teubner, Leipzig).

[124] Liptser, R. Sh. and Shiryaev, A. N. (1989). *Theory of Martingales* (Kluwer, Dordrecht).

[125] Liptser, R. Sh. and Shiryaev, A. N. (2001). *Statistics of Random Processes. Vol.* I: *General Theory; Vol.* II: *Applications*, 2nd edn., rev. and expanded (Springer-Verlag, Berlin).

[126] Lundberg F. I. (1903). Approximerad Framställning Sannolikhetsfunktionen. II. Återförsäkering av Kollectivrisker. Akad. Afhandling. (Almqvist & Wiksell, Uppsala).

[127] Madan, D. B., Carr, P., and Chang, E. (1998). The variance gamma process and option pricing, *Europ. Finance Rev.* **2**, pp. 79–105.

[128] Madan, D. B. and Seneta, E. (1990). The variance gamma model for share market returns, *J. Bus.* **63**, pp. 511–524.

[129] Mandelbrot, B. B. and Taylor, H. M. (1967). On the distribution of stock price difference, *Operations Research* **15**, 6, pp. 1057–1062.

[130] Merton, R. C. (1973). Theory of rational option pricing, *Bell J. Econ. and Management Sci.* no. 4, pp. 141–183.

[131] Monroe, I. (1978). Processes that can be embedded in Brownian motion, *Ann. Probab.* **6**, 1, pp. 42–56.

[132] Musiela, M. and Rutkowski, M. (1997). *Martingale Methods in Financial Modelling*, Appl. Math., Vol. 36 (Springer-Verlag, Berlin–Heidelberg).

[133] Nikiforov, A. F. and Uvarov, V. B. (1974). *Osnovy teorii spetsial'nykh functsij* (Russian) [Foundations of the Theory of Special Functions] (Nauka, Moskva).

[134] Obukhov, A. M. (1941). On the energy distribution in the spectrum of a turbulent flow (Russian), *Dokl. Akad. Nauk SSSR* **32**:1, pp. 22–24.

[135] Obukhov, A. M. (1941). On the energy distribution in the spectrum of a turbulent flow (Russian), *Izv. Akad. Nauk SSSR, Geography and Geophysics ser.* **5**:4/5, pp. 453–466.

[136] Ocone, D. L. (1993). A symmetry characterization of conditionally independent increments martingales, *Progr. Probab.* **32**, pp. 147–167.

[137] Øksendal, B. (2000). *Stochastic Differential Equations. An Introduction with Applications*, 5th edn. (Springer, Berlin–Heidelberg).

[138] *Karl Pearson's Early Statistical Papers* (Cambridge Univ. Press, Cambridge), 1948.

[139] Peskir, G. and Shiryaev, A. (2006). *Optimal Stopping and Free-Boundary Problems*, Lectures in Math. ETH Zürich (Birkhäuser, Basel).

[140] Pillai, R. N. (1990). On Mittag-Leffler functions and related distributions, *Ann. Inst. Statist. Math.* **42**, pp. 157–161.

[141] Prause, K. (1999). The generalized hyperbolic model: Estimation, financial derivatives and risk measures, Dissertation, Albert-Ludwigs-Universität, Freiburg i. Br.

[142] Protter, P. (1990). *Stochastic Integration and Differential Equations* Appl. Math. (New York), Vol. 21 (Springer-Verlag, Berlin).

[143] Rachev, S. T. (ed.) (2003). *Handbook of Heavy Tailed Distributions in Finance* (Elsevier/North-Holland, Amsterdam).

[144] Raible, S. (2000). Lévy processes in finance: Theory, numerics, empirical facts, Dissertation, Albert-Ludwigs-Universität, Freiburg i. Br.

[145] Renault, E. and Touzi, N. (1996). Option hedging and implied volatilities in a stochastic volatility model, *Math. Finance* **6**, pp. 279–302.

[146] Revuz, D. and Yor, M. (1991). *Continuous Martingales and Brownian Motion*, Grundlehren Math. Wiss., Vol. 293 (Springer-Verlag, Berlin); 3rd edn., 1999.

[147] Rogers, L. C. G. (1995). Equivalent martingale measures and no-arbitrage, *Stochastics Stochastics Rep.* **51**, pp. 41–50.

[148] Rogers, L. C. G. and Williams, D. (1987, 1994). *Diffusions, Markov Processes, and Martingales. Vol.* 1: *Foundations*; *Vol.* 2: *Itô calculus* (Wiley, Chichester–New York).

[149] Rydberg, T. H. (1997). A note on the existence of unique equivalent martingale measures in a Markovian setting, *Finance Stoch.* **1**, 3, pp. 251–257.

[150] Rydberg, T. H. (1999). Generalized hyperbolic diffusion processes with applications in finance, *Math. Finance* **9**, 2, pp. 183–201.

[151] Samorodnitsky, G. and Taqqu, M. S. (1994). *Stable Non-Gaussian Random Processes* (Chapman & Hall, New York).

[152] Samuelson, P. A. (1965). Rational theory of warrant pricing, *Industrial Management Rev.* **6**, pp. 13–31.

[153] Sato K. (1999). *Lévy Processes and Infinitely Divisible Distributions* (Cambridge Univ. Press, Cambridge).

[154] Sato K. (2000). Density transformation in Lévy processes, Lecture Notes 7, Centre for Mathematical Physics and Stochastics, Aarhus University, Aarhus.

[155] Sato, K. (2001). Subordination and self-decomposability, *Statist. Probab. Lett.* **54**, 3, pp. 317–324.

[156] Schweizer, M. (1996). Approximation pricing and the variance-optimal martingale measure, *Ann. Probab.* **24**, 1, pp. 206–236.

[157] Selivanov, A. V. (2005). On martingale measures in exponential Lévy models, *Theory Probab. Appl.* **49**, 2, pp. 261–274.

[158] Selivanov, A. V. (2004). *Filtration of volatility and martingale mesures in the exponential Lévy models*, Ph.D. thesis, Moscow State University, Moscow.

[159] Shiryaev, A. N. (1978). *Optimal Stopping Rules* (Springer-Verlag, New York); 2nd English edn., 2008.

[160] Shiryaev, A. N. (1996). *Probability*, Grad. Texts in Math., Vol. 95 (Springer-Verlag, New York).

[161] Shiryaev, A. N. (1999). *Essentials of Stochastic Finance. Facts, Models, Theory*, Adv. Ser. Statist. Sci. Appl. Probab., Vol. 3 (World Scientific, River Edge, NJ).

[162] Shiryaev, A. N. and Cherny, A. S. (2000). Vector stochastic integrals and the fundamental theorems of asset pricing, *Proc. Steklov Inst. Math.* **237**, pp. 6–49.

[163] Shiryaev, A. N. and Yor, M. (2004). On stochastic integral representations of functionals of Brownian motion. I, *Theory Probab. Appl.* **48**, 2, pp. 304–313.

[164] Shiryaev, A. N. (2007). On the classical, statistical, and stochastic approaches to the hydrodynamic turbulence, *Reserch report* 02,Thiele Centre for Applied Mathematics in Natural Science, Aarhus, 62 pp.

[165] Shreve, S. E. (2004). *Stochastic Calculus for Finance.* II: *Continuous-time Models* (Springer-Verlag, New York).

[166] Skorokhod, A. V. (1965). *Studies in the Theory of Random Processes* (Addison-Wesley, Reading).

[167] Stricker, C. (1990). Arbitrage et lois de martingale, *Ann. Inst. H. Poincaré* **26**, 2, pp. 451–460.

[168] Stroock, D. W. and Varadhan S. R. S. (1979). *Multidimensional Diffusion Processes* (Springer-Verlag, Berlin–Heidelberg).

[169] Tsirel'son, B. S. (1975). An example of a stochastic equation having no strong solution, *Theory Probab. Appl.* **20**, 2, pp. 427–430.

[170] Vajda, I. (1989). *Theory of Statistical Inference and Information* (Kluwer, Dordrecht).

[171] Vostrikova, L. and Yor, M. (2000). Some invariance properties (of the laws) of Ocone's martingales, *Lecture Notes in Math.* Vol. 1729, pp. 417–431.

[172] Wald, A. (1947). *Sequential Analysis* (Wiley, New York).

[173] Watson, G. N. (1944). *A Treatise on the Theory of Bessel Functions* (Cambridge, Cambridge Univ. Press).

[174] Winkel, M. (2001). The recovery problem for time-changed Lévy processes, MaPhySto Research Report 38, Aarhus University, MaPhySto, Aarhus.

[175] Zanzotto, P. A. (1998). Representation of a class of semimartingales as stable integrals, *Theory Probab. Appl.* **43**, 4, pp. 666–676.

[176] Zvonkin, A. K. (1974). A transformation of the phase space of a diffusion process that removes the drift, *Math. USSR-Sb.* **22**, 1, pp. 129–149.

Index

(B, S)-market, 196
X-representation, 212
$\widetilde{\mathcal{O}}$-optional function, 49

American option, 225
  buyer price, 241
  lower price, 241
  seller price, 241
  upper price, 241
angular-bracket process, 48
arbitrage, 198
arbitrage opportunity, 198
arbitrage-free market, 198
arbitrage-free model, 216
asymmetry parameter, 111
asymptotic arbitrage, 217
at-the-money, 262
autocorrelation function, 262

Bachelier
  formula, 231
  model, 9
backward differential equation, 75
Bernstein result, 162
Black–Scholes formulae, 238
BNS model, 285
Brownian motion
  economic, 234
  geometric, 234
buyer price
  of American option, 241
  of European option, 203

càdlàg process, 43
call-put parity formula, 228
canonical representation, 45
capital, 209
capital of a portfolio, 197
Carr–Geman–Madan–Yor
  distribution, 286
Cauchy
  distribution, 112
  process, 120
CGMY distribution, 286
changes of time
  dual, 4
  mutually inverse, 4
clustering effect, 263
compensator, 47
complete (B, S)-market, 201
complete model of a market, 212
completely monotone function, 160
compound interest formula, 93
compound Poisson process, 116, 130
compound return representation, 136
contiguity, 218
conversion formula for conditional
  expectations, 124
Cox–Ross–Rubinstein model, 201
CRR-model, 201
cumulant process, 97
cumulant (Fourier) function
  local, 98

Dambis–Dubins–Schwarz theorem, 17

de Finetti–Kolmogorov–Lévy–
Khinchin formula,
109
density process, 122
differential characteristics of a
semimartingale, 103
diffusion
coefficient, 74
process, 78
discrete intervention of chance, 208
distribution
$\mathbb{G}$H, 273
$\mathbb{G}$IG, 180
$\mathbb{N}$ormal Gamma, 191
$\mathbb{N} \circ$ Gamma, 279
$\mathbb{N} \circ \mathrm{H}^+$, 276
$\mathbb{N} \circ$ IG, 276
$\mathbb{N} \circ$ PH, 276
$\mathbb{R}$Gamma, 184
$\mathbb{R}\mathrm{H}^+$, 184
$\mathbb{R}$IG, 184
$\mathbb{V}$ariance Gamma, 191
$\mathbb{V} \circ$ G, 279
$\mathbb{V}$Gamma, 277
$\mathbb{V}$G, 277
$\mathbb{N} \circ \mathbb{G}$IG, 193
H^+, 13, 184
IG, 13, 179
PH, 13, 184
Carr–Geman–Madan–Yor, 286
Cauchy, 112
CGMY, 286
Gamma, 271
generalized hyperbolic, 273
generalized inverse Gaussian, 180
hyperbolic, 187, 273, 276
infinitely divisible, 109
inverse Gaussian, 13, 179
Lévy–Smirnov, 112
negative Gaussian, 179
normal, 112
normal Gamma, 279
normal generalized inverse
Gaussian, 193
normal inverse Gaussian, 276
normal positive hyperbolic, 276
one-sided stable, 112
positive hyperbolic, 13, 184
reciprocal Gamma, 184
reciprocal inverse Gaussian, 184
reciprocal positive hyperbolic, 184
Variance Gamma, 277, 279
Doob decomposition, 47, 167
drift coefficient, 74
driving process, 123
dual
model, 245
process, 245

economic Brownian motion, 234
Esscher
change of measure, 100
measure, 134
transform, 133
European option, 225
buyer price, 203
lower price, 203
seller price, 203
upper price, 203
evanescent set, 43
exercise price, 202

fair price, 204
filtered probability space, 1
extension, 18
finite change of time, 2
formula
of change of variables in the
Lebesgue–Stieltjes integral,
8
of integration by substitution, 8
forward differential equation, 75
Fourier cumulant process, 97
modified, 97
fractional Brownian motion
self-similarity, 265
free lunch
with bounded risk, 210
with vanishing risk, 210
function
$\widetilde{\mathcal{P}}$-predictable, 49
simple, 7

Gamma process, 14
generalized
  Bayes formula, 124
  hyperbolic Lévy process, 158
  inverse Gaussian distribution, 180
  tempered stable process, 286
geometric Brownian motion, 234
Girsanov theorem
  discrete version, 126
  original formulation, 128
  semimartingale version, 127
global Lipschitz condition, 81
growth condition, 77

Halphen law, 185
harmonic law, 185
  generalized, 186
Hellinger
  distance, 219
  integral, 218
homogeneous Poisson measure, 51
Hurst exponent, 113
hyperbolic distribution, 187, 273, 276
  generalized, 273
hyperbolic Lévy process, 13, 157, 275

IG, 179
implied volatility, 261
in-the-money, 262
indistinguishable processes, 43
infinite divisibility, 109
innovation
  process, 152
  representation, 152
integrability w.r.t. a semimartingale, 73
inverse
  Gaussian distribution, 179
inverse Gaussian
  distribution, 13
  Lévy subordinator, 273
    generalized, 273

jump measure, 49

Knight result, 19
Kolmogorov–Chapman equation, 74
Kolmogorov–Lévy–Khinchin's formula, 98

Lévy
  measure, 109
Lévy martingale, 28
Lévy process, 108
  generalized hyperbolic, 158
  hyperbolic, 157, 275
  normal inverse Gaussian, 157
  strictly α-stable, 113
  Variance Gamma, 275
Lévy–Khinchin formula, 106, 111
Lévy–Smirnov distribution, 112
Laplace cumulant process, 99
  modified, 99
large market, 217
leverage parameter, 285
leverage-effect, 269
life time of the process, 2
Lipschitz condition, 77
  global, 81
local characteristics of a semimartingale, 103
local Laplace cumulant function, 100
local martingale, 44
  purely discontinuous, 45
localizing sequence, 44
location parameter, 111
long memory, 265
lower price
  basic formula, 204
  of American option, 241
  of European option, 203

Markov time, 43
martingale, 43
  problem, 87
  transform, 44
mean-reversion, 266, 269
measure
  absolutely continuous, 121
  locally absolutely continuous, 122
  of jumps, 168
measures

equivalent, 121
locally equivalent, 122
Mittag-Leffler
function, 15
process, 15
monotone class, 68
Monroe theorem, 154
multivariate point process, 115

natural filtration, 151
negative Gaussian distribution, 179
no arbitrage, 198
no free lunch with vanishing risk, 210
normal
distribution, 112
Gamma distribution, 191, 279
inverse Gaussian Lévy process, 13, 157, 282
positive hyperbolic distribution, 276
Novikov condition, 86, 124

Ocone result, 155
optional
function, 49
process, 44
random measure, 50
sigma-algebra, 44, 49
out-of-the-money, 262

Pearson system, 171
perfect hedge, 214
Poisson
process
compound, 116
random measure, 51, 107
extended, 107
portfolio, 197
positive hyperbolic distribution, 13, 184
predictable
function, 49
process, 44
quadratic covariation, 48
random measure, 50
sigma-algebra, 49
price
quasi-rational, 254
process
$\mathbb{V}$G, 14
L($\mathbb{G}$H), 158
L($\mathbb{N} \circ$ IG), 157
L($\mathbb{H}$), 157
Cauchy, 120
generalized hyperbolic Lévy, 158
hyperbolic Lévy, 157
Lévy, 108
Lévy hyperbolic, 275
Lévy strictly α-stable, 113
Lévy Variance Gamma, 275
normal inverse Gaussian Lévy, 157
of bounded variation, 43
optional, 44
predictable, 44
progressively measurable, 61
stable, 112, 113
with independent increments, 105
with stationary independent increments, 108
progressively measurable
process, 61
sets, 60

quadratic
characteristic, 16, 48
covariation, 46
variation, 46
quasi-rational price, 254

random change of time, 1
random measure, 50
$\widetilde{\mathcal{P}}$-σ-finite, 50
optional, 50
predictable, 50
random variable
infinitely divisible, 109
stable, 110, 112
rational price, 204
reflection principle, 256
replication, 214
return, 209

Samuelson model, 9
scale parameter, 111
scheme of series, 217
self-financing portfolio, 198
self-similarity, 113
seller price
  of American option, 241
  of European option, 203
semi-strong representation, xiii
semimartingale, 44
  exponentially special, 245
  locally square-integrable, 55
short selling, 215
sigma-martingale, 211
simple interest formula, 93
simple point process, 51
simple random function, 63
simple return representation, 136
skewness parameter, 111
smile effect, 261
special semimartingale, 44
stable process, 112, 113
stochastic basis, 1
stochastic differential equation, 76
stochastic exponential, 92
stochastic integral, 70
  with respect to a Wiener process, 59
stochastic logarithm, 92
stochastic volatility model, 123
stopping time, 44
strategy, 209
strike price, 202
strong aftereffect, 265
strong intermittency, 266
strong solution, 77
strong uniqueness, 77
submartingale, 43
subordinator, 2, 117, 272
  in the strong sense, 2

Tanaka formula, 80, 170
triplet of predictable characteristics, 54

upper price
  basic formula, 204, 242
  of American option, 241
  of European option, 203
usual conditions on filtration, 42
usual conditions on stochastic basis, 1

Variance Gamma
  distribution, 191, 277, 279
  Lévy process, 275
  process, 14
vector stochastic integral, 73

weak representation, xiii
weak solution, 84, 86
Weber function, 194
Wold decomposition, xi